ADVANCES IN CELL AGING AND GERONTOLOGY

VOLUME 13

Basic Biology and Clinical Impact of Immunosenescence

ADVANCES IN CELL AGING AND GERONTOLOGY
VOLUME 13

Basic Biology and Clinical Impact of Immunosenescence

Volume Editor:

Graham Pawelec,
University of Tübingen,
ZMF,
Tübingen,
Germany

Series Editor:

Mark P. Mattson,
National Institute on Aging,
Baltimore,
USA

2003

ELSEVIER

Amsterdam – Boston – London – New York – Oxford – Paris
San Diego – San Francisco – Singapore – Sydney – Tokyo

ELSEVIER SCIENCE B.V.
Sara Burgerhartstraat 25
P.O. Box 211, 1000 AE Amsterdam, The Netherlands

First edition 2003

Library of Congress Cataloging in Publication Data
A catalog record from the Library of Congress has been applied for.

British Library Cataloguing in Publication Data
Basic biology and clinical impact of immunosenescence. -
  (Advances in cell aging and gerontology ; v. 13)
  1.Geriatrics - Immunological aspects 2.Aging -
  Immunological aspects
  I. Pawelec, G. (Graham)
  618.9'76079

ISBN: 0-444-51316-7
ISSN: 1566-3124 (Series)

Transferred to digital printing 2007

TABLE OF CONTENTS

PREFACE

Like other organ systems, the immune system does not remain unaffected by age. Work begun over three decades ago already hinted at the importance of immunity for the achievement of healthy old age. The real clinical relevance of age-related alterations in the immune system is, however, still difficult to document and remains controversial. The first part of this volume of ACAG contains contributions describing studies which are helping to resolve these difficulties, focussing almost exclusively on humans. The first chapter, by Wikby et al., stresses the importance of longitudinal studies and summarises results from ongoing studies which have now been monitoring elderly Swedes from one small area of the country for 11 years. The results suggest that at least in these very elderly (i.e. the small fraction of the entire population selected for greater than average longevity), clusters of immune parameters can be defined which predict morbidity and mortality (an "immunological risk phenotype", IRP). Importantly, this seems to be independent of the overt health status of the individuals at the beginning of the study. However, awareness of a critical role for persistent viruses, particularly CMV, appears to be emerging from their, and several others', studies. Whether the concept of the IRP can be extended to the younger elderly urgently needs to be determined in further longitudinal studies. The second chapter in this volume begins to approach this question, wherein Huppert et al. examine two larger unselected populations at younger starting ages (65 or over), and assess predictive ability of immune markers for 9-year survival. There are some intriguing similarities as well as differences in the results of these two approaches of Wikby et al. and Huppert et al. In the next chapter, Candore et al. review what is known about changes in T cells and NK cells with ageing, and how these might impact on immunosenescence. They go on to consider the role of cytokines and cytokine receptors, and, importantly, the impact of genetic factors (polymorphisms) on these parameters, in the context of human health and longevity. This chapter introduces the concept of a critical role for the balance between inflammatory and anti-inflammatory activities within the immune system, and the role of chronic antigenic stress in determining immune parameters and overall longevity.

The next section of this volume includes contributions focussing more on the mechanisms responsible for those observed changes in the human immune system shown above to be clinically relevant. Although all cells of the immune system show age-associated changes, it is the T cell compartment which is thought to be most severely affected and which has been studied most intensively. As T cells are generated primarily in the thymus, this organ must be the first focus of attention when considering T cell immunosenescence. Two chapters by Globerson and by Pido-Lopez and Aspinall consider not only age-related changes to the thymus and thymic function, but how these changes might be prevented or reversed (and whether this would in fact be desirable). Peripheral T cell ageing is considered in the next chapter by Fülöp et al. in terms of alterations in signal transduction pathways required for activation of immune responses. The T cell arm is here contrasted with an important compartment of the innate immune system, the neutrophil. Next, Casado et al. focus on costimulatory requirements for T cell activation in their contribution and introduce the important recent idea of negative as well as positive T cell costimulation mediated by families of receptors originally discovered on natural killer (NK) cells. The status and importance of NK cells themselves in the elderly, and their sensitive reflection of other physiological ageing parameters, is discussed in the following chapter by Mariani and Facchini.

All these changes in the immune system may contribute to morbidity and mortality due to decreased resistance to infection and conceivably also at least some cancers in the aged. The next chapter, by McLeod, reviews how immunosurveillance against tumours may be influenced by changes in the immune system. One facet of particular importance in this context may be the control of apoptosis in the immune system (and its exploitation by tumours) which is mentioned by McLeod and others here, and discussed in detail in the next chapter by Gupta.

The general question of DNA damage and repair, not only in the context of apoptosis, is the subject of the next chapter by Ross et al., who consider both nuclear and mitochondrial damage and how to prevent or reverse this. Improving DNA repair capacity might offer opportunities for amelioration so that a greater understanding of repair mechanisms is important in this respect, as discussed in the next chapter by Frasca. In the next contribution, by Sourlingas and Sekeri-Pataryas, the role of changes in histone deacetylation is discussed, a topic of increasing importance since the realisation that the *SIR2* gene, which is required for the lifespan extension caused by caloric restriction in yeast and C. elegans, is a histone deacetylase.

The importance of zinc, required by many enzymes including several transcription factors, for proper function of the immune system has been recognized for some time, and is reviewed in the next chapter by Ibs et al. A hypothesis regarding the dynamics of zinc transport and zinc sinks is then developed in the next chapter, by Mocchegiani et al.

Finally, Effros discusses the role of replicative senescence and the impact of telomere length control on human immunosenescence, especially of CD8 cells in vitro and in vivo; and Pawelec dissects in vitro models of CD4 cell senescence in clonal cultures in vitro.

While this volume has not covered all areas of immunity known to be affected by ageing – with B cells and dendritic cells getting particularly short-shrift, and the innate immune system in general being under-represented – it has focussed on the most important and susceptible compartment, the T cell. As changes in antigen presentation by dendritic cells appear to be minimally altered by ageing, and many at least of the changes to B cells are contingent upon T cell alterations, I hope that this may not be too grave an omission. I would like to thank the authors for their excellent contributions and to express my gratitude to the European Union for supporting the Thematic Network "Immunology and Ageing in Europe" (ImAginE, QLK6-CT-1999-02031), under the aegis of which this volume was assembled.

Finally, I thank the series editor for enthusiastically agreeing to the idea of a volume on immunosenescence, and Ashley Knights, MSc, for his assistance in putting this together.

GRAHAM PAWELEC

Advances in
Cell Aging and
Gerontology

The OCTO and NONA immune longitudinal studies: a review of 11 years studies of Swedish very old humans

Anders Wikby[a], Boo Johansson[b,c] and Frederick Ferguson[d]

[a]Department of Natural Science and Biomedicine, School of Health Sciences, Jönköping University, Box 1026, 551 11 Jönköping, Sweden
Correspondence address: Tel.: +46-381-35101; fax: +46-381-17620
E-mail: anders.wikby@ltjkpg.se
[b]Institute of Gerontology, School of Health Sciences, Jönköping University, Box 1026, 551 11 Jönköping, Sweden
[c]Department of Psychology, Göteborg University, Box 500, 405 30 Göteborg, Sweden
Tel.: +46-31-773-1656; fax: +46-31-773-4628. E-mail: boo.johansson@psy.gu.se
[d]The Pennsylvania State University, Department of Veterinary Science, College of Agricultural Sciences, 101 Centralized Biological Laboratory, University Park, PA 16802-4803, USA
Tel.: +1-814-865-1495; fax: +1-814-865-3685. E-mail: fgf@psu.edu

Contents

Advances in Cell Aging and Gerontology, vol. 13, 1–16

1. Introduction

From a societal and population perspective, the very old constitute the fastest growing age segment with compromised health and significant requirements for service and health care. Immune studies of elderly populations have mainly been conducted on individuals in their 60s and 70s. Few have focused on samples over 80 years of age and employed longitudinal designs. In the Swedish OCTO (Wikby et al., 1994; Ferguson et al., 1995; Olsson et al., 2000) and NONA (Wikby et al., 2002) immune studies, however, octo- and nonagenarians were deliberately focused upon using the longitudinal design. The main reason for the inclusion of very old individuals in these studies of the immune system is that oldest-old samples provide the potential for a model useful for detecting intra-individual change in a period in life with high probability for compromised health and morbidity. Changes in the immune system may provide presumptive predictors for subsequent mortality and clinical parameters related to the substantial morbidity/co-morbidity seen in late life. From a practical perspective, detection of predictive changes may enable clinical interventions that could assist in improvement of the quality of life for the individuals in this expanding population segment.

The OCTO immune longitudinal study is a population-based study of ageing and the immune system in a sample of Swedish octogenarians. It began in 1989 and ended in 1997. In 1999, the subsequent NONA immune longitudinal study of nonagenarians was initiated. This review summarises results from the OCTO study and the initial wave of the NONA. First, however, we address the significant design and sampling considerations that directed this research.

2. Methodological design and sampling considerations in ageing studies

2.1. Design considerations

The two methods used in studies of ageing in a population are the cross-sectional and longitudinal designs. The most common design is the cross-sectional, a method by which two or more age groups are compared at a single time of examination. Age changes cannot be measured directly but are inferred from the difference in mean values, observed in the different age groups. Caution is necessary in this interpretation, however, since age differences also reflect the fact that birth-cohorts may have been exposed to various environmental exposures and socio-cultural influences. Another confounding effect in cross-sectional studies is that of selective mortality. As a study population ages it becomes more and more selected, since deaths do not occur randomly in a population. If, for example, a low value in a variable is deleterious, death might occur first in individuals with low values and last in individuals with high values. In a cross-sectional study an observed difference in mean values between age groups may falsely be interpreted as a real age change rather than as an effect of selective mortality. Many studies have characterised changes in the immune system with age, but a number of these have yielded conflicting results. This may partly relate to the fact that a majority of these studies are cross-sectional, limited by single measurements on different individuals.

In a longitudinal design, individuals are followed across time, usually with a number of years in between measurement occasions. This allows the detection of intra-individual change that minimises or overcomes many of the confounding arte-facts likely to emerge from a cross-sectional design. Although the longitudinal design represents the superior alternative for conducting ageing research, the use of this design has been very limited, particularly in studies of the immune system. The main reason for objection to the longitudinal design is that such studies are expensive and require sustained effort, financial support, and commitment of personnel. In addition, longitudinal studies require careful coordination, standardised procedures, and control of studied panels to avoid dropouts. Also, there is a risk of confounding between age and time of measurement effects. Time of measurement confounding involves numerous factors, such as the motivation and interest of the subjects as well as experimenter effects including changes in personnel, in their motivation, and in the methods and techniques used across time. Many of these problems can be compensated for by including a younger group for comparisons across measurement occasions, since immune system changes across the limited time periods between these occasions will be negligible in healthy young people compared to the very old. Also, restricted time periods between the measurements and the use of identical methods will prevent time of measurement effects.

2.2. Sampling considerations

Advancing age brings increased disease problems. This is one of the primary problems in the selection and definition of a sample in population studies of ageing. To overcome this, most studies have used various selection schemes to exclude individuals with underlying diseases from participation in studies of the immune system. The stringent *SENIEUR* Protocol (Ligthart et al., 1984) represents an excellent example of a widespread application of a set of exclusion criteria used to select individuals in perfect health, to be able to distinguish between age changes caused by *primary* and *secondary ageing*, i.e. by diseases. The exclusion of *non-SENIEUR* individuals, however, will result in the study of less than 10% of all individuals aged over 80 years in a representative population (Pawelec et al., 2001). Another way to diminish confusion between ageing and disease has been to employ exclusion criteria tailored to the experimental situation (Hallgren et al., 1988; Wikby et al., 1994), i.e. in immune studies to exclude individuals that have immune-related diseases or who use drugs that affect the immune system. Such a strategy, however, will also generate a highly selected sample, since it will exclude about 50% of the individuals in a population aged over 80 years (Wikby et al., 1994).

A way to overcome these selection problems is to examine a population-based sample, combined with careful continuous evaluation of individual health parameters. The clinical variables needed for the evaluation of individual health and morbidity are then of considerable value in the comparison of findings from the application of various protocols and in the categorisation of individuals into subgroups according to their health status (Nilsson et al., 2003). Thus, the significance of change in health status, an important consideration in these ageing studies, is included rather than excluded.

The establishment of specific markers for ageing, that are independent of morbidity and valid for the population, would be of great assistance in ageing studies. Such markers of ageing, being predictive of longevity, would represent the current status of the immune system independent of disease and, thereby, replace the need for the elaborate exclusion procedures using *SENIEUR* Protocol assessments (Pawelec et al., 2001).

3. The OCTO and NONA immune longitudinal studies

3.1. The OCTO immune study

The OCTO immune longitudinal study was an integrated part of the OCTO longitudinal study on bio-behavioural ageing, in Jönköping, Sweden. The municipality of Jönköping has 110,000 inhabitants and is situated in the south-central part of Sweden. The aim of the OCTO immune study was to explore age changes in the immune system in Swedish octogenarians relative to an array of medical, bio-behavioural, and social variables measured in the OCTO (Wikby et al., 1994).

3.1.1. Sample and design

Census data was used to identify octogenarians living in Jönköping and born in 1897, 1899, 1901, and 1903. A non-proportional sample composed of 100 persons in each of the birth-cohorts was recruited. From these 400 individuals, 324 were examined in the first wave in 1987/1988 of the OCTO study. The persons were then at the ages of 84, 86, 88, and 90 years. At the second wave of the study, the OCTO immune longitudinal study was initiated. Of the 324 examined at baseline of the OCTO, 96 were deceased before the start of the second wave of this study. Another 15 declined to participate, giving a total number of potential participants of 213 for the OCTO immune.

Exclusion criteria were set to diminish confound between ageing, disease, and medications and to secure reliable psychosocial self-reports. Potential candidates were included if they:

- were non-institutionalised;
- had normal cognition according to neuropsychological tests (Johansson et al., 1992);
- were not on a drug regimen that may influence the immune system.

These exclusion criteria were similar to those of Hallgren et al. (1988). Of the potential 213 individuals, 110 met the inclusion criteria. Of these, 102 individuals participated in the first wave. Sixty-nine individuals were available throughout the three waves in the longitudinal analysis and 23 participated in the longitudinal analysis over all four time-points, T1 (1989), T2 (1990), T3 (1991), and T4 (1997) (Table 1). Non-participation at the various measurement occasions was mainly due to mortality in the sample.

Table 1
Characteristics of the subjects included in the OCTO Immune Longitudinal Study

Occasion (time)	Year	Number of subjects investigated	Age (years)	
			Mean	Range
1	1989	102	88	86-92
2	1990	83	89	87-93
3	1991	69	90	88-94
4	1997	23	95	94-100

Fourteen healthy middle-aged volunteers (39 years SD ± 5.8) of men and women working in the laboratories at Ryhov Hospital in Jönköping were included across the measurement occasions for comparative reasons.

3.1.2. Immune components

The very old individuals were examined in their place of residence. Blood samples were drawn in the morning between 08:00 and 10:00 h. The following immune system parameters were investigated:

- Complete blood cell count.
- Differential WBC count.
- Antibody defined T and B cell surface molecules using three colour flow cytometry.
- Proliferative response of PBMC using a mitogen stimulation assay with ConA in cell culture.
- Interleukin 2 production.
- Cytomegalovirus (CMV) and Herpes simplex serology.

3.2. The NONA immune study

Findings in the longitudinal OCTO immune study constituted the background for the subsequent ongoing NONA immune study of nonagenarian individuals living in the municipality of Jönköping. The NONA immune is an integrated part of the NONA longitudinal study initiated to examine the disablement process in late life. The overall aim in the NONA immune is to examine predictive factors for longevity in the very old and to further investigate in greater depth the immune risk profile identified in the OCTO immune. The aim is also to consider immune data in the context of functional and disability parameters examined in the overall NONA. The overall study includes measurements of the following functional and disability domains:

- physical and mental health,
- cognitive functioning,
- personal control/coping,
- social networks,
- provision of service,
- care and everyday functioning capacity.

The NONA immune is also part of the EU-supported program *Immunology and Ageing in Europe*, with collaborations between the NONA immune project and several laboratories participating in this Thematic Network (Pawelec, 2000).

3.2.1. Sample and design

NONA immune is a population-based random sample with no exclusion criteria. Individuals were drawn from the population (census) register of Jönköping. A non-proportional random sampling procedure was employed, including all individuals permanently residing in the municipality, with the goal to have individuals aged 86, 90, and 94 years old. The sampling frame was defined from the census information available in September 1999. As the number of available subjects in the oldest birth cohort was limited, a few subjects were also included from the birth cohorts of 1904 and 1905. Blood samples for the immune system analysis were drawn in 138 individuals, of whom 42 belonged to the oldest birth cohort, 47 were 90-years old, and 49 86-years old.

The mean age of the sample was 90.3 years with a total proportion of women of 70%. While about 60% of them lived in ordinary housing, 40% resided in sheltered housing or in an institution. A comparison between individuals who participated in the in-person testing part of the NONA study ($n = 157$), and those who agreed to have blood was drawn ($n = 138$), indicated no significant differences for demographics or overall ratings of physical and mental health.

3.2.2. Immune components

Subjects were examined in their place of residence. The blood samples were drawn in the morning between 09:00 and 10:00 h. The following immune and clinical components are studied in the first wave of the NONA immune longitudinal study:

- Complete blood cell count.
- Differential WBC count.
- Proteins, albumin, transthyretin, C-reactive protein, orosomucoid, haptoglobulin, IgG, IgM, IgA, urea, cystatinC, creatinine as indicators of malnutrition, inflammation or kidney disease.
- Antibody defined T cell surface molecules of T, NKT, NK cell populations, using three colour flow cytometry.
- Secretion of cytokines.
- TCR clonotype mapping with Denaturing Gradient Gel Electrophoresis (DGGE).
- CMV and Herpes simplex serology.
- DNA damage and defence in PBMC, comet assay to detect damage, Ferric Reducing Ability of Plasma (FRAP) to measure antioxidant capacity.
- Mt-DNA damage, PCR methodology to detect and quantify levels of mt-DNA^{4977} and heteroduplex Reference Strand Conformational Analysis (RSCA) to analyse the accumulation of point mutations.
- MHC/peptide tetramers to analyse the number of CMV- and EBV-specific CD8+ cells.

4. Results and discussion

4.1. Immune parameters and mortality

Deterioration of the immune system with ageing is believed to contribute to morbidity in humans due to a greater incidence of infectious, autoimmune disease and cancer (Pawelec et al., 1999). The relationship between immune parameters and mortality in the OCTO immune study has been reported in two studies (Ferguson et al., 1995; Wikby et al., 1998), indicating that no single immune parameter could predict 2-year survival. Multivariate cluster analysis, on the other hand, allowed the identification of a cluster of immune parameters constituting an "immune risk" phenotype (IRP). This was composed of a combination of a low number of CD4 cells, a high number of CD8 cells, poor T cell proliferation and low IL2 production. The IRP predicted subsequent 2-year mortality using immune data both at baseline in 1989 and two years later (Table 2). Whether these findings can be generalised also to other populations will be tested in the broader-defined NONA immune sample, although the IRP was shown to be independent of the individual's health condition (Nilsson et al., 2003).

The relationship between a reduced immune functional response and mortality has also been described in several previous studies. In humans, Murasko et al. (1987) reported poor responses to three T cell mitogens, Con A, phytohemagglutinin and pokeweed, to be associated with increased mortality in old individuals. In a study of octogenarians, Roberts-Thomson et al. (1974) found that anergy was predictive of 2-year mortality. Old subjects that were anergic had a 2-year mortality rate of 80% as compared with 35% in subjects that were not anergic. An association between anergy and morbidity might have confounded these results, however, which was not taken into account. Wayne et al. (1990) reported, on the other hand, that anergy is associated with higher mortality rates even in healthy people older than 60 years.

4.2. Immune parameters and morbidity

The prevalence and incidence of diseases were examined in the NONA immune by a detailed evaluation of health and morbidity information (Nilsson et al., 2003).

Table 2
Chi square analysis of 2-year survival/non-survival in IRP/non-IRP octogenarians

Survival	IRP	Non-IRP	$P <$
1989–1991			
Survivors	5	59	0.001
Non-survivors	9	16	
1991–1993			
Survivors	9	34	0.05
Non-survivors	9	11	

The examination allowed a comparison of findings from the application of a modified SENIEUR protocol (Ligthart et al., 1984) with results using the exclusion criteria of the OCTO immune study (Wikby et al., 1994; Hallgren et al., 1988). The protocols were applied by the use of medical records, self-reports, laboratory data and information about medication usage. The modified SENIEUR protocol excluded more than 90% of the NONA immune sample. The use of the original protocol, suggesting additional laboratory analysis for exclusion, would probably have excluded even more individuals, demonstrating the need for use of broader criteria in studies of the immune system. The OCTO immune criteria excluded almost 65% of the initial sample compared with 48% in the previous OCTO immune study by use of these criteria (Nilsson et al., 2003).

A step-by-step exclusion procedure was applied beginning with the criterion that reduced the sample most, continuing with the next and so on in decreasing order (Nilsson et al., 2003). Applying the five most common exclusion criteria, cardiac insufficiency, medication, laboratory data, urea and malignancy, the modified SENIEUR protocol excluded 120 of the original sample (87%). When the OCTO immune protocol was applied, medication was found to be the most common criterion, excluding 59 (43%), institutionalisation the second, excluding 54 (39%), and cognitive dysfunction the third, excluding 20 (14%).

The application of the protocols allowed us to define three independent subgroups, very healthy (SENIEUR, $n = 13$), moderately healthy (non-SENIEUR/ OCTO immune, $n = 38$), and frail (non-SENIEUR/non-OCTO immune, $n = 87$). Noteworthy, a comparison of the number of T cells across these subgroups indicated no group differences for the IRP (Nilsson et al., 2003), previously identified in octogenarians (Ferguson et al., 1995; Wikby et al., 1998). The IRP might thus serve as a marker of ageing, independent of the individuals' state of health. This is compatible with results in non-inbred mouse populations, showing that clusters of immune markers can predict longevity in old as well as middle-aged individuals independently of the health condition of the animals (Miller, 2001).

4.3. Changes in T-lymphocyte sub-populations

A primary advantage of the longitudinal design is that it enables us to estimate age changes over a specific time period for any specific variable. In the OCTO immune, octogenarians (102 at baseline) were followed for 8 years from T1 in 1989, though T2 1990, T3 1992 and T4 1997. At baseline 14 individuals (14%) with an IRP pattern were identified using cluster analysis. Two years later another 12 individuals (12%) with this immune profile were identified by cluster analysis (Wikby et al., 1998). The new individuals were recruited into the subgroup at-risk due to a significant increase in the CD8+ T lymphocyte levels and simultaneous decreases in the levels of CD4+ cells. Only minor changes in the total number of T cells (CD3+) were found across the 2-year period of time indicating that there were progressive homeostatic changes occurring in the T cell compartment of these octogenarians.

At T4 in the OCTO immune an inverted CD4/CD8 ratio was used to identify individuals at risk (Olsson et al., 2000). The inverted ratio is closely associated with the IRP. The number of individuals in the sample ($n = 23$) was too small at this time to allow the use of multivariate cluster analysis. The CD4/CD8 ratio can also be used in an evaluation of the immune system status in a number of clinical situations

An increasing fraction of individuals with a CD4/CD8 ratio less than 1 was continuously recruited over time. At T1 16%, at T2 17%, at T3 23% and at T4 27% of the individuals had an inverted CD4/CD8 ratio. Of the octogenarians followed longitudinally from T1 in 1989 to T4 in 1997 32% originally had (14% prevalent cases at baseline) or developed (18% incidental cases) an inverted CD4/CD8 ratio (Olsson et al., 2000). All the individuals that once moved into the category with an inverted CD4/CD8, never moved back to normal values. They remained in the IRP category until their death.

At T1 of the NONA immune the fraction of individuals with an inverted CD4/CD8 ratio was about 20% (Fig. 1), which is compatible with the proportions found on different occasions in the OCTO immune (Wikby et al., 2002).

Studies of variations in the CD4/CD8 ratio among healthy adults have indicated that about 5% of subjects have an inverted ratio (Amadori et al., 1995). As much as 57% of the variation was under genetic control. Prevalence of 14–27% at different measurement occasions in the OCTO and NONA studies are therefore too large to be explained only by means of normal genetic distributions of these Swedish populations. The findings of 18% incidental cases across the 8 years in the OCTO support this view. Rather, the results suggest other factors affecting very old individuals and influencing the balance between the CD4 and CD8 T cell subsets. While this change in balance involved significant increases in the CD8+ level with CD4+ level being concurrently deceased, the total T cell level (CD3+) remained unchanged. This suggests the T cell homeostasis model is operating in very old individuals.

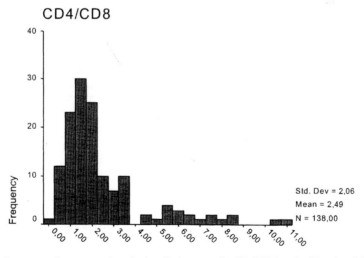

Fig. 1. The frequency of nonagenarians (at baseline) versus the CD4/CD8 ratio. (Reprinted from Wikby et al. (2002), with permission from Elsevier Science.)

According to this model an increase in the number of CD8+ cells is accompanied by a decrease in the number of CD4+ cells keeping the total number of T lymphocytes at a relatively constant level (Mittler et al., 1996). Such a loss in the T cell homeostasis may be associated with age and thymic involution.

4.4. T lymphocyte sub-populations and chronic viral infection

More detailed analysis of presumptive factors affecting the CD4/CD8 balance in octo- and nonagenarians indicated that nutritional disorders, medications, inflammatory diseases or cancer were not associated with IRP (Pawelec, 2000). At T4, 8 years after the initiation of the OCTO immune study, however, an association was found between CD4/CD8 ratio, lymphocyte activation and the level of CMV-IgG antibodies in plasma, indicating evidence of a persistent CMV infection (Olsson et al., 2000). No association with persistent Herpes simplex virus infection was found, similar to the findings of Looney et al. (1999). The results were confirmed and even more evident at baseline in the larger NONA sample (Wikby et al., 2002). The prevalence of CMV-IgG antibodies among octo- and nonagenarians was about 90%, significantly greater than for middle-aged (about 60%). The results also indicated that the very old individuals with an inverted CD4/CD8 ratio are unique, having a CMV prevalence of 100%, while those in the range of 1–4 had a prevalence of 90% and those with ratios greater than 4 only 55% (Wikby et al., 2002). Very old individuals with an inverted CD4/CD8 ratio were also characterised by profound T cell changes apparent in a number of T cell subsets, particularly by increases in the number of CD8+CD57+CD28− and CD8+CD45RA+CD27− cells, indicating T cell activation (Table 3). These changes were comparable with findings by Merino et al. (1998), reporting that the expansion of CD8+ CD57+CD28− T cells in the elderly is dependent both on age and CMV status. Similar results have been found by others (Fagnoni et al., 1996; Nociari et al., 1999),

Table 3
CD3+CD8+CD4−, CD3+CD8−CD4+ and CD8+ lymphocyte subsets (numbers of cells/mm^3) in nonagenarians at baseline categorised by their CD4/CD8 ratios and compared with middle-aged[a]

Subset	Very old with a CD4/CD8		Middle-aged (n = 18)
	Less than 1 (n = 24)	Greater than 1 (n = 114)	
CD3+CD8−CD4+	551 ± 41[b,d]	729 ± 32	863 ± 78
CD3+CD8+CD4−	850 ± 68[c]	338 ± 19	387 ± 56
CD8+CD57+CD28−	494 ± 50[c]	212 ± 14	138 ± 37
CD8+CD45RA+CD27−	462 ± 66[c]	197 ± 12	140 ± 43
CD8+CD45RA+CDRO+	330 ± 41[c]	145 ± 9	132 ± 24
CD8+CD57+CD56+	221 ± 45[c]	105 ± 9	53 ± 14

[a] Reprinted from Wikby et al. (2002), with permission from Elsevier Science.
[b] Mean ± SE.
[c] P < 0.001 compared to other subgroups.
[d] P < 0.01 compared to middle-aged.

demonstrating at a single cell level that these cells proliferate poorly and produce mainly IL4, IL10 and IFN-gamma associated with ageing.

A significant increase in the CD8+ cell compartment in very old individuals with an inverted CD4/CD8 ratio may reflect an attempt to counteract disease problems, such as the CMV infectious process, by generation of clones of protective effector cytotoxic cells. Exposed to a recurrent or persistent viral antigen, like CMV, CD8+ cells lose their CD28 expression with up-regulation of the CD57 cell activation marker. Several authors have reported that clonal expansions in the CD8+CD57+ and CD8+CD28− subsets occur and persist in old individuals (Wang et al., 1995; Schwab et al., 1997; Mugniani et al., 1999; Posnett et al., 1999; Weekes et al., 1999a,b).

Preliminary T cell clonotype mapping, using the DGGE procedure was done in 11 NONA immune individuals by thor Straten et al. (unpublished results) using specific primers covering the TCRBV 1–24 variable regions. The map covers the vast majority of T cells (thor Straten et al., 1998). Significantly greater numbers of distinct bands on the gel, indicating clonal expansion, were obtained in individuals with an inverted CD4/CD8 ratio compared to those with a normal ratio, for whom distinct bands were found to be rare. Interestingly, the number of bands correlated with the number of CD8+CD57+CD28− cells (Wikby et al., 2001).

In vitro studies of the role of the CD28 marker on CD8+ cells have convincingly demonstrated that repeated antigen-induced T cell division leads to a state of T cell senescence with irreversible cell cycle arrest, shortened telomeres, undetectable telomerase, and down-regulation of CD28 expression. These results indicate immune exhaustion (Effros, 1997, 2000). These "senescent" CD8+CD28− T cells, however, show an increased resistance to apoptosis and to retain good functional antigen-specific cytotoxicity (Spaulding et al., 1999).

4.5. The CD3+CD8+ phenotype associated with IRP

In the NONA immune study an extended panel of surface antigen markers was used which enabled simultaneous analysis of several T cell phenotypes. It was found that the predominant phenotypes of the CD3+CD8+ cells, associated with the inverted CD4/CD8 ratio as well as with persistent CMV infection, were CD27−, CD28−, CD56+, CD57+, CD45RA+ and partly double-labelled CD45RA+RO+ cells, representing late differentiation stages of highly overlapping populations (Wikby et al., 2002). The high frequency of the CD45RA+ population is not surprising since the use of the CD45RA/RO marker system seems inadequate to differentiate naive and memory cells. It has been shown that in the differentiation of the CD45RO+ memory marker, cells revert to CD45RA+ (Okumura et al., 1993; Roederer, 1995). The presence of a double-labelled population suggests a transitional double-positive state that may relate to the CD8+ population changes and cell activation.

Recent data suggest that human effector CD8+ express a CD27−CD28− CD45RA+CD57+ phenotype (Pittet et al., 2000). The data also indicated that effector function correlates even better with CD56+ surface expression. Dramatic clonal expansion was confined to this CD56+ subset and found to be associated with CTL effector function. This conclusion was based on findings of high amounts of intracellular perforin and granzyme B (Pittet et al., 2000). It has also been proposed that the majority of these "effector/senescent" cells are the results of chronic activation (Tarazona et al., 2000). In addition, Pittet et al. (2000) reported a reduced lytic capacity of these cells in vivo, which prevents damage to tissue cells by the presence of specific NK receptors capable of inducing inhibitory signals. Baars et al. (2000) showed that these NKT cells express both Ig-super family and C-type lectin classes of NKR. They suggest that binding may cause inhibition of the activation in the cytolytic machinery. Future studies will examine the possibility whether this suppression is important in individuals with a reduced CD4/CD8 ratio.

4.6. MtDNA damage, DNA damage, antioxidant capacity

DNA, mt-DNA damage and antioxidant capacity of T cells have the potential to be significant contributors to the age-related homeostatic CD4/CD8 changes in the T cell subsets and functions found in the OCTO and NONA studies. To address this question, two studies were performed in the NONA immune sample, comparing nonagenarian subjects with the middle-aged control samples (Hyland et al., 2002; Ross et al., 2002).

Hyland et al. (2002) investigated the antioxidant capacity of plasma using the FRAP assay as well as the levels and types of DNA damage using the alkaline comet assay in peripheral blood mononuclear cells (PBMCs). An increase in the levels of oxidative DNA damage was previously demonstrated when T cells grow older in cell culture (Barnett et al., 1999; Hyland et al., 2000). An increased DNA damage accumulation in the T cells will result in T cell cycle arrest and the prevention of T cell replication, contributing to the progression of immunosenescence. Results from NONA immune indicated significantly higher plasma antioxidant capacity in NONA subjects and similar levels of DNA damage of PBMC, as compared with the middle-aged controls (Hyland et al., 2002). This suggests a relationship between longevity and an intact immune function of NONA T cells, underpinned by an elevated antioxidant defence. There was no association, however, between these results and the homeostatic T cell changes previously identified in the OCTO and NONA immune studies. This supports the view that the T cells from NONA subjects, including those with substantial increases in the CD8+, CD27−, CD28−, CD56+, CD57+, CD45RA+ NKT cell phenotype, represent "non-senescent" cells that maintain proper function.

Studies of mitochondrial DNA damage of lymphocytes of the NONA immune study were performed by Ross et al. (2002). The mitochondria play significant roles in apoptosis and energy production processes and any changes in their functions are, therefore, believed to be of importance for the T cell function and, thereby, immunosenescence. A competitive polymerase chain reaction (PCR) methodology was used to evaluate the level of mtDNA^{4977} in addition to a novel heteroduplex RSCA technique to study the accumulation of point mutations with age. The mtDNA^{4977} was detected at very low concentrations in all NONA samples, independently of the individual's age. No accumulation of point mutation was detected.

The low level of mt-DNA damage and absence of age association, support the idea that a vast majority of T cells are still able to replicate rather than being senescent (Ross et al., 2002).

5. Conclusions and future direction

Large clones of CD8 T cells are frequently found with age in both humans and mice. In apparently healthy adults these clonal expansions increase linearly with age (Ricalton et al., 1998). In mice most cells in the clones are in continuous slow division independent of antigenic stimulation (Ku et al., 2001). It has been suggested that the clonally expanded T cells compete with normal CD8 cells, using IL-15 more effectively or resist the inhibitory effects of IL-2, thereby affecting the immune response. Along with these increased clonal expansions, immune senescence has been characterised by thymic involution and a decrease in naïve T cell production resulting in a progression that may compromise the immune capabilities of the elderly. As suggested by Lemaoult et al. (2000), repeated cycles of clonal selection and expansion in a non-renewing T cell population expectedly would be predicted to alter the peripheral lymphocyte population, including its structural integrity and functional responsiveness. Ultimately the consequences are a loss of diversity and immune protective capabilities in the aged. The results of the OCTO and NONA immune studies suggest that the latter, in fact, is occurring selectively and dramatically in the expanded CD8 T cell population(s) described in the individuals with the IRP.

The ongoing NONA longitudinal immune study is examining factors underlying the dysregulation of homeostasis in the peripheral T cell compartments. Establishment of the clonality of the CD8 cells is expected to better characterise the T cell changes and to provide pathways to examine the factors driving this expansion in the IRP individuals. This along with future cytokine profiling will provide important information on the nature of the expanded cells and their functional capabilities. The analyses of relative relationship to pathogens, particularly of CD8+ T cell responses to latent pathogens such as CMV using tetramer staining (Ouyang et al., 2003), should provide supporting evidence that previously controlled factors could drive the observed T cell changes in the increasingly compromised immune system of the elderly.

In addition to cutting edge immunology studies, future research, also, must be multidisciplinary and include provision of more detailed psychosocial and medical clinical evaluation of the affected individuals in comparison to those elderly individuals with lower risk profiles. The immune alterations may only be a reflection of other significant stress-related factors that may be different, but ultimately result in similar immune system changes. Only carefully controlled longitudinal studies which are not selectively exclusive will provide the information necessary to elucidate the mechanism or mechanisms underlying these important age-associated changes, as well as their ultimate health related effects. Omitting this important additional information will result in ineffective practical approaches to not only understanding, but also improving the health and well-being of ageing individuals. The components of the ongoing NONA study have been designed to consider, on a broader basis, potential factors impacting upon the latter.

Acknowledgements

The authors acknowledge the support from the Research Board in the County Council of Jönköping and the Research Council in the Southeast of Sweden (FORSS) for funding these projects. We also acknowledge Länsjukhuset Ryhov for provision of laboratory resources for the completion of these studies. The authors are also indebted to our co-workers Sture Löfgren, Bengt-Olof Nilsson, Jan Ernerudh, Jadwiga Olsson, Jan Strindhall and Per-Eric Evrin for their important contributions to these studies. We particularly would like to thank the nursing staff including Annica Andersson, Inga Boström, Gerd Martinsson, Agneta Carholt, Lene Ahlbäck, Lena Blom, Monica Janeblad and Lena Svensson for their efforts in obtaining the blood samples used. We are also indebted to Roberta Valeski, Florence Confer, Margaret Kensinger, Penn State University, United States, and Andrea Tompa, Gunilla Isaksson, Inger Johansson, Cecilia Ottosson, Helen Olsson, Lisa Stark Jönköping, Sweden, for secretarial and technical assistance. We finally acknowledge our ImAginE collaborators, Graham Pawelec and Qin Ouyang, University of Tubingen, Germany, Yvonne Barnett, Paul Hyland, Owen Ross and colleagues, University of Ulster, Northern Ireland, Rosalyn Forsey, Jonathan Powell and Julie Thompson, Unilever, UK, and Per thor Straten, Danish Cancer Society, Copenhagen, Denmark, for successful co-operation.

References

Amadori, A., Zamarchi, R., DeSilvestro, G. et al., 1995. Genetic control of the CD4/CD8 T-cell ratio in humans. Nat. Med. 1, 531–541.

Baars, P.A., Ribeiro do Couto, L.M., Leusen, J.H.W., Hooibrink, B., Kuijpers, T.W., Lens, S.M.A., van Lier, R.A.W., 2000. Cytolytic mechanisms and expression of activation-regulating receptors on effector-type CD8+CD45RA+CD27− human T cells. J. Immunol. 165, 1910–1917.

Barnett, Y.A., King, C., Bristow-Craig, H. et al., 1999. Age-related increases in DNA damage and mutations in T cells in vivo and in vitro: contributors to alterations in T cell mediated immune responses? In: Pawelec, G. (Ed.), EUCAMBIS: Immunology and Ageing in Europe. IOS Press, Berlin, pp. 54–66.

Effros, R.B., 1997. Loss of CD28 expression on T lymphocytes: a marker of replicative senescence. Dev. Comp. Immunol. 21(6), 471–478.

Effros, R.B., 2000. Costimulatory mechanisms in the elderly. Vaccine 18(16), 1661–1665.

Fagnoni, F.F., Vescovini, R., Mazzola, M. et al., 1996. Expansion of cytotoxic CD8+CD28− T cells in healthy ageing people, including centenarians. Immunology 88(4), 501–507.

Ferguson, F.G., Wikby, A., Maxson, P., Olsson, J., Johansson, B., 1995. Immune parameters in a longitudinal study of a very old population of Swedish people: a comparison between survivors and nonsurvivors. J. Gerontol. Biol. Sci. 50A, B378–B382.

Hallgren, H.M., Berg, N., Rodysill, K.J., O'Leary, J.J., 1988. Lymphocyte proliferative response to PHA and anti-CD3/Ti monoclonal antibodies, T cell surface marker expression, and serum IL-2 receptor levels as biomarkers of age and health. Mech. Ageing Dev. 43, 175–185.

Hyland, P., Duggan, O., Hipkiss, A., Barnett, C., Barnett, Y., 2000. The effects of carnosine on oxidative DNA damage levels and in vitro lifespan in human peripheral derived CD4+ T cell clones. Mech. Ageing Dev. 121, 203–215.

Hyland, P., Duggan, O., Turbitt, J. et al., 2002. Nonagenarians from the Swedish NONA immune study have increased plasma antioxidant capacity and similar levels of DNA damage in peripheral blood mononuclear cells compared to younger control subjects. Exp. Gerontol. 37, 465–473.

Johansson, B., Zarit, S.H., Berg, S., 1992. Changes in cognitive functioning of the oldest old. J. Gerontol. Psychiatry Sci. 47, P75–P80.

Ku, C.C., Kappler, J., Marrack, P., 2001. The growth of the very large CD8+ T cell clones in older mice is controlled by cytokines. J. Immunol. 166(4), 2186–2193.

Lemaoult, J., Messaoudi, I., Manavalan, J.S. et al., 2000. Age-related dysregulation in CD8 T cell homeostasis: kinetics of a diversity loss. J. Immunol. 165(5), 2367–2373.

Ligthart, G.J., Corberand, J.X., Fournier, C. et al., 1984. Admission criteria for immuno-gerontological studies in man: the SENIEUR protocol. Mech. Ageing Dev. 28, 47–55.

Looney, R.J., Falsey, A., Campbell, D. et al., 1999. Role of cytomegalovirus in the T-cell changes seen in elderly individuals. Clin. Immunol. 90, 213–219.

Merino, J., Martinez-Gonzalez, M.A., Rubio, M., Inoges, S., Sanchez-Ibarrola, A., Subira, M.L., 1998. Progressive decrease of CD8 high+CD8+CD57− cells with ageing. Clin. Exp. Immunol. 112: 48–51.

Miller, R.A., 2001. Biomarkers of aging: prediction of longevity by using age-sensitive T-cell subset determinations in a middle-aged genetically heterogeneous mouse population. J. Gerontol. Biol. Sci. 56A, B180–B186.

Mittler, J.E., Levin, B.R., Antia, R., 1996. T-cell homeostasis, competition and drift: AIDS as HIV-accelerated senescence of the immune repertoire. J. Acquir. Immune Defic. Syndr. Hum. Retrovirol. 12, 233–248.

Mugniani, E.N., Egeland, T., Spurkland, A., Brinchmann, J.E., 1999. The T-cell receptor repertoire of CD8+CD28− T-lymphocytes is dominated by expanded clones that persist overtime. Clin. Exp. Immunol. 117(2), 298–303.

Murasko, D.M., Weiner, P., Kaye, D., 1987. Decline in mitogen induced proliferation of lympocytes with increasing age. Clin. Exp. Immunol. 70, 440–448.

Nilsson, B.-O., Ernerudh, J., Johansson, B., Evrin, P.-E., Löfgren, S., Ferguson, F., Wikby, A., 2003. Morbidity does not influence the T-cell immune risk phenotype in the elderly: findings in the Swedish NONA immune study using sample selection protocols. Mech. Ageing Dev. (in press).

Nociari, M.M., Telford, W., Russo, C., 1999. Postthymic development of CD28−CD8+ T cell subset: age-associated expansion and shift from memory to naive phenotype. J. Immunol. 162, 3327–3335.

Okumura, M., Fujii, Y., Takeuchi, Y., Inada, K., Nakahara, K., Matsuda, H., 1993. Age related accumulation of LFA-I[high] cells in a CD8+CD45RA[high] T cell population. Eur. J. Immunol. 23, 1057–1063.

Olsson, J., Wikby, A., Johansson, B., Löfgren, S., Nilsson, B.-O., Ferguson, F.G., 2000. Age-related change in peripheral blood T-lymphocyte subpopulations and cytomegalovirus infection in the very old: the Swedish longitudinal OCTO immune study. Mech. Ageing Dev. 121, 187–201.

Ouyang, Q., Wagner, W., Wikby, A., et al., 2003. Large numbers of dysfunctional CD8+ T lymphocytes bearing receptors for a single dominant CMV epitope in the very old. J. Clin. Immunol. (submitted).

Pawelec, G., Effros, R.B., Caruso, C., Remarque, E., Barnett, Y., Solana, R., 1999. T cells and aging. Front. Biosci. 1(4), D216–D269.

Pawelec, G., 2000. Meeting report: first conference of the EU-supported thematic network on immunology and ageing in Europe (ImAginE), Schloss Hohentubingen, April, 2000. Exp. Gerontol. 35, 1095–1103.

Pawelec, G., Ferguson, F.G., Wikby, A., 2001. The SENIEUR protocol after 16 years. Mech. Ageing Dev. 122(2), 132–134.

Pittet, M.J., Speiser, D.E., Valmori, D., Cerottini, J.C., Romero, P., 2000. Cutting edge: cytolytic effector function in human circulating CD8+ T cells closely correlates with CD56 surface expression. J. Immunol. 164(3), 1148–1152.

Posnett, D.N., Edinger, J.W., Manavalan, J.S., Irwin, C., Marodon, G., 1999. Differentiation of human CD8 T cells: implications for *in vivo* persistence of CD8+CD28− cytotoxic effector clones. Int. Immunol. 11(2), 229–241.

Ricalton, N.S., Roberton, C., Norris, J.M., Rewers, M., Hamman, R.F., Kotzin, B.L., 1998. Prevalence of CD8+ T cell expansion in relation to age in healthy individuals. J. Gerontol. A. Biol. Sci. Med. Sci. 53(3), B196–B203.

Roberts-Thomson, I.C., Wittingham, S., Youngchaiyud, U., Mackay, I.R., 1974. Ageing, immune response and mortality. Lancet 2(7877), 368–370.

Roederer, M., 1995. T-cell dynamics of immunodeficiency. Nat. Med. 1(7), 621–622.

Ross, O.A., Hyland, P., Curran, M.D. et al., 2002. Mitochondrial DNA damage in lymphocytes: a role in immunosenescence? Exp. Gerontol. 37, 329–340.

Schwab, R., Szabo, P., Manavalan, J.S. et al., 1997. Expanded CD4$^+$ and CD8$^+$ T cell clones in elderly humans. J. Immunol. 158, 4493.

Spaulding, C., Guo, W., Effros, R.B., 1999. Resistance to apoptosis in human CD8+ T cells that reach replicative senescence after multiple rounds of antigen-specific proliferation. Exp. Gerontol. 34(5), 633–644.

thor Straten, P., Barfoed, A., Seremet, T., Saeterdal, I., Zeuthen, J., Guldberg, P., 1998. Detection and characterisation of alpha-beta-T-cell clonality by denaturing gradient gel electrophoresis (DGGE). Biotechniques 25(2), 244–250.

Tarazona, R., Delarosa, O., Alonso, C., Ostos, B., Espejo, J., Pena, J., Solana, R., 2000. Increased expression of NK cell markers on T lymphocytes in aging and chronic activation of the immune system reflects the accumulation of effector/senescent T cells. Mech. Ageing Dev. 121(1–3), 77–88.

Wang, E.C.Y., Moss, P.A.H., Frodsham, P., Lehner, P.J., Bell, J.I., Borysiewicz, L.K., 1995. CD8^{high}CD57$^+$ T lymphocytes in normal, healthy individuals are oligoclonal and respond to human cytomegalovirus. J. Immunol. 155, 5046–5056.

Wayne, S.J., Rhyne, R.L., Garry, P.J., Goodwin, J.S., 1990. Cell mediated immunity as a predictor of morbidity and mortality in subjects over 60. J. Gerontol. Med. Sci. 45, M45–M48.

Weekes, M.P., Wills, M.R., Mynard, K., Carmichael, A.J., Sissons, J.G.P., 1999a. The memory cytotoxic T lymphocyte (CTL) response to human cytomegalovirus infection contains individual peptide-specific CTL clones that have undergone extensive expansion in vivo. J. Virol. 73(3), 2099–2108.

Weekes, M.P., Carmichael, A.J., Wills, M.R., Mynard, K., Sissons, J.G.P., 1999b. Human CD28−CD8+ T-cells contain greatly expanded functional virus-specific memory CTL clones. J. Immunol. 162, 7569–7577.

Wikby, A., Johansson, B., Ferguson, F., Olsson, J., 1994. Age-related changes in immune parameters in a very old population of Swedish people: a longitudinal study. Exp. Gerontol. 29(5), 531–541.

Wikby, A., Maxson, P., Olsson, J., Johansson, B., Ferguson, F.G., 1998. Changes in CD8 and CD4 lymphocyte subsets, T cell proliferation responses and non-survival in the very old: the Swedish longitudinal OCTO-immune study. Mech. Ageing Dev. 102, 187–198.

Wikby, A., Strindhall, J., Johansson, B., Löfgren, S., Ferguson, F., 2001. Age changes in T cell homeostasis associated with cytomegalovirus infection: the Swedish longitudinal NONA immune study. Presented at Immunology and Ageing in Europe, 2nd Conference on basic biology and clinical impact of immunosenescence, Cordoba, Spain, 22–26 March, 2001.

Wikby, A., Johansson, B., Olsson, J., Löfgren, S., Nilsson, B.-O., Ferguson, F., 2002. Expansions of peripheral blood CD8 T-lymphocyte subpopulations and an association with cytomegalovirus seropositivity in the elderly: the Swedish NONA immune study. Exp. Gerontol. 37, 445–453.

Advances in
Cell Aging and
Gerontology

Immune measures which predict 9-year survival in an elderly population sample

Felicia A. Huppert[a], Eleanor M. Pinto[b], Kevin Morgan[c],
MRC CFAS and Carol Brayne[d]

[a]*Department of Psychiatry, University of Cambridge, Cambridge, UK*
[b]*Centre for Applied Medical Statistics, University of Cambridge, Cambridge, UK*
[c]*Department of Human Sciences, Loughborough University, Leicestershire, UK*
[d]*Department of Public Health and Primary Care, University of Cambridge, Cambridge, UK*
Correspondence address: Tel.: +44-1223-330-334; fax: +44-1223-330-330.
E-mail: carol.brayne@medschl.cam.ac.uk

Contents

Abbreviations

MRC CFAS: Medical Research Council Cognitive Function and Ageing Study; MMSE: Mini-Mental State Examination; ONS: Office of National Statistics; HR: hazard ratio; CI: confidence interval.

1. Introduction

The immune system has evolved as a highly complex and sophisticated network of defences to facilitate survival in a hostile environment. From this we can make two important inferences: first that individuals who survive into advanced old age should have an immune system which is functioning well, and second that among older people in general, how well the immune system is functioning should be a predictor of survival. In view of the sizeable differences in male and female longevity, it would also be expected that the immune system in general functions better in females than in males.

Recent years have seen a great deal of research activity into immunosenescence. Studies analysing changes in T-cell subsets in peripheral blood show age-related changes in T-cell ratios and their ability to proliferate (see Ginaldi et al., 1999 for a review). Studies of healthy centenarians, who are the finest examples of successful ageing, show no change in subset ratios and, furthermore, no loss in the proliferative vigour of T-cells. There appears to be a decrease in the absolute number of T-cells in the peripheral blood of these exceptional individuals (Franceschi et al., 1995; Sansoni et al., 1997).

Such findings suggest that age-related changes in immune function, which are often reported, may be the consequence of age-related disease. However, most studies use very selected samples of individuals who are either extremely healthy (Ligthart et al., 1990) or hospital-based (Lehtonen et al., 1990). To avoid the possible biases inherent in using selected groups, data need to be obtained from population-based studies, using random sampling methods. In a cross-sectional analysis of the cohort considered in this paper, age differences were found on all the lymphocyte measures examined, and the age-related decrease was statistically significant for CD3, CD4, CD8, CD19 and total lymphocytes (Huppert et al., 1998). The decrease was most pronounced for CD3, with individuals aged 64–69 years having a mean count of 1.55 per 10^{-9} L (standard error 0.003) compared to a mean of 1.07 amongst those aged 80 years and over (standard error 0.007). However, no significant differences in lymphocyte total or subsets were found when the subgroup that reported some health problems was compared with the much smaller group ($n = 99$) who reported no health problems and took no medication. Those who reported no health problems tended to have slightly lower counts than the reminder; the strongest association was for CD3 where absolute counts per 10^{-9} L were 0.12 lower amongst those reporting no health problems (95% CI 0.25 lower to 0.011 higher, $P = 0.07$) (Huppert et al., 1998). The only individuals not included in this study were those who were cognitively impaired or physically frail and dependent on others. The sample can therefore be regarded as reasonably representative of non-demented and independent older adults (aged 65+ years). The same study showed the predicted gender differences in immune markers, which would be expected if immune function plays a part in gender differential survival. On all measures, women had higher values than men, significantly so for CD3, CD4, CD19 and the CD4:CD8 ratio. The means and standard errors per 10^{-9} L were 1.41 (0.034) for men vs. 1.52 (0.036) for women on CD3, 0.95 (0.023) vs. 1.06 (0.027) on CD4, 0.22 (0.0089) vs. 0.26 (0.0086) on CD19 and 1.80 (0.072) vs. 2.03 (0.070) for the CD4:CD8 ratio. An earlier population-based study of 266 healthy non-smoking adults also reported a higher percentage of CD4 cells and a higher CD4:CD8 ratio in women compared with men (Tollerud et al., 1989).

The relationship between T-cells and survival has been examined in a number of longitudinal studies. In a study of 102 very old individuals (86–92 years at baseline), it was found that the 27 who died within 2 years were characterised at baseline by low CD4 and CD19 percentages and a high CD8 percentage (Ferguson et al., 1995). Further follow-up (Wikby et al., 1998; Olsson et al., 2000) confirmed that the combination of low CD4 and high CD8, leading to an inverted CD4:CD8 ratio, was predictive of death over the 8-year follow-up.

ııı ıııııeı ııı eⱴpıⱱıⱳ ıⱨe ⱳⱥııııⱳⱥⱨⱨıⱤ ⱨⱥⱤⱨⱨⱥⱨ ıⱨⱨⱨⱨⱨ ⱨⱷⱨⱷⱨⱷ and ⱨⱨⱤⱨⱨⱨⱨ ın a representative population, we sampled over 1000 individuals aged 65 years and older in two centres in the UK. These centres represent an urban and a rural population and have different lifestyles and life expectancies. This study was part of a larger study which examined dementia and cognitive decline (Medical Research Council Cognitive Function and Ageing Study, MRC CFAS). Based on this study, we have already published normative values for lymphocyte subsets as described above (Huppert et al., 1998). We now report on the relationship of immune markers to survival.

2. Methods

2.1. Sample for interview and venepuncture

The sample was taken from the Healthy Ageing Study, which has examined 2041 elderly community residents participating in the Cambridge and Nottingham centres of the MRC CFAS. The study began in 1991 and has been described in detail elsewhere (MRC CFAS, 1998). Briefly, population samples in each area were selected from Family Health Services Authorities, who hold the general practice sample lists for remuneration purposes. General practices in the UK include cover the population in a comprehensive manner, and have almost 100% coverage including institutions. This frame was used to randomly select individuals aged 65 years and over, stratified by age, to take part in the parent study. In each centre a sufficient number was approached to achieve an interviewed sample of 2500 with half aged 65–74 and half aged 75 and over. A screening interview was administered which, among other measures, collected information on cognitive function and dependency. Individuals with physical frailty and/or cognitive impairment were selected for another study; for the Healthy Ageing Study we sampled from those who did not have these problems, until a sample size of at least 1000 was achieved in each of the two participating centres. Individuals were asked whether they would be willing to provide a blood sample, and were recruited until at least 500 had been collected in each centre. Funding limitation precluded taking blood samples from all participants. The outcome measure for the present study is survival time from recruitment (between June 1991 and December 1992) up to 31/12/2000, derived from death certificates from the Office of National Statistics (ONS). The follow-up time varies from 3 to 115 months, with a total of 7624 person-years of observation.

2.2. Blood samples

Individuals who consented to giving a blood sample were visited at home by a nurse or phlebotomist, a median of 17 weeks after the screening interview. At the time of venepuncture, current health and medication were again recorded.

The nurse visited respondents between 09:30 and 15:10h in order that blood samples could be assayed within 6h (fresh K2 EDTA peripheral blood samples). The blood measures obtained for both Cambridge and Nottingham were total white blood cells, neutrophils, monocytes, eosinophils, basophils and lymphocytes. In addition, CD3, CD4, CD8, CD19 and CD57 counts were measured for Cambridge, from which the CD4:CD8 ratio was calculated. All samples were subjected to full blood count (FBC) with Coulter ST KS (Coulter Electronics Ltd., Luton). FBC results falling outside normal ranges were examined microscopically and followed up as appropriate.

For the Cambridge group, measurements were made of CD19 (B cells), CD3 (functional T cells), CD4 (helper T cells), CD8 (suppressor T cells), and CD57 (preferentially expressed on NK cells). Dual-colour immunoflourescence was carried out using a panel of monoclonal antibodies as follows: CD3 Flourescien Isothiocyanate (FITC)/CD4 Phycoerythrin (PE), CD3 FITC/CD8 PE, CD5 FITC/CD19 PE, and dual-colour FITC/PE isotype-matched control (Dako Ltd., UK), CD57 FITC (Immunotech/The Binding Site Ltd., Birmingham, UK), and FacsLyse reagent (Becton-Dickinson, Oxford, UK).

Immunophenotyping was carried out using a Direct Immunoflourescence technique. Fifty microlitres of well-mixed whole blood was added to 10 µL of monoclonal antibody. The mixture was incubated at room temperature, in the dark for 20 min. The samples were then treated with FacsLyse for 5 min to lyse the red blood cells (RBC). The samples were centrifuged at 1500 rpm for 5 min and the supernatant was discarded. The samples received two more washes with PBS/0.5% albumin. The cell platelets were resuspended in 400 µL of 0.5% formaldehyde fixative solution, and stored at 4 °C in the dark until analysed.

Flow cytometry was carried out with a Coulter Epics Elite (Coulter Electronics Ltd., Luton, UK) flow cytometer that has an air-cooled 488 mm Argon Laser. The instrument was set up and aligned daily using DNA-Check and Standard-Brite fluorospheres (Coulter Electronics Ltd.). The samples were analysed using the Coulter software, 5000.

Lymphocyte subsets were counted and the listmode data were saved. Lymphocytes were gated on side scatter (SSC) and forward scatter (FSC) properties. Each patient's sample was checked individually for optimum instrument setting and minor adjustments made as necessary. The marked settings were displayed as a dual-parametric histogram, and the results expressed as percent positive. The percent positive results were converted into absolute counts using the automated lymphocyte count obtained from the Coulter ST KS and expressed as $\times 10^9/L$.

2.3. Statistical analysis

Socio-demographic variables used in the analyses are based on data collected at the baseline screening interview. Within each centre each blood measure was categorised into tertiles in order to assess its association with survival, since measurements were not standardised between centres. Results are reported as hazard ratios (HRs) and 95% CI for the second and third tertiles compared to the first, and are derived from proportional hazards regression analysis. The effect of centre (Cambridge vs. Nottingham), and interactions between centre and immune measures, were assessed using the likelihood ratio statistics of the relevant models. All *p*-values for the relationship of an immune measure to survival relate to the Wald statistic of the relevant covariate in the regression model. In cases where there appeared to be a trend in hazard across the tertiles, this was quantified by forcing the HR for the third tertile vs. the first to be the square of that for the second tertile vs. the first. The resulting change in fit of the model was assessed by comparing the likelihood ratio statistic for this restricted model to that for the model allowing unrestricted hazards in the second and third tertiles. The proportional hazards assumption was tested by examining the scaled Schoenfeld residuals (Grambsch and Therneau, 1994). Survivor curves were derived from the Kaplan–Meier estimator.

Analyses were initially unadjusted. Subsequent adjustment for known predictors of survival was carried out through a sequence of nested models, as follows:

1. Adjusting for age and sex. Age at baseline is included in the model as a continuous covariate, and an age group variable (< 75 or ≥ 75) was added to allow for a non-linear effect of age on log hazard.
2. Adding smoking (never/past/current).
3. Adding chronic disease: indicator variable for presence of the chronic diseases most commonly associated with death, i.e. vascular disease (diagnosed angina, diagnosed intermittent claudication, self-reported heart attacks and strokes) and respiratory disease (self-reported asthma or bronchitis, excluding conditions limited to childhood).
4. Adding cognitive function: Mini-Mental State Examination (MMSE) score, since this has been shown to predict survival independently of other variables (categorised into ≤ 27 or > 27) (Neale et al., 2001; Dewey and Saz, 2001).

We did not adjust for education, as it was neither significantly associated with survival in this sample, nor in the MRC CFAS sample as a whole (Neale et al., 2001).

In cases where the HR of interest appeared to be time-dependent, this association was further explored by undertaking separate survival analyses for years 0–2, 3–5 and 6–8. Each time interval included only those individuals who were alive at the beginning of the interval, and those who survived beyond the end of the interval were censored at the end of the interval. There were roughly equal numbers of deaths in each 3-year time period.

A Bonferroni correction was applied in order to maintain the study's type I error rate at 0.05, resulting in a *P*-value of 0.0042 ($= 0.05/12$) or smaller being considered to indicate statistical significance. Whilst there were many more than 12 hypothesis tests, many of these were associated and hence we correct for the number of immune measures rather than the number of tests.

There were only small amounts of missing data for most of the blood measures, with the exception of CD3 counts, which were not available for 84 of the 513 participants. While this is a high percentage of missing data, the indicator variable specifying whether or not the CD3 measurement was missing did not predict survival.

3. Results

In total, blood samples were obtained from 1046 individuals, 513 from Cambridge and 533 from Nottingham. Over the course of the 8 to 9-year follow-up, there were 395 deaths, 189 in Cambridge and 206 in Nottingham (Table 1). There were no important differences in survival of these samples between the two centres.

The effects of the immune measures on survival in the two centres were similar, and the results are reported for the two centres combined. Individuals in the top tertile of white blood cells had an 81% greater hazard of death than those the bottom tertile (95% CI 41–132%). Similarly, when compared to the bottom tertile, the top tertile of monocyte count was associated with an 84% increase in hazard (95% CI 43–136%), and the top tertile of neutrophil count carried a 135% greater hazard (95% CI 81–204%). A test for trend across tertiles suggested an increase in hazard of 53% (95% CI 35–73%) per neutrophil tertile, and did not lead to important loss of fit in the model ($P = 0.75$). There was an increase of 36% in hazard (95% CI; 20–55%) per monocyte tertile; again, this did not lead to loss of fit ($P = 1.00$). The results for white blood cells were almost identical to those for monocytes (36% increase in hazard per white blood cell tertile, 95% CI 20–54%), with no significant loss of fit ($P = 0.061$).

These effects diminished on adjusting for age and sex and although their significance was reduced, it was nonetheless retained. White blood cells, neutrophils and monocytes had respective hazard increases of 32% (95% CI 16–50%), 41% (95% CI 24–61%) and 23% (95% CI 9–40%) per tertile. The effects of tertiles 2 and 3 relative to the first are shown in Table 2. Additional adjustment for smoking brought some of the HRs closer to unity, but the changes were small and did not affect the significance of the immune measures (results not shown).

On adjusting for age, sex, chronic disease, smoking and cognition, the comparisons of the top tertiles of white blood cells and neutrophils to the bottom tertiles remained significant, and were associated with 51% (95% CI 17–94%) and 70% (95% CI 30–122%) increases in hazard, respectively. The hazard for neutrophils increased by 31% (95% CI 15–49%) per tertile, whilst the hazard for white blood cells increased by 25% (95% CI 10–42%) per tertile. The effect of monocytes both in terms of the top tertile relative to the bottom tertile and as a trend became non-significant on including chronic disease. The fully adjusted effect of monocyte count was a 38% increase (95% CI 7–78%) for the top tertile compared to the bottom, or an increase in hazard of 18% (95% CI 4–35%) per tertile.

Table 1
Survivors and non-survivors by age group, sex and centre

Sex and age		Men < 75	Men ≥ 75	Women < 75	Women ≥ 75	Total
Cambridge	Alive	102	33	134	55	324
	Dead	56 (35%)	60 (65%)	22 (14%)	51 (48%)	189 (37%)
Nottingham	Alive	91	46	125	65	327
	Dead	45 (33%)	81 (64%)	29 (19%)	51 (44%)	206 (39%)

Table 2
Hazard Ratios and 95% Confidence Intervals[a]

Variable (sample size)	Unadjusted		Adjusted for age and sex		Fully adjusted[b]	
	HR	(95% CI)	HR	(95% CI)	HR	95% CI
White blood cells (n = 993)	1.09	(0.83, 1.42)	1.00	(0.76, 1.31)	0.96	(0.73, 1.23)
	1.81	(1.41, 2.32)	1.68	(1.31, 2.15)	1.51	(1.17, 1.94)
Neutrophils (n = 1014)	1.58	(1.20, 2.07)	1.36	(1.03, 1.78)	1.27	(0.96, 1.67)
	2.35	(1.81, 3.04)	1.98	(1.52, 2.57)	1.70	(1.30, 2.22)
Lymphocytes (n = 1014)	0.74	(0.58, 0.95)	0.78	(0.61, 1.00)	0.77	(0.60, 0.98)
	0.72	(0.56, 0.92)	0.90	(0.70, 1.15)	0.87	(0.62, 1.11)
Monocytes (n = 1014)	1.35	(1.04, 1.75)	1.18	(0.91, 1.54)	1.06	(0.81, 1.38)
	1.84	(1.43, 2.36)	1.51	(1.17, 1.95)	1.38	(1.07, 1.78)
Eosinophils (n = 1008)	0.84	(0.65, 1.08)	0.86	(0.67, 1.11)	0.82	(0.64, 1.05)
	1.06	(0.84, 1.35)	0.90	(0.71, 1.15)	0.83	(0.65, 1.05)
Basophils (n = 993)	0.83	(0.65, 1.06)	0.86	(0.67, 1.10)	0.86	(0.67, 1.09)
	0.82	(0.64, 1.05)	0.86	(0.67, 1.10)	0.84	(0.66, 1.08)
CD3[c] (n = 429)	0.52	(0.35, 0.75)	0.60	(0.41, 0.88)	0.55	(0.38, 0.81)
	0.55	(0.37, 0.81)	0.73	(0.49, 1.09)	0.62	(0.41, 0.93)
CD4[c] (n = 497)	0.65	(0.46, 0.92)	0.74	(0.52, 1.05)	0.63	(0.44, 0.90)
	0.59	(0.41, 0.84)	0.75	(0.52, 1.08)	0.68	(0.47, 0.97)
CD8[c] (n = 499)	0.97	(0.68, 1.38)	1.01	(0.71, 1.44)	0.97	(0.68, 1.40)
	0.94	(0.66, 1.34)	0.99	(0.69, 1.42)	0.94	(0.65, 1.35)
CD19[c] (n = 499)	0.54	(0.38, 0.76)	0.61	(0.43, 0.88)	0.66	(0.46, 0.94)
	0.47	(0.32, 0.67)	0.62	(0.43, 0.89)	0.69	(0.48, 1.00)
CD4:8[c] (n = 496)	0.79	(0.56, 1.13)	0.85	(0.60, 1.21)	0.78	(0.55, 1.11)
	0.77	(0.54, 1.10)	0.89	(0.62, 1.27)	0.87	(0.61, 1.25)
CD57[c] (n = 487)	0.85	(0.58, 1.23)	0.95	(0.65, 1.38)	0.94	(0.65, 1.38)
	1.05	(0.74, 1.48)	1.03	(0.73, 1.46)	0.97	(0.68, 1.38)

[a] Comparison between tertiles (middle tertile vs. top and top vs. bottom).
[b] Adjusted for age, sex, smoking, chronic disease and MMSE score.
[c] Data available for Cambridge only (513 individuals in total).

3.1. Lymphocyte subpopulations

Measures of total lymphocytes, CD3, CD4 and CD8 were all very closely related to one another, with Spearman correlations above 0.7, and the correlation between total lymphocytes and CD3 was 0.9. In the absence of any adjustment CD3, CD4 and CD19 were significant predictors of survival, the lowest tertile of each being associated with poor survival while the top two tertiles carried similar prognoses. Compared to the lowest tertile, the middle tertile of CD3 carried a 48% decrease in hazard (95% CI 25–52%), while the middle tertile of CD19 carried a 46% decrease in hazard (95% CI 24–62%). CD4 showed similar variation, with the top tertile carrying a 41% lower hazard than the bottom tertile (95% CI 16–39%).

On adjusting for age and sex, these values decreased slightly, and all became non-significant at the 0.0042 level: the middle tertiles compared to the lowest tertiles were associated with a 40% decrease in hazard (95% CI 12–59%) for CD3, a 26% decrease (95% CI 48% decrease to 5% increase) for CD4, and a 39% decrease (95% CI 12–57%) for CD19. On adjusting for chronic disease, the effects of CD3 and CD4 became more pronounced than the age and sex adjusted effects, and the middle tertile of CD3 regained its significance: the fully adjusted effect (relative to the lowest tertile) was a 45% decrease (95% CI 19–62%). CD8 and CD57 showed no significant association with survival. The top tertile of the CD4 : CD8 ratio had 23% smaller hazard than the bottom tertile, but this was not significant (95% CI 46% lower to 10% higher).

The hazards for several of the CD counts (CD3, CD8, CD57) showed evidence of departures from the proportional hazards model; that is, the HR appeared to vary over time. This was investigated as described in Section 2.3 above. Although the effects of CD8 and CD57 were time-dependent, the actual HRs involved were small and not significantly associated with survival in any of the three time periods examined. CD3 was a very strong predictor of survival in the first two 3-year periods, but this effect did not appear to carry over beyond 6 years. In the unadjusted model, the second tertile of CD3 carried a 74% lower hazard than the first tertile (95% CI 42–88%) in the first 3 years, a 76% lower hazard (95% CI 49–89%) in the following 3 years and a 62% greater hazard (95% CI 17% lower to 219% higher) in the final 3 years. On adjusting for age and sex, the effects were weaker but still significant: the hazard for the middle tertile compared to the first was 69% lower (95% CI 32–88%) in years 0–2, 73% lower (95% CI 43–87%) in years 3–5 and 91% higher (95% CI; 3% lower to 278% higher) in years 6–8. The 95% CI were wide but even the 99.6% CI (corresponding to 0.4% significance, see statistical methods) excluded unity in the first two time periods. There were too few deaths in each time period to adjust reliably for smoking, chronic disease and cognition.

The Kaplan–Meier plot in Fig. 1 shows that survival in the second and third tertiles of CD3 reached the same value by year 8, despite having diverged in years 3–6. Figure 2 is a complementary log–log plot of the survivor curves; if the hazards were constant over time the three lines would be parallel.

4. Discussion

This study has examined whether immune status predicts 9-year survival in older adults. Notable features of the study design are: (a) the use of a representative sample of community-dwelling elders, excluding only those who screened positive for probable dementia or physical frailty (i.e. were dependent on the help of others to manage the activities of daily living); (b) the participation of individuals from two distinct geographical locations (rural Cambridgeshire, urban Nottingham) to establish the consistency of the findings; (c) the large sample size ($n = 1046$ for white blood cell measures; $n = 513$ for lymphocyte subsets); and (d) a follow-up interval at least as long as any currently published (8–9 years).

Kaplan-Meier Survival Curves for Tertiles of CD3

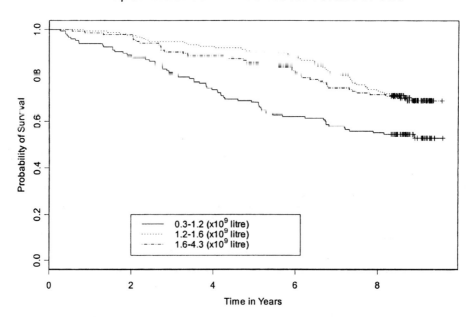

Fig. 1

Complementary Log-Log Plot for Tertiles of CD3

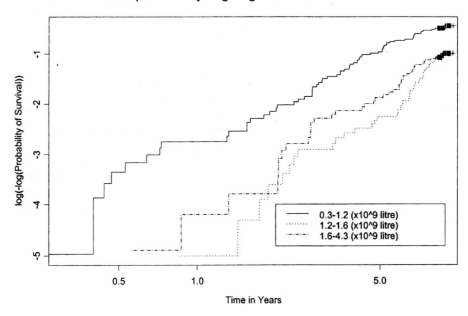

Fig. 2

Studies of immune function in elderly people, particularly studies with a longitudinal component, are often undertaken on small samples. The combination of small sample size and the large number of immune measures typically being investigated may lead to results that cannot be replicated. For this reason, we examined a large sample of elderly people and used a high level of statistical significance to avoid spurious results. Although Bonferroni corrections have been criticised (Perneger, 1998), since our hypothesis was very general we felt that the traditional 5% significance level (itself arbitrary) was not sufficiently stringent. We have, however, presented 95% CI for HRs, as a range of plausible effect sizes for a single measure may be more useful than a simultaneous confidence region for the set of 12 HRs.

The principal findings were that white blood cells, neutrophils and monocytes were significant predictors of 9-year survival in the unadjusted model. In all cases, high values were associated with an increased risk of death while low values were protective. In the fully adjusted model (adjusted for age, sex, chronic disease, smoking and cognitive function), white blood cells and neutrophils remained significant predictors, whilst monocytes became non-significant. Individuals with high neutrophil counts had a 70% greater hazard compared to those with low neutrophil counts.

Examination of lymphocyte subsets (Cambridge only, $n = 513$) showed that CD3, CD4 and CD19 were significant predictors of 9-year survival in the unadjusted model. CD3 remained significant in the fully adjusted model while CD4 and CD19 became non-significant. We found no association between survival and CD8 or CD57, and estimated that those in the top tertile of the CD4:CD8 ratio had a 23% decrease in hazard compared to those in the bottom tertile. However, our confidence interval was wide (46% decrease to 10% increase) and the association did not achieve significance even at the 5% level. This contrasts with previous studies (Ferguson et al., 1995; Wikby et al., 1998) and may be due to differences in sampling frames. We are further investigating inversion of the CD4:CD8 ratio in this sample and Huppert et al. (2003) report on it separately.

Nine years is a long time period over which to expect a relationship between baseline immune measures and survival. Nevertheless, for most measures we found no significant evidence that the effects varied over time, the one exception being CD3, for which the second tertile carried a 70% smaller hazard than the first in years 0–6, but had a non-significantly greater hazard (90% greater) in years 7–9.

Many studies in the field of immunosenescence are undertaken using the SENIEUR protocol (Ligthart et al., 1990), which excludes individuals who are not in excellent health. This procedure may be appropriate when the aim of the study is to examine the optimal state of the immune system in old age. However, for other purposes, such as examining predictors of survival, we believe it is important to investigate representative samples of the elderly. A strength of the present study is that it is based on a representative population sample (aged 65+ years) and excludes only those individuals who have probable dementia or who are physically frail. Otherwise it is a typical sample of community-dwelling older people, with health problems and medication use which are common in this age group. In an earlier study (Huppert et al., 1998) we found no significant differences in lymphocyte subpopulations between members of the sample whose health was excellent and who were taking no medication (around 20% of the sample) compared with the remainder of the sample. Nevertheless, in the present study, our findings were adjusted for common disorders and smoking behaviour, amongst other potential confounders.

Lymphocyte subpopulations that show no evidence of an association with survival were CD8 (T suppressor cells), CD57 (natural killer cells) and the CD4:CD8 ratio. The absence of a relationship between NK cells and survival is consistent with the absence of age differences in NK cell numbers (Huppert et al., 1998) or NK cell function (Ligthart et al., 1989). Likewise the absence of an association between CD8 and survival is consistent with the failure to find cross-sectional age differences in CD8 numbers in either a typical or a very healthy population sample (Huppert et al., 1998). Further studies are required for a detailed understanding of the nature of these relationships and the processes that determine their association (or lack of association) with age and survival.

Acknowledgements

This study was a bolt-on to the MRC CFA Study, which is funded by the Medical Research Council and the Department of Health. It was supported by the Economic and Social Research Council. We would like to thank Valerie Jackson for her work in the administration of this study, Sheila O'Connor for blood measurements, Pamela Sussams for study co-ordination and Wendy Solomou for database management. We are grateful to the general practitioners, their staff, the interviewers and nurses, and particularly the respondents for their support of the study.

References

Dewey, M.E., Saz, P., 2001. Dementia, cognitive impairment and mortality in persons aged 65 and over living in the community: a systematic review of the literature. International Journal of Geriatric Psychiatry, 16, 751–761.

Ferguson, F.G., Wikby, A., Maxson, P., Olsson, J., Johansson, B., 1995. Immune parameters in a longitudinal study of a very old population of Swedish people: a comparison between survivors and nonsurvivors. Journal of Gerontology. A Biological Science and Medical Science, 50(6), B378–B382.

Franceschi, C., Monti, D., Sansoni, P., Cossarizza, A., 1995. The immunology of exceptional individuals: the lesson of centenarians (see comments). Immunology Today, 16, 12–16.

Ginaldi, L., De Martinis, M., D'Ostilio, A., Marini, L., Loreto, M.F., Quaglino, D., 1999. Immunological changes in the elderly. Aging (Milano), Oct. 11(5), 281–286.

Grambsch, P., Therneau, T., 1994. Proportional hazards tests and diagnostics based on weighted residuals. Biometrika, 81, 515–526.

Huppert, F.A., Solomou, W., O'Connor, S., Morgan, K., Sussams, P., Brayne, C., 1998. Aging and lymphocyte subpopulation: whole blood analysis of immune makers in a large population of elderly individuals. Experimental Gerontology, 33(6), 593–600.

Huppert, F.A., Pinto, E.M., Morgan, K., MRC CFAS, Brayne, C., 2003. Survival in a population sample is predicted by proportions of lymphocyte subsets, in press with Mechanisms of Ageing and Development, editors Pawelec, G. and Caruso, C., Special issue on immunosenescence.

Lehtonen, L., Eskola, J., Vainio, O., Lehtonen, A., 1990. Changes in lymphocyte subsets and immune competence in very advanced age. Journal of Gerontology, 45(3), M108–M112.

Ligthart, G.J., Schuit, H.R., Hijmans, W., 1989. Natural killer cell function is not diminished in the healthy aged and is proportional to the number of NK cells in the peripheral blood. Immunology, 68(3), 396–402.

Ligthart, G.J., Corberand, J.X., Geertzen, H.G., Meinders, A.E., Knook, D.L., Hijmans, W., 1990. Necessity of the assessment of health status in human immunogerontological studies: evaluation of the SENIEUR protocol. Mechanisms of Ageing and Development, 55(1), 89–105.

MRC CFAS, 1998. Cognitive function and dementia in six areas of England and Wales: the distribution of MMSE and prevalence of GMS organicity level in the MRC-CFA Study. Psychological Medicine 28, 319–335.

Neale, R., Brayne, C., Johnson, A.L., 2001. Cognition and survival: an exploration in a large multicentre study of the population aged 65 years and over. International Journal of Epidemiology, 30(6), 1383–1388.

Olsson, J., Wikby, A., Johansson, B., Lofgren, S., Nielsson, B.O., Ferguson, F.G., 2000. Age-related change in peripheral blood T-lymphocyte subpopulations and cytomegalovirus infection in the very old: the Swedish longitundinal OCTO immune study. Mechanisms of Ageing and Development, 121(1–3), 187–201.

Perneger, T.V., 1998. What's wrong with Bonferroni adjustments. British Medical Journal, 316, 1236–1238.

Sansoni, P., Fagnoni, F., Vescovini, R., Mazzola, M., Brianti, V., Bologna, G., Nigro, E., Lavagetto, G., Cossarizza, A., Monti, D., Franceschi, C., Passeri, M., 1997. T lymphocyte proliferative capability to defined stimuli and costimulatory CD28 pathway is not impaired in healthy centenarians. Mechanisms of Ageing and Development, 96, 127–136.

Tollerud, D.J., Clark, J.W., Brown, L.M., Neuland, C.Y., Pankiw-Trost, L.K., Blattner, W.A., Hoover, R.N., 1989. The influence of age, race, and gender on peripheral blood mononuclear-cell subsets in healthy nonsmokers. Journal of Clinical Immunology, 9(3), 214–222.

Wikby, A., Maxson, P., Olsson, J., Johansson, B., Ferguson, F.G., 1998. Changes in CD8 and CD4 lymphocyte subsets, T cell proliferation responses and non-survival in the very old: the Swedish longitudinal OCTO-immune study. Mechanisms of Ageing and Development, 102(2–3), 187–198.

Advances in
Cell Aging and
Gerontology

Immunological and immunogenetic markers of successful and unsuccessful ageing

Giuseppina Candore, Giuseppina Colonna-Romano,
Domenico Lio and Calogero Caruso

*Gruppo di Studio sull'Immunosenescenza, Dipartimento di Biopatologia e Metodologie Biomediche,
Università di Palermo, Corso Tukory 211, 90134 Palermo, Italy
Correspondence address: Tel.: +39-091-655-5911; fax: +39-091-655-5933.
E-mail: marcoc@unipa.it (C.C.)*

Contents

Abbreviations

AD: Alzheimer's disease; AH: ancestral haplotype; CD: cluster of differentiation; CRP: C-reactive protein; FAD: familial AD; HFE: hemochromatosis; HLA: human leukocyte antigens; IFN: interferon; IL: interleukin; NK: natural killer; SNP: single nucleotide polymorphism; TCR: T cell receptor; TNF: tumor necrosis factor.

1. Introduction

Some people live in good health to great ages while others die relatively young, though we do not understand why this is so. However, several studies show that longevity may be correlated with optimal functioning of the immune system. In fact, both longitudinal and cross-sectional studies performed in the last few years have indicated that several functional markers of the immune system may be used either as markers of successful ageing or conversely as markers of unsuccessful ageing.

2. Immunological markers

2.1. T lymphocytes

Many years ago, it was established that several tests of T cell function gave different results in the elderly than in the young, and this was interpreted as indicating depressed responses in elderly individuals (Thoman and Weigle, 1989; Candore et al., 1992). In particular, the ability of T cells to undergo clonal expansion when stimulated in vitro is repeatedly identified as an age-associated decrease in immune function. Various mechanisms have been proposed to account for such suboptimal proliferation, but they probably still offer only an incomplete explanation of this phenomenon. Nonetheless, several early studies already suggested a positive association between "good" T cell function, as it could be measured in vitro, and individual longevity (Roberts-Thomson et al., 1974; Murasko et al., 1988; Wayne et al., 1990).

Higher absolute lymphocyte counts have been also associated with longevity (Bender et al., 1986). A progressive decrease of B lymphocytes (Franceschi et al., 1995; Potestio et al., 1999) with oligoclonal expansion of CD5+ B lymphocytes, accompanied by increased serum concentration of autoantibodies, (Weksler, 2000) has been reported with age, but these changes seem not to be significantly associated with longevity because they are also observed in the best example of successful ageing, i.e. the centenarians (Candore et al., 1997, 2002b; Potestio et al., 1999; Colonna-Romano et al., 2002b). A large number of studies have focused on changes within the T lymphocyte compartment. A decrease in mature CD3+ T cells (both CD4 and CD8) with age has been described, and among CD3+ lymphocytes, also T cells bearing the T cell receptor (TCR) γδ are reduced in number (Cossarizza et al., 1997; McNerlan et al., 1999; Colonna-Romano et al., 2002a,b) but again these changes are also observed in centenarians (Colonna-Romano et al., 2002a,b). Strictly linked to investigations on numerical changes, within the main lymphocyte subsets, is the Swedish longitudinal OCTO immune study (Wikby et al., 1998), in which it has been demonstrated that a combination of high CD8, low CD4 and poor T cell proliferation is associated with a higher 2-year mortality in very old subjects.

A well-known phenomenon in the elderly is the increase of the T cell memory phenotype, which is better represented in CD4+ cells than in CD8+ cells (Cossarizza et al., 1996). The relevance of the immunological experience over the years is considered to be the cause of the rapid increase of memory T lymphocytes during the first decades of life (and the concomitant loss of "virgin" cells), with maintenance of these cells until the centenarian group. The expression of the apoptosis-related molecule CD95 has been proposed to be a good marker to discriminate between "virgin" and previously activated cells (Potestio et al., 1999; Fagnoni et al., 2000). In fact, it increases both with age "ex vivo" and after "in vitro" stimulation with mitogens. It has also been shown that a decrease of CD95-negative cells (virgin cells) within the CD4+ subset, and an almost complete loss of CD95-negativity within the CD8+ subset, is common in old people, particularly in centenarians (Fagnoni et al., 2000). This report could explain the reduced ability to respond to new intracellular pathogens, which occurs during senescence. In fact it has been described that a reduction in naïve CD4, and a great reduction of CD8+ cells, is not accompanied by a reduction of CD95+ (memory) cells. It has been hypothesized that the reduction of "virgin" T cells renders the immune system incapable of defending the body against new antigens, rendering old people highly susceptible to infections, cancer and autoimmune diseases. All together these data support the hypothesis that the loss of CD8+ "virgin" cells (and the reciprocal increase of CD8+CD95+ lymphocytes) is related to an increased mortality. Because a greater prevalence of CD8/CD95+ cells is commonly found in very old subjects, it may be hypothesized that younger people with increased CD8/CD95+ cells may also be at greater risk of death. Another possibility is that in each different subject the increased expression of CD95 on CD8+ cells may be considered a neutral marker of age, whereas the increase of CD95 on CD4+ T cells could be considered as a marker of senescence. This idea comes from our study (Potestio et al., 1999) in which an increase of the percentage of CD4+/CD95+ cells is shown until 74 years, thereafter a slight decrease is observed until the last group of 85–102 years of age. A longitudinal study might evaluate the individual variations of this parameter and indicate whether CD95 may also have predictor significance on CD4+ T lymphocytes. On the other hand, the expression of CD95 was evaluated on lymphocytes bearing the costimulatory molecule CD28 and it was demonstrated that, whereas the percentage of CD28+ cells decrease with age, the percentage of CD28+CD95+ cells increases (Potestio et al., 1999). Therefore, it may be possible to presume that the high expression of CD95 is related to the loss of CD28+ lymphocytes. Regarding CD28 in the elderly, there is a large consensus that this molecule is reduced on T lymphocytes both "ex vivo" and after "in vitro" culture; moreover, the loss of CD28+ cells is more evident within the CD8 subset (Boucher et al., 1998). It is well known that CD28 acts as a costimulatory receptor on T cells, which, together with the TCR, is required for activation and induction of cytokine secretion. The decreased expression of this molecule has also been reported within "in vitro" cultures (Effros et al., 1994), and in centenarians the CD8+CD28− lymphocytes are more numerous (Boucher et al., 1998), suggesting their role in the attainment of longevity.

2.2. Natural killer cells

In any case, it is clear that T cells are an important parameter, but a combination of both T cell and natural killer (NK) assays might be more critical. In fact there was a reported correlation in the elderly between a high NK cell activity and the T cell proliferative responses (Mysliwski et al., 1993). Several studies have reported that NK and NK-T cell numbers increase with age (Franceschi et al., 1995; Cossarizza et al., 1997; McNerlan et al., 1999; Solana et al., 1999). Recently, a population of T cells that share some characteristics with NK cells has been observed, termed NKT cells; these cells produce high levels of IL-4 and IFN-γ when stimulated. In addition to cytokine production, NKT cells can exhibit potent lytic activity. They constitutively express Fas-L and can kill Fas+ (i.e. CD95+) target cells, including double positive thymocytes, and can also kill tumor targets in a perforin-dependent manner (Solana and Mariani, 2000). There is a major shift within the lymphocyte population, from conventional T cells to NK cells and NKT cells, with senescence (Solana and Mariani, 2000; Godfrey et al., 2000). Miyaji et al. (2000) have demonstrated that the proportion and the absolute number of NK and NKT cells (especially CD57+ T cells) were highly increased in the blood of centenarians. Moreover, it has been demonstrated that the proportion of IFN-γ (rather than IL-4)-producing cells increased in NK and NKT cells in centenarians. Since IFN-γ is important in maintaining the cytotoxic function of NK, NKT and conventional T cells, the pattern of cytokine production might be essential for maintaining the function of lymphocytes in centenarians (Huang et al., 1993; Chen et al., 1997).

In addition, human NK cells can be divided into two subsets based on their cell surface density of CD56: namely, CD56^{bright} and CD56^{dim}, each with distinct phenotypic properties. These subsets have unique functional attributes and, therefore, distinct roles in the human immune response. The CD56^{dim} NK cell subset is more naturally cytotoxic and expresses higher levels of Ig-like NK receptors and CD16 than the CD56^{bright} NK cell subset. By contrast, the CD56^{bright} subset has the capacity to produce abundant cytokines following activation by monocytes, but has low natural cytotoxicity and is CD16^{dim} or CD16 negative (Cooper et al., 2001). As regards analysis in healthy ageing of NK subpopulations, defined by the relative expression of CD56, phenotypic changes in the CD56^{bright} subset are not observed between young and elderly whereas, CD56^{dim} NK cells from elderly donors expressed higher levels of HLA-DR and CD95 and lower levels of CD69 than CD56^{dim} NK from young donors. Furthermore, the increased percentage of NK cells observed in the elderly is mainly due to an increase in the CD56^{dim} subset. As the CD56^{dim} subset represent the mature NK cell subset, these results support the hypothesis that a phenotypic and functional shift in the maturity status of NK cells occurs in ageing (Baume et al., 1992; Krishnaraj and Svanborg, 1992; Borrego et al., 1999).

However, it has been hypothesized that the number of NK cells on the whole might represent a good marker of successful ageing. In fact, in a follow-up study in Leiden, it has been demonstrated that elderly people (aged >85 years) with low numbers of NK cells have a three times higher mortality risk in the first 2 years of follow-up than those with high NK cell levels (Remarque and Pawelec, 1998). Further supporting this hypothesis, centenarians display a high number of NK cells (Franceschi et al., 1995; Colonna-Romano et al., 2002b).

Functional NK activity, assessed by the ^{51}Cr release assay using K562 cells, a human erythromyeloid leukaemia-derived cell line, as targets, has been extensively analyzed in elderly people. Studies of age-associated alterations in the level of NK cell mediated-cytotoxicity have resulted in conflicting data i.e. no significant change in the basal NK activity or increase or decrease in the magnitude of response of NK cells (see Di Lorenzo et al., 1999). As regards the possible explanations for the different results reported by various authors, it is worth noting that these conflicting reports may be due to differences in donor selection criteria, in sample size, or in the use of different experimental procedures and different parameters of activity (Kutza et al., 1995; Di Lorenzo et al., 1999). Although results of NK cells are discordant in the elderly, the few studies performed in centenarians are concordant and, in particular, in these subjects NK cell activity was preserved, thus suggesting that NK cell activity might be considered a good marker of successful ageing. In fact, an emphasis on the significance of NK cells in healthy ageing comes from studies of centenarians. Centenarians have a very well-preserved NK cell cytotoxicity (Sansoni et al., 1993; Franceschi et al., 1995). Besides, when the NK cells are studied in this group of "very old" people, the results show that those individuals with higher NK-cell numbers and NK cytolytic activity had a better ability to maintain an autonomous life style, had higher serum vitamin D levels, were well nourished, and had a balanced basal metabolism (Mariani et al., 1999).

Recent data also suggest that leading causes of death in very old people are common infectious diseases (reviewed in Pawelec et al., 2002). In fact, NK cell-mediated innate immunity is important for fighting all kinds of disease and the in vivo importance of NK cells for fighting infections in humans is supported by reports that rare patients who almost totally lacked NK cell activity, but had no derangements of other immunological functions, suffered repeated viral and bacterial infectious (Fleisher et al., 1982; Biron et al., 1989). NK cell activity varies significantly among normal subjects. However, there is no evidence that differences in NK cell activity among normal subjects influence their ability to fight infections. Based on these findings, it is postulated that well-preserved NK cell activity is important for human longevity, at least in part, because of its anti-infectious effect. Ogata et al. (2001) have found that low NK cell cytotoxicity, considered either on absolute values or on a per cell basis, was the only parameter that correlated with a past history of severe infection or death due to infection in the elderly group.

Finally, a recent study has demonstrated an important role of NK cells in human survival, suggesting an effective role as "survival" biomarkers. In fact Imai et al. (2000), in an 11-year follow-up study on a general Japanese population, have observed that medium and high cytotoxic activity of NK cells is associated with a reduced cancer risk, whereas low activity is associated with an increased cancer risk, suggesting a role for natural immunological host defense mechanisms against cancer.

3. Immunogenetic markers

As discussed by Pawelec et al. (2002), in these and very many other studies, the "chicken and egg" question must be asked; i.e. do people live longer because of "good" immune function, or do they possess good immune function because other factors have enabled them to survive longer? Thus, to better understand the role of the immune system in longevity, we have to search for immunogenetic markers of longevity. In other words, if the immune system plays a key role in the attainment of successful ageing, then genetic determinants of longevity should reside within polymorphisms of the immune system genes that regulate immune responses, such as the HLA and cytokine genes. Thus, these polymorphic markers could be markers of successful ageing or conversely of unsuccessful ageing.

3.1. HLA

The studies performed on the association between longevity and HLA are generally difficult to interpret, owing to major methodological problems. However, as reported by Caruso et al. (2000, 2001) some of them, well designed and performed, suggest an HLA effect on longevity. In studies performed in Caucasoids, an increase in HLA-DR11 (that is a HLA-DR5 split) in Dutch women over 85 years was observed (Lagaay et al., 1991). The same laboratory performed a further study, and by using a 'birth-place-restricted comparison' in which the origin of all the subjects was ascertained, the authors were able to confirm that ageing in women was positively associated with HLA-DR5 (Izaks et al., 2000). Two French studies confirmed the relevance of HLA-DR11 to longevity in aged populations (Ivanova et al., 1998; Henon et al., 1999). This increase is consistent with the protective effects of this allele in viral diseases, as HLA-DR5, or its subtype HLA-DR11, frequencies have been shown to be decreased in some viral diseases (Caruso et al., 2000, 2001). Finally, an association between longevity and the ancestral haplotype (AH) 8.1 (or part of this haplotype, i.e. HLA-B8, DR3) apparently emerges. In fact, an excess of this AH in the 'oldest' old men has been reported in both French and in Northern Irish populations (Proust et al., 1982; Rea and Middleton, 1994). This association appears to be sex-specific. In fact, a Greek study showed a significant decrease of 8.1 AH in aged women (Papasteriades et al., 1997). Immune dysfunctions of the 8.1 AH appear to contribute to early morbidity and mortality in elderly women, more susceptible to autoimmune diseases than men, but to longevity in men (Candore et al., 2002a). These associations are gender-related, but that is not unexpected based on the available data on the genetics of longevity, showing that the association of longevity with particular alleles may be found only in one gender (Franceschi et al., 2000b). Thus, these studies seem to suggest that HLA-DR11 in women and HLA-B8, DR3 in men may be considered markers of successful ageing. However, it is noteworthy that in a longitudinal study, in which a total of 919 subjects aged 85 years and older were HLA-typed and followed up for at least 5 years, no HLA-association with mortality was found (Izaks et al., 1997).

Hemochromatosis (HFE), the most telomeric HLA class I gene, codes for a class I α chain, which seemingly no longer participates in immunity because its has lost its ability to bind peptides, due to a definitive closure of the antigen-binding cleft that prevents peptide binding and presentation. Nonetheless, the HFE protein, by regulating iron uptake, indirectly regulates immune responses because iron availability plays a role in specific and non-specific immune responses. In fact, iron deficiency may be associated with reversible abnormalities of immune-inflammatory responses, although it is difficult to demonstrate the severity and relevance of these in observational studies (Klein and Sato, 2000; Lio et al., 2002a). Our recent results show that possession of the C282Y HFE allele, known to be associated with an increase in iron uptake, significantly increases the possibility for women to live long, since its frequency was higher in older women than in the younger controls (Lio et al., 2002a). It is intriguing that HFE mutations have been suggested to be involved in unsuccessful ageing too. In fact, Alzheimer's disease (AD) patients carrying the H63D allele had a mean age of onset at 72 vs. 77 years for those who were homozygous for the wild-type allele (Sampietro et al., 2001). Thus, it seems that H63D mutations may anticipate sporadic late-onset of AD clinical presentation in susceptible individuals, although we have not been able to confirm these results (Balistreri et al., 2002). In another study, in patients with familial AD (FAD), C282Y and H63D mutations were over-represented in men and under-represented in women with FAD (Moalem et al., 2000). Thus, with the possibility that HFE mutations are important, new genetic risk factors for AD should be pursued further.

3.2. Cytokine genes

Many aspects of ageing involve inflammatory processes. In fact, ageing is associated with chronic, low-grade inflammatory activity leading to long-term tissue damage, and systemic chronic inflammation has been found to be related to all-cause mortality risk in older persons (Bruunsgaard et al., 2001). Additionally, the organ-specific effects of systemic inflammation may be influenced by the genetic constitution of the individual. Age-related diseases such as AD, Parkinson disease, atherosclerosis, type 2 diabetes, sarcopenia and osteoporosis, are initiated or worsened by systemic inflammation, thus suggesting the critical importance of unregulated systemic inflammation in the shortening of survival in humans (Brod, 2000; Bruunsgaard et al., 2001; Pawelec et al., 2002). Accordingly, pro-inflammatory cytokines are believed to play a pathogenetic role in age-related diseases, and genetic variations located within their promoter regions have been shown to influence susceptibility to age-related diseases, by increasing gene transcription and therefore cytokine production (Bidwell et al., 1999, 2001; Pawelec et al., 2002).

3.2.1. IL-6
IL-6 is a pleiotropic cytokine capable of regulating proliferation, differentiation and activity of a variety of cell types, and plays a major role in bone remodeling, neuro-endocrine homeostasis, hemopoiesis and immune system regulation. In particular, IL-6 plays a pivotal role in the acute-phase response and in the balancing of the pro-inflammatory/anti-inflammatory pathways (Ershler and Keller, 2000). For some years, it has been known that circulating levels of acute-phase proteins are associated with, and can predict, the onset of cardiovascular events. The predictive value of these markers, including C-reactive protein (CRP), has been demonstrated for subjects with pre-existing coronary heart disease as well as for apparently healthy subjects. More recently, increased levels of IL-6 and CRP have been also associated with a high risk of all-cause mortality in older people. Many chronic conditions that are common causes of death in older persons may stimulate and sustain a systemic inflammatory state, which can be measured by increased levels of CRP, IL-6, or other pro-inflammatory cytokines. Only a few population-based studies have investigated the association between circulating levels of IL-6 or CRP and the risk of all-cause

mortality. Results from these studies suggest that markers of inflammation may be helpful in identifying high-risk subjects. However, the influence of clinical and sub-clinical chronic diseases on this association has not been fully investigated, and consequently, the prognostic value of these markers of inflammation, independent of disease status, remain uncertain. These findings are in agreement with recent studies in older populations, and they suggest that chronic inflammation is probably involved in the pathophysiology of many conditions; accordingly, the magnitude of IL-6 plasma levels has been associated with the degree of functional disability and mortality (Harris et al., 1999; Ferrucci et al., 1999; Gussekloo et al., 2000; Volpato et al., 2001). Despite the high sensitivity of IL-6 plasma levels to acute and chronic infections, as well as to other environmental conditions, there is a strong genetic control of IL-6 plasma levels that suggest different reactivity in ageing people. In fact, a C/G polymorphism 5'-upstream of IL-6 has been identified (Fishman et al., 1998). This single nucleotide polymorphism (SNP) influences the rate of IL-6 gene transcription and is associated with different IL-6 plasma levels. As far as the func-tional implications of this SNP are concerned, carriers of the G allele at −174 locus appear to be prone to develop lipid abnormalities and have a worse glucose handling capacity, higher blood glycosylated hemoglobin, higher fasting insulin levels and a higher sensitivity to insulin (Fernandez-Real et al., 2000). Accordingly, recent data suggest that C+ men may be protected from cardiovascular diseases. In particular, an under-representation of homozygosity at the C allele at the −174 locus has been found in a group of Swedish men having a myocardial infarction under the age of 40 (Yudkin et al., 2000). Moreover, middle-aged men homozygous for the G allele at the −174 C/G locus, have an increased artery intima-media thickness in the carotid bifurcation, a predilection site for atherosclerosis (Rauramaa et al., 2000). Recent studies also showed that another IL-6 polymorphism, i.e. a variable number of tandem repeat polymorphisms in the 3' flanking region of the IL-6 gene, plays a complex role in AD. In fact, it was demonstrated that the C allele of this polymor-phism was associated with a reduced risk of sporadic AD and a delayed initial onset (Papassotiropoulos et al., 1999). Recently, in a study performed on centenarians (Bonafe et al., 2001), it has been reported that those individuals who are genetically predisposed to produce high levels of IL-6 during ageing, i.e. men who are C− at the IL-6 −174 C/G locus, are less likely to reach the extreme limits of human longevity. On the other hand, low level IL-6 production throughout the life span (C+ indi-viduals) appears to be beneficial for longevity, at least in men. On the whole, these data suggest that in an ageing population, those people who have the tendency to produce elevated IL-6 quantities are less likely to achieve maximum longevity, prob-ably because of their increased susceptibility to age-related inflammatory diseases.

3.2.2. IL-1 cluster

The "prototypic" inflammatory cytokine IL-1 is a primary mediator of systemic inflammatory responses, such as hypophagia, slow-wave sleep, sickness behavior and neuroendocrine changes (Dantzer, 2001). It is a family of at least three closely-related proteins, two agonists IL-1α, IL-1β and one antagonist. The IL-1 receptor antagonist is a highly selective, competitive receptor antagonist that binds to the receptor, but fails to trigger signal transduction, thereby blocking all the actions of IL-1α or IL-β. A number of biallelic and multiallelic markers in the surrounding region of IL-1 genes have been identified. Several reports have provided evidence that these polymorphisms influence cytokine regulation. Accordingly, genetic poly- morphisms of the IL-1 cluster influence the susceptibility and the severity of a variety of disorders, including age-related diseases (Nicklin et al., 1994; Hurme et al., 1998; Rosenwasser, 1998). IL-1 levels are elevated in AD patients' brains, and over-expres- sion of IL-1 is associated with β amyloid plaque progression. IL-1 interacts with the gene products of several other known or suspected genetic risk factors for AD, mostly ApoE ε allele 4. IL-1 over-expression is also associated with environmental risk factors for AD, including normal ageing and head trauma (Akiyama et al., 2000). These observations suggest an important pathogenic role for IL-1, and for IL-1-driven cascades, in the pathogenesis of AD. A large number of studies clearly demonstrates an involvement of IL-1α −889 SNP (a C to T transition, also desig- nated as allele 2) (Franceschi et al., 2001) in terms of association and early onset. An association with the age at onset was also shown for the IL-1β and IL-1 receptor antagonist genes. Plasma levels of IL-1β were found to be increased in patients with AD and the high levels of the cytokine were linked to IL-1β TT genotype. This genotype in the presence of a particular α1-ACT genotype increased the risk of AD and decreased the age at onset of the disease (Murphy et al., 2001). On the whole, these results clearly demonstrate that IL-1 polymorphisms play a significant role in AD. In two recent studies (Wang et al., 2001; Cavallone et al., 2002), no significant genotypic or allelic differences were observed between young and oldest old for all the polymorphisms examined, by analyzing data on the whole or sepa- rately in men and women. This apparently negative result suggests that no single polymorphism of the IL-1 gene cluster confers an advantage for survival in the last decade of life and is compatible with the hypothesis that these polymorphisms constitute an evolutionarily conserved shield against pathogens which remain a threat over the entire lifespan of the individual.

3.2.3. TNF cluster

The tumor necrosis factor (TNF) cluster genes, located in the HLA region on chromosome 6, encode three inflammation-related proteins: TNF-α, TNF-β and lymphotoxin-β. All three are important mediators of the immune response with multiple biologic activities. Several polymorphic areas are documented within the TNFα gene, coding for a central player in inflammatory age-related diseases. Notably, some two-allele polymorphisms for the TNF promoter region and several microsatellite polymorphic sites have been described. Polymorphisms in the TNF promoter region have been observed to result in differences in the rate of gene transcription and in the rate of protein production. Many associations with immune-mediated diseases have been described, both with TNF gene promoter polymorphisms, and TNF region microsatellites (Candore et al., 2002a). TNF may be involved in the pathogenesis of AD based on observations that senile plaques have been found to upregulate proinflammatory cytokines (Akiyama et al., 2000) and a collaborative genome-wide scan for AD genes in 266 late-onset families implicated a 20 centimorgan region at chromosome 6p21.3 that includes the TNF gene (Collins et al., 2000). However, discrepant results have been obtained indicating that increased intrathecal production of TNF-α in AD is preferentially controlled by environmental stimuli rather than genetic makeup (Tarkowski et al., 2000). On the other hand, results from three TNF polymorphism typings, (a −308 TNF promoter polymorphism, whose TNF2 allele is associated with autoimmune inflammatory diseases and strong transcriptional activity; the −238 TNF promoter polymorphism; and the microsatellite TNFa, whose so called allele 2 is associated with high TNF secretion) indicate an increased frequency of a "high secretion" haplotype among AD patients using the sibling disequilibrium test (Collins et al., 2000). A further polymorphism in the regulatory region of the TNF-α gene was analyzed in a case-control study. The polymorphism (C850T) was typed in 242 patients with sporadic AD, 81 patients with vascular dementia, 61 stroke patients without dementia, and 235 normal controls. Possession of the TNF-α T allele significantly increases the risk of vascular dementia, and increases the risk of AD associated with ApoE (McCusker et al., 2001).

There is ample evidence to support a role of TNF-α in the development of cardiovascular disease. TNF-α is expressed in atherosclerotic plaques but not in healthy vessels. In atherosclerotic plaques, TNF-α may contribute to foam cell formation, to T-lymphocyte activation and to the expression of matrix metalloproteinases that may destabilize the plaque by degrading the extra-cellular matrix (Lee and Libby, 1997). It has been demonstrated that there are significant positive associations between the TNF-α −308 A allele and levels of extracellular superoxide dismutase and homocysteine, implicated in oxidative stress relevant to atherogenesis. However, the TNF-α polymorphism was not associated directly with the occurrence or severity of atherosclerosis as documented angiographically and in a retrospective study performed on an autopsy series that comprised 700 Finnish men, aged 33–70 years (Wang and Oosterhof, 2000; Keso et al., 2001), where coronary stenosis and surface area of atherosclerotic changes were measured and the presence of myocardial infarction and coronary thrombosis recorded. Only a weak association with the TNFA and TNFB polymorphisms and atherosclerotic morphometric changes in coronary arteries was found, but not with coronary stenosis and frequency of old or recent myocardial infarction or coronary thrombosis. Thus TNFA and TNFB polymorphisms are unlikely to contribute to progression of atherosclerosis in a clinically important manner. The genotypic frequencies of the TNF-α promoter SNPs 308G and 308A, suggested to be associated with low and high TNF-α production, respectively, were not significantly different between centenarians and controls (Crivello et al., 2002).

3.2.4. IFN-γ

It has been reported recently that the 12 CA repeat microsatellite allele at the first intron of the IFN-γ gene, one of the pivotal cytokines in the induction of immune-mediated inflammatory responses, is associated with a higher level of cytokine production in vitro. This is correlated with the presence of the T allele at a SNP located at the +874 position (+874T→A) from the translation start site coinciding with a putative NF-κB binding site (Pravica et al., 2000). No data have yet been presented on any associations between age-related inflammatory diseases and IFN-γ polymorphisms. However, both in experimental animals and in humans, IFN-γ is involved in the development of atherosclerotic plaques (Laurat et al., 2001). Thus, in AD patients, an increased spontaneous and IL-2-induced release of IFN-γ and TNF-α from NK cells compared to healthy subjects was found. Furthermore, significant negative correlations between the spontaneous release of IFN-γ and TNF-α from NK cells and a decrease of cognitive function score were found in these patients (Solerte et al., 2000). Thus, it is conceivable that IFN-γ polymorphisms can be involved in these diseases. On the other hand, Lio et al. (2002b) have studied the distribution of the +874T→A IFN-γ polymorphism in a large number of Italian centenarians. The +874T allele was found less frequently in centenarian women than in control women, whereas allele frequencies in centenarian men were not found significantly different from those in control men. This suggests that the possession of the +874T allele, known to be associated with low IFN-γ production, may have a gender-related positive effect on longevity.

3.2.5. IL-10

IL-10 is a major immunoregulatory cytokine, usually considered to mediate potent downregulation of the inflammatory response (Moore et al., 2001). IL-10 production is genetically controlled by polymorphisms in the IL-10 promoter sequence. Several polymorphisms located close to or within the IL-10 gene are potentially associated with transcription levels. The best-documented is the promoter polymorphism −1082G→A. Possession of this allele −1082A, results in lower IL-10 production after stimulation of lymphocytes with concanavalin A than in allele −1082A-negative cells (Turner et al., 1997). A recent extensive study suggests that IL-10 polymorphisms are not associated with an increased risk of myocardial infarction (Donger et al., 2001), although serum levels of IL-10 are decreased in patients with unstable angina, in keeping with previous data from animal model studies suggesting that IL-10 has a protective role in atherosclerosis (Smith et al., 2001). Concerning AD, in a recent study, whole blood samples from AD and controls were stimulated ex vivo with endotoxin under standard conditions and cytokine levels were assessed (Remarque et al., 2001). The data suggest that reduced production of IL-10 may contribute to the development of AD, but no study has been performed on the association between AD and IL-10 polymorphisms. A recent study has demonstrated a role of IL-10 polymorphisms in successful ageing. In fact, the 1082G homozygous genotype was increased in centenarian men but not in centenarian women. No difference was found between centenarians and control subjects regarding the other two polymorphisms. The presence of the −1082GG genotype, suggested to be associated with high IL-10 production, significantly increases the possibility to reach the extreme limit of human lifespan in men (Lio et al. 2002c). To further strengthen these results, in a recent report we have evaluated combined IL-10 and TNF-α genotypes showing that there was a significant increase of the "anti-inflammatory" (IL-10 −1082GG/TNF-α −308GG) genotype in centenarians with respect to controls (Crivello et al., 2002).

4. Conclusions

On the whole, data on cytokine polymorphisms suggest that polymorphic alleles of inflammatory cytokines, involved in high cytokine production, play an important role in age-related inflammatory diseases, i.e. in unsuccessful ageing. Reciprocally, they suggest that controlling the inflammatory status may facilitate achievement of successful ageing. In fact, the major findings reported in the recent papers on cytokine polymorphisms and longevity, discussed here, suggest that those individuals who are genetically predisposed to produce low levels of inflammatory cytokines, or high levels of anti-inflammatory cytokines, have an increased chance of reaching the extreme limits of the human life-span. Further immunogenetic analyses will need to investigate the genetic control of acute-phase proteins such as CRP, whose significance as predictive markers of disability or mortality was briefly discussed above. Other genetic parameters not yet investigated in this context, for example, the control of platelet surface antigens suggested to play a role in the evolution of atherosclerosis (Sperr et al., 1998) also require study. However, it is noteworthy that no longitudinal study, such as that previously described concerning HLA, has been performed for cytokine gene polymorphisms; all the discussed studies are merely association studies.

In any case, further progress will require studies correlating immunological and immunogenetic markers. For example, concerning the change in T cell subpopulations assessed as markers of unsuccessful ageing, the matter should be to assess how much these changes depend on the immunogenetic background and how much they depend on the natural history of the individual, i.e. on an extra burden of antigenic load such as chronic infections (Franceschi et al., 2000a). However, it has to be pointed out that an efficient control of chronic infections also depends on the immunogenetic background of the host (Bidwell et al., 1999, 2001; Caruso et al., 2000, 2001; Lio et al., 2002d).

Acknowledgements

The "Gruppo di Studio sull'immunosenescenza" coordinated by Professor C. Caruso is funded by grants from MIUR, Rome (ex 40%, to CC and DL; ex 60% to CC, DL, GC and GCR), and from Ministry of Health Projects "Immunological parameters age-related" and "Pharmacogenomics of Alzheimer's Disease".

References

Akiyama, H., Barger, S., Barnum, S., Bradt, B., Bauer, J., Cole, G.M., Cooper, N.R., Eikelenboom, P., Emmerling, M., Fiebich, B.L., Finch, C.E., Frautschy, S., Griffin, W.S., Hampel, H., Hull, M., Landreth, G., Lue, L., Mrak, R., Mackenzie, I.R., McGeer, P.L., O'Banion, M.K., Pachter, J., Pasinetti, G., Plata-Salaman, C., Rogers, J., Rydel, R., Shen, Y., Streit, W., Strohmeyer, R., Tooyoma, I., Van Muiswinkel, F.L., Veerhuis, R., Walker, D., Webster, S., Wegrzyniak, B., Wenk, G., Wyss-Coray, T., 2000. Inflammation and Alzheimer's disease. Neurobiol. Aging 21, 383–421.
Balistreri, C.R., Licastro, F., Chiappelli, M., Franceschi, C., Piazza, G., Lio, D., Colonna-Romano, G., Candore, G., Caruso, C., 2002. HFE polymorphisms in unsuccessful ageing: a study in Alzheimer's disease patients from Northern Italy, Proc. 3rd Imagine Congr. Palermo (abs).

Baume, D.M., Robertson, M.J., Levine, H., Manley, T.J., Schow, P.W., Ritz, J., 1992. Differential responses to interleukin 2 define functionally distinct subsets of human natural killer cells. Eur. J. Immunol. 22, 1–6.

Bender, B.S., Nagel, J.E., Adler, W.H., Andres, R., 1986. Absolute peripheral blood lymphocyte count and subsequent mortality of elderly men. The Baltimore Longitudinal Study of Aging. J. Am. Geriatr. Soc. 34, 649–654.

Bidwell, J., Keen, L., Gallagher, G., Kimberly, R., Huizinga, T., McDermott, M.F., Oksenberg, J., McNicholl, J., Pociot, F., Hardt, C., D'Alfonso, S., 1999. Cytokine gene polymorphism in human disease, on-line databases, Genes Immun. 1, 3–19.

Bidwell, J., Keen, L., Gallagher, G., Kimberly, R., Huizinga, T., McDermott, M.F., Oksenberg, J., McNicholl, J., Pociot, F., Hardt, C., D'Alfonso, S., 2001. Cytokine gene polymorphism in human disease: on-line databases, supplement 1. Genes Immun. 2, 61–70.

Biron, C.A., Byron, K.S., Sullivan, J.L., 1989. Severe herpes virus infections in an adolescent without natural killer cells. N. Engl. J. Med. 320, 1731–1735.

Bonafe, M., Olivieri, F., Cavallone, L., Giovanetti, S., Mayegiani, F., Cardelli, M., Pieri, C., Marra, M., Antonicelli, R., Lisa, R., Rizzo, M.R., Prolisso, G., Monti, D., Franceschi, C., 2001. A gender-dependent genetic predisposition to produce high levels of IL-6 is detrimental for longevity. Eur. J. Immunol. 31, 2357–2361.

Borrego, F., Alonso, C., Galiani, M., Carracedo, J., Ramirez, R., Ostos, B., Pena, J., Solana, R., 1999. NK phenotypic markers and IL-2 response in NK cells from elderly people. Exp. Gerontol. 34, 253–265.

Boucher, N., Dufeu-Duchesne, T., Vicaut, E., Farge, D., Effros, R.B., Schachter, F., 1998. CD28 expression in T cell aging and human longevity. Exp. Gerontol. 33, 267–282.

Brod, S.A., 2000. Unregulated inflammation shortens human functional longevity. Inflamm. Res. 49, 561–570.

Bruunsgaard, H., Pedersen, M., Pedersen, B.K., 2001. Aging and proinflammatory cytokines. Curr. Opin. Hematol. 8, 131–136.

Candore, G., Di Lorenzo, G., Caruso, C., Modica, M.A., Colucci, A.T., Crescimanno, G., Ingrassia, A., Barbagallo Sangiorgi, G., Salerno, A., 1992. The effect of age on mitogen responsive T cell precursors in human beings is completely restored by interleukin-2. Mech. Ageing Dev. 63, 297–307.

Candore, G., Di Lorenzo, G., Mansueto, P., Melluso, M., Fradà, G., Li Vecchi, M., Esposito Pellitteri, M., Drago, A., Di Salvo, A., Caruso, C., 1997. Prevalence of organ-specific and non organ-specific autoantibodies in healthy centenarians. Mech. Ageing Dev. 94, 183–190.

Candore, G., Lio, D., Colonna-Romano, G., Caruso, C., 2002a. Pathogenesis of autoimmune diseases associated with 8.1 ancestral haplotype, effect of multiple gene interactions. Autoimmun. Rev. 1, 29–35.

Candore, G., Grimaldi, M.P., Listì, F., Ferlazzo, V., Colonna Romano, G., Motta, M., Malaguarnera, M., Fradà, G., Lio, D., Caruso, C., 2002b. Prevalence of non organ-specific autoantibodies in healthy centenarians, Arch. Gerontol. Geriatrics. 35, 75–80.

Caruso, C., Candore, G., Colonna Romano, G., Lio, D., Bonafè, M., Valensin, S., Franceschi, C., 2000. HLA, aging and longevity: a critical reappraisal. Hum. Immunol. 61, 942–949.

Caruso, C., Candore, G., Colonna Romano, G., Lio, D., Bonafè, M., Valensin, S., Franceschi, C., 2001. Immunogenetics of longevity. Is major histocompatibility complex polymorphism relevant to the control of human longevity? A review of literature data. Mech. Ageing Dev. 122, 445–462.

Cavallone, L., Lio, D., Colonna-Romano, G., Candore, G., Bonafè, M., Cardelli, M., Giovanetti, S., Giampieri, C., Mugianesi, E., Stecconi, R., Marchegiani, F., Oliveri, F., Franceschi, C., Caruso, C., 2002. Molecular typing of IL-1 gene cluster: a study in Italian centenarians, Proc. 3rd Imagine Congr. Palermo (abs).

Chen, Y.H., Chiu, N.M., Mandal, M., Wang, N., Wang, C.R., 1997. Impaired NK1+ T cell development and early IL-4 production in CD1-deficient mice. Immunity 6, 459–467.

Collins, J.S., Perry, R.T., Watson, B., Jr., Harrell, L.E., Acton, R.T., Blacker, D., Albert, M.S., Tanzi, R. E., Bassett, S.S., McInnis, M.G., Campbell, R.D., Go, R.C., 2000. Association of a haplotype for tumor necrosis factor in siblings with late-onset Alzheimer disease, the NIMH Alzheimer disease genetics initiative. Am. J. Med. Genet. 96, 823–830.

Colonna-Romano, G., Potestio, M., Aquino, A., Candore, G., Lio, D., Caruso, C., 2002a. Gamma/delta T lymphocytes are affected in the elderly. Exp. Gerontol. 37, 205–211.

Colonna-Romano, G., Cossarizza, A., Aquino, A., Scialabba, G., Bulati, M., Lio, D., Candore, G., Di Lorenzo, G., Caruso, C., 2002b. Age- and sex-related values of lymphocyte subsets in subjects from North and South Italy. Arch. Gerontol. Geriatrics. 35, 99–107.

Cooper, M., Fehniger, T.A., Caligiuri, A., 2001. The biology of natural killer cell subsets. Trends Immunol. 22, 633–640.

Cossarizza, A., Ortolani, C., Paganelli, R., Barbieri, D., Monti, D., Sansoni, P., Fagiolo, U., Castellani, G., Bersani, F., Londei, M., Franceschi, C., 1996. CD45 isoforms expression on CD4+ and CD8+ T cells throughout life, from newborns to centenarians: implications for T cell memory. Mech. Ageing Dev. 86, 173–195.

Cossarizza, A., Ortolani, C., Monti, D., Franceschi, C., 1997. Cytometric analysis of immunosenescence. Cytometry 27, 297–313.

Crivello, A., Lio, D., Scola, L., Colonna-Romano, G., Candore, G., Bonafè, M., Cavallone, L., Marchegiani, F., Oliveri, F., Franceschi, C., Caruso, C., 2002. Inflammation and longevity: further studies on the protective effects in men of cytokine SNPs. Proc. 3rd Imagine Congr. Palermo (abs).

Dantzer, R., 2001. Cytokine-induced sickness behavior: where do we stand? Brain Behav. Immun. 15, 7–24.

Di Lorenzo, G., Balistreri, C.R., Candore, G., Cigna, D., Colombo, A., Colonna Romano, G., Colucci, A. T., Gervasi, G., Listì, F., Potestio, M., Caruso, C., 1999. Granulocyte and natural killer activity in the elderly. Mech. Ageing Dev. 108, 25–38.

Donger, C., Georges, J.L., Nicaud, V., Morrison, C., Evans, A., Kee, F., Arveiler, D., Tiret, L., Cambien, F., 2001. New polymorphisms in the interleukin-10 gene-relationships to myocardial infarction. Eur. J. Clin. Invest. 31, 9–14.

Effros, R.B., Boucher, N., Porter, V., Zhu, X.M., Spaulding, C., Walford, R.L., Kronenberg, M., Cohen, D., Schachter, F., 1994. Decline in CD28 (+) T cells in centenarians and in long-term T cell cultures: a possible cause for both in vivo and in vitro immunosenescence. Exp. Gerontol. 29, 601–609.

Ershler, W.B., Keller, E.T., 2000. Age-associated increased interleukin-6 gene expression, late-life diseases, and frailty. Annu. Rev. Med. 51, 245–270.

Fagnoni, F.F., Vescovini, R., Passeri, G., Bologna, G., Pedrazzoni, M., Lavagetto, G., Cast, A., Franceschi, C., Passeri, C., Sansoni, P., 2000. Shortage of circulating naïve CD8+ T cells provides new insights on immunodeficiency in aging. Blood 95, 2860–2866.

Fernandez-Real, J.M., Broch, M., Vendrell, J., Gutierrez, C., Casamitjana, R., Pugeat, M., Richart, C., Ricart, W., 2000. Interleukin-6 gene polymorphism and insulin sensitivity. Diabetes 49, 517–520.

Ferrucci, L., Harris, T.B., Guralnik, J.M., Tracy, R.P., Corti, M.C., Cohen, H.J., Penninx, B., Pahor, M., Wallace, R., Havlik, R.J., 1999. Serum IL-6 level and the development of disability in older persons. J. Am. Geriatr. Soc. 47, 639–646.

Fishman, D., Faulds, G., Jeffery, R., Mohamed-Ali, V., Yudkin, J.S., Humphries, S., Woo, P., 1998. The effect of novel polymorphisms in the interleukin-6 (IL-6) gene on IL-6 transcription and plasma IL-6 levels, and an association with systemic-onset juvenile chronic arthritis. J. Clin. Invest. 102, 1369–1376.

Fleisher, G., Starr, S., Koven, N., Kamiya, H., Douglas, S.D., Henle, W., 1982. A non X-linked syndrome with susceptibility to severe Epstain-Barr virus infections. J. Pediatr. 100, 727–730.

Franceschi, C., Monti, D., Sansoni, P., Cossarizza, A., 1995. The immunology of exceptional individuals: the lesson of centenarians. Immunol. Today 16, 12–16.

Franceschi, C., Valensin, S., Bonafè, M., Paolisso, G., Yashin, A.I., Monti, D., De Benedictis, G., 2000a. The network and the remodeling theories of aging: historical background and new perspectives. Exp. Gerontol. 35, 879–896.

Franceschi, C., Motta, L., Valensin, S., Rapisarda, R., Frantone, A., Berardelli, M., Motta, M., Monti, D., Bonafe, M., Ferrucci, L., Deiana, L., Pes, G.M., Carru, C., Desole, M.S., Barbi, C., Sartoni, G., Gemelli, C., Lescai, F., Olivieri, F., Marchigiani, F., Cardelli, M., Cavallone, L., Gueresi, P., Cossarizza, A., Troiano, L., Pini, G., Sansoni, P., Passeri, G., Lisa, R., Spazzafumo, L., Amadio, L., Giunta, S., Stecconi, R., Morresi, R., Viticchi, C., Mattace, R., De Benedictis, G., Baggio, G., 2000b. Do men and women follow different trajectories to reach extreme longevity? Italian Multicenter Study on Centenarians. Aging Clin. Exp. Res. 12, 77–84.

Franceschi, C., Valensin, S., Lescai, F., Olivieri, F., Licastro, F., Grimaldi, L.M., Monti, D., De Benedictis, G., Bonafe, M., 2001. Neuroinflammation and the genetics of Alzheimer's disease, the search for a pro-inflammatory phenotype. Aging (Milan) 13, 163–170.

Godfrey, D.I., Hammond, K.J.L., Poulton, L.D., Smyth, M.J., Baxter, A.G., 2000. NKT cells: facts, functions and fallacies. Immunol. Today 21, 573–583.

Gussekloo, J., Schaap, M.C., Frolich, M., Blauw, G.J., Westendorp, R.G., 2000. C-reactive protein is a strong but nonspecific risk factor of fatal stroke in elderly persons. Arterioscler. Thromb. Vasc. Biol. 20, 1047–1051.

Harris, T.B., Ferrucci, L., Tracy, R.P., Corti, M.C., Wacholder, S., Ettinger, W.H., Jr., Heimovitz, H., Cohen, H.J., Wallace, R., 1999. Associations of elevated interleukin-6 and C-reactive protein levels with mortality in the elderly. Am. J. Med. 106, 506–512.

Henon, N., Busson, M., Dehay Martuchou, C., Charron, D., Hors, J., 1999. Familial versus sporadic longevity and MHC markers. J. Biol. Regulat. Homeost. Agent. 13, 27–31.

Huang, S., Hendriks, W., Althage, A., Hemmi, S., Bluethmann, H., Kamijo, R., Vilcek, J., Zinkernagel, R.M., Aguet, M., 1993. Immune response in mice that lack the interferon-γ receptor. Science 259, 1742–1745.

Hurme, M., Lahdenpohja, N., Santtila, S., 1998. Gene polymorphisms of interleukins 1 and 10 in infectious and autoimmune diseases. Ann. Med. 30, 469–473.

Imai, K., Matsuyama, S., Miyake, S., Suga, K., Nakachi, K., 2000. Natural cytotoxic activity of peripheral blood lymphocytes and cancer incidence: an 11-year follow up study of a general population. Lancet 356, 1795–1799.

Ivanova, R., Hénon, N., Lepage, V., Charron, D., Vicaut, E., Schächter, F., 1998. HLA-DR alleles display sex-dependent effects on survival and discriminate between individual and familial longevity. Hum. Mol. Genet. 7, 187–194.

Izaks, G.J., van Houwelingen, H.C., Schreuder, G.M., Ligthart, G.J., 1997. The association between human leucocyte antigens (HLA) and mortality in community residents aged 85 and older. J. Am. Geriatr. Soc. 45, 56–60.

Izaks, G.J., Remarque, E.J., Schreuder, G.M.T., Westendorp, R.G.J., Ligthart, G.J., 2000. The effect of geographic origin on the frequency of HLA antigens and their association with ageing. Eur. J. Immunogenet. 27, 87–92.

Keso, T., Perola, M., Laippala, P., Ilveskoski, E., Kunnas, T.A., Mikkelsson, J., Penttila, A., Hurme, M., Karhunen, P.J., 2001. Polymorphisms within the tumor necrosis factor locus and prevalence of coronaryartery disease in middle-aged men. Atherosclerosis 154, 691–697.

Klein, J., Sato, A., 2000. The HLA system. Second of two parts. N. Engl. J. Med. 343, 782–790.

Krishnaraj, R., Svanborg, A., 1992. Preferential accumulation of mature NK cells during human immunosenescence. J. Cell. Biochem. 50, 386–391.

Kutza, J., Kaye, D., Murasko, D.M., 1995. Basal natural killer cell activity of young versus elderly humans. J. Geront. 50, 110–116.

Lagaay, A.M., D'Amaro, J., Ligthart, G.J., Schreuder, G.M., van Rood, J.J., Hijmans, W., 1991. Longevity and heredity in humans Association with the human leukocyte antigen phenotype. Ann. N. Y. Acad. Sci. 621, 78–89.

Laurat, E., Poirier, B., Tupin, E., Caligiuri, G., Hansson, G.K., Bariety, J., Nicoletti, A., 2001. In vivo downregulation of T helper cell 1 immune responses reduces atherogenesis in apolipoprotein E-knockout mice. Circulation 104, 197–202.

Lee, R.T., Libby, P., 1997. The unstable atheroma. Arterioscler. Thromb. Vasc. Biol. 17, 1859–1867.

Lio, D., Balisteri, C.R., Colonna-Romano, G., Motta, M., Franceschi, C., Malaguarnera, M., Candore, G., Caruso, C., 2002a. Association between the MHC class I gene HFE polymorphisms and longevity: a study in Sicilian population. Genes Immun. 3, 20–24.

Lio, D., Scola, L., Crivello, A., Bonafè, M., Franceschi, C., Olivieri, F., Colonna-Romano, G., Candore, G., Caruso, C., 2002b. Allele frequencies of +874T→A single nucleotide polymorphism at the first intron of interferon-γ gene in a group of Italian centenarians. Exp. Gerontol. 37, 317–321.

Lio, D., Scola, L., Crivello, A., Colonna-Romano, G., Candore, G., Bonafè, M., Cavallone, L., Franceschi, C., Caruso, C., 2002c. Gender specific association between −1082 IL-10 promoter polymorphism and longevity. Genes Immun. 3, 30–33.

Lio, D., Marino, V., Gioia, V., Scola, L., Crivello, A., Forte, G.I., Colonna-Romano, G., Candore, G., Caruso, C., 2002d. Genotype frequencies of +874T→A single nucleotide polymorphism at the first intron of interferon-γ gene in a sample of Sicilian patients affected by tuberculosis. Eur. J. Immunogenet. 229, 371–375.

Mariani, E., Ravaglia, G., Forti, P., Meneghetti, A., Tarozzi, A., Maioli, F., Boschi, F., Pratelli, L., Pizzoferrato, A., Piras, F., Facchini, A., 1999. Vitamin D, thyroid hormones and muscle mass influence natural killer (NK) innate immunity in healthy nonagenarians and centenarians. Clin. Exp. Immunol. 116, 19–27.

McCusker, S.M., Curran, M.D., Dynan, K.B., McCullagh, C.D., Urquhart, D.D., Middleton, D., Patterson, C.C., McIlroy, S.P., Passmore, A.P., 2001. Association between polymorphism in regulatory region of gene encoding tumour necrosis factor alpha and risk of Alzheimer's disease and vascular dementia, a case-control study. Lancet 357, 436–439.

McNerlan, S.E., Alexander, H.D., Rea, I.M., 1999. Age-related reference intervals for lymphocyte subsets in whole blood of healthy individuals. Scand. J. Clin. Lab. Invest. 59, 89–92.

Miyaji, C., Watanabe, H., Toma, H., Akisaka, M., Tomiyama, K., Sato, Y., Abo, T., 2000. Functional alteration of granulocytes, NK cells, and natural killer T cells in centenarians. Hum. Immunol. 61, 908–916.

Moalem, S., Percy, M.E., Andrews, D.F., Kruck, T.P., Wong, S., Dalton, A.J., Mehta, P., Fedor, B., Warren, A.C., 2000. Are hereditary hemochromatosis mutations involved in Alzheimer disease? Am. J. Med. Genet. 393, 58–66.

Moore, K.W., de Waal, Malefyt, R., Coffman, R.L., O'Garra, A., 2001. Interleukin-10 and the interleukin-10 receptor. Annu. Rev. Immunol. 19, 683–765.

Murasko, D.M., Weiner, P., Kaye, D., 1988. Association of lack of mitogen induced lymphocyte proliferation with increased mortality in the elderly. Aging: Immunol. Inf. Dis. 1, 1–23.

Murphy, G.M., Jr., Claassen, J.D., DeVoss, J.J., Pascoe, N., Taylor, J., Tinklenberg, J.R., Yesavage, J.A., 2001. Rate of cognitive decline in AD is accelerated by the interleukin-1 alpha −889 *1 allele. Neurology 56, 1595–1597.

Mysliwski, A., Mysliwska, T., Chodnik, T., Bigda, E., Bryl, E., Foerster, J., 1993. Elderly high NK responders are characterized by intensive proliferative response to PHA and Con-A and optimal health status. Arch. Gerontol. Geriatr. 16, 199–205.

Nicklin, M.J., Weith, A., Duff, G.W., 1994. A physical map of the region encompassing the human interleukin-1 alpha, interleukin-1 beta, and interleukin-1 receptor antagonist genes. Genomics 19, 382–384.

Ogata, K., An, E., Shioi, Y., Nakamura, K., Luo, S., Yokose, N., Minami, S., Dan, K., 2001. Association between natural killer cell activity and infection in immunologically normal elderly people. Clin. Exp. Immunol. 124, 392–397.

Papassotiropoulos, A., Bagli, M., Jessen, F., Bayer, T.A., Maier, W., Rao, M.L., Heun, R., 1999. A genetic variation of the inflammatory cytokine interleukin-6 delays the initial onset and reduces the risk for sporadic Alzheimer's disease. Ann. Neurol. 45, 666–668.

Papasteriades, C., Boki, K., Pappa, H., Aedonopoulos, S., Papasteriadis, E., Economidou, J., 1997. HLA phenotypes in healthy aged subjects. Gerontology 43, 176–181.

Pawelec, G., Barnett, Y., Forsey, R., Frasca, D., Globerson, A., McLeod, J., Caruso, C., Franceschi, C., Fülöp, T., Gupta, S., Mariani, E., Mocchegiani, E., Solana, R., 2002. T cells and aging. 2002 update. Front. Biosci. 7, d1056–d1183.

Potestio, M., Pawelec, G., Di Lorenzo, G., Candore, G., D'Anna, C., Gervasi, F., Lio, D., Tranchida, G., Caruso, C., Colonna-Romano, G., 1999. Age-related changes in the expression of CD95 (APO1/FAS) on blood lymphocytes. Exp. Gerontol. 34, 659–673.

Pravica, V., Perrey, C., Stevens, A., Lee, J.H., Hutchinson, I.V., 2000. A single Nucleotide polymorphism in the first intron of the human IFN-γ gene: absolute correlation with a polymorphic CA microsatellite marker of high IFN-γ gene. Hum. Immunol. 61, 863–866.

Proust, J., Moulias, R., Fumeron, F., Bekkhoucha, F., Busson, M., Schmid, M., Hors, J., 1982. HLA and longevity. Tissue Antigens 19, 168–173.

Rauramaa, R., Vaisanen, S.B., Luong, L.A., Schmidt-Trucksass, A., Penttila, I.M., Bouchard, C., Toyry, J., Humphries, S.E., 2000. Stromelysin-1 and interleukin-6 gene promoter polymorphisms are determinants of asymptomatic carotid artery atherosclerosis. Arterioscler. Thromb. Vasc. Biol. 20, 2657–2662.

Rea, I.M., Middleton, D., 1994. Is the phenotypic combination A1B8Cw7DR3 a marker for male longevity? J. Am. Geriatr. Soc. 42, 978–983.

Remarque, E., Pawelec, G., 1998. T cell immunosenescence and its clinical relevance in man. Rev. Clin. Gerontol. 8, 5–14.

Remarque, E.J., Bollen, E.L., Weverling-Rijnsburger, A.W., Laterveer, J.C., Blauw, G.J., Westendorp, R.G., 2001. Patients with Alzheimer's disease display a pro-inflammatory phenotype. Exp. Gerontol. 36, 171–176

Roberts-Thomson, L.C., Whittingham, S., Youngchaiyud, U., Mackay, I.R., 1974. Ageing, immune response and mortality. Lancet 2, 368–370.

Rosenwasser, L.J., 1998. Biologic activities of IL-1 and its role in human disease. J. Allergy Clin. Immunol. 102, 344–350.

Sampietro, M., Caputo, L., Casatta, A., Meregalli, M., Pellagatti, A., Tagliabue, J., Annoni, G., Vergani, C., 2001. The hemochromatosis gene affects the age of onset of sporadic Alzheimer's disease. Neurobiol. Aging 22, 563–568.

Sansoni, P., Cossarizza, A., Brianti, V., Fagnoni, F., Snelli, G., Monti, D., Marcato, A., Passeri, G., Ortolani, C., Forti, E., 1993. Lymphocyte subsets and natural killer activity in healthy old people and centenarians. Blood 82, 2767–2773.

Smith, D.A., Irving, S.D., Sheldon, J., Cole, D., Kaski, J.C., 2001. Serum levels of the anti-inflammatory cytokine interleukin-10 are decreased in patients with unstable angina. Circulation 104, 746–749.

Solana, R., Alonso, M.C., Pena, A.J., 1999. Natural killer cells in healthy aging. Exp. Gerontol. 34, 435–443.

Solana, R., Mariani, E., 2000. NK and NK/T cells in human senescence. Vaccine 18, 1613–1620.

Solerte, S.B., Cravello, L., Ferrari, E., Fioravanti, M., 2000. Overproduction of IFN-gamma and TNF-alpha from natural killer (NK) cells is associated with abnormal NK reactivity and cognitive derangement in Alzheimer's disease. Ann. N.Y. Acad. Sci. 917, 331–340.

Sperr, W.R., Huber, K., Roden, M., Janisiw, M., Lang, T., Grat, S., Maurer, G., Mayr, W.R., Panzer, S., 1998. Inherited platelet glycoprotein polymorphisms and a risk for coronary heart disease in young central europeans. Thrombosis Res. 90, 117–123.

Tarkowski, E., Liljeroth, A.M., Nilsson, A., Ricksten, A., Davidsson, P., Minthon, L., Blennow, K., 2000. TNF gene polymorphism and its relation to intracerebral production of TNFalpha and TNFbeta in AD. Neurology 54, 2077–2081.

Thoman, M.L., Weigle, W.O., 1989. The cellular and subcellular bases of immunosenescence. Adv. Immunol. 46, 221–262.

Turner, D.M., Williams, D.M., Sankaran, D., Lazarus, M., Sinnott, P.J., Hutchinson, I.V., 1997. An investigation of polymorphism in the interleukin-10 gene promoter. Eur. J. Immunogenet. 24, 1–8.

Volpato, S., Guralnik, J.M., Ferrucci, L., Balfour, J., Chaves, P., Fried, L.P., Harris, T.B., 2001. Cardiovascular disease, interleukin-6 and risk of mortality in older women. The women's health and aging study. Circulation 20, 947–953.

Wang, X.L., Oosterhof, J., 2000. Tumour necrosis factor alpha G-308→A polymorphism and risk for coronary artery disease. Clin. Sci. (London) 98, 435–437.

Wang, X.Y., Hurme, M., Jylha, M., Hervonen, A., 2001. Lack of association between human longevity and polymorphisms of IL-1 cluster, IL-6, IL-10 and TNF-alpha genes in Finnish nonagenarians. Mech. Ageing Dev. 123, 29–38.

Wayne, S.J., Rhyne, R.L., Garry, P.J., Goodwin, J.S., 1990. Cell-mediated immunity as a predictor of morbidity and mortality in subjects over 60. J. Gerontol. 45, 45–48.

Weksler, M.E., 2000. Changes in the B-cell repertoire with age. Vaccine 18, 1624–1628.

Wikby, A., Maxson, P., Olsson, J., Johansson, B., Ferguson, F.G., 1998. Changes in CD8 and CD4 lymphocyte subsets, T-cell proliferation responses and non-survival in the very old: the Swedish longitudinal OCTO immune study. Mech. Ageing Dev. 102, 187–198.

Yudkin, J.S., Kumari, M., Humphries, S.E., Mohamed-Ali, V., 2000. Inflammation, obesity, stress and coronary heart disease, is interleukin-6 the link? Atherosclerosis 148, 209–214.

Advances in
Cell Aging and
Gerontology

Developmental aspects of the thymus in aging

Amiela Globerson

Department of Immunology, Weizmann Institute of Science, Rehovot, Israel
Correspondence address: Fax: +972-8-934-4141. E-mail: amiela.globerson@weizmann.ac.il

Contents

Abbreviations

5-HT: serotonin; AchE: acetylcholinesterase; BM: bone marrow; CCK-8s: sulfated cholecystokinin octapeptide; CK: cytokeratin; DN: double negative

Advances in Cell Aging and Gerontology, vol. 13, 47–78

(CD4−CD8−); DP: double positive (CD4+CD8+); E2: 17 beta estradiol; ER: estrogen receptor; FTOC: fetal thymus organ culture; GC: glucocorticoids; GH: growth hormone; GRP: gastrin-releasing peptide; Gx: gonadectomized; IGF: insulin-like growth factor; KO: knockout; LH−RH: luteinizing hormone–releasing hormone; mAb: monoclonal antibody; NA: noradrenaline; NK: natural killer; NPY: neuropeptide Y; PBL: peripheral blood lymphocytes; RAG: recombinase activating gene; SDF: stromal derived factor; SPG: sucrose phosphate glyoxylic acid; SS: somatostatin; SSR: somatostatin receptor; TCR: T cell receptor; TEC: thymic epithelial cells; TREC: T cell receptor excision circles; WT: wild type.

1. Introduction

The thymus resumes its maximal size at puberty in the human and at the age of about 1 month in mice. Thereafter the thymus declines, and this process has a significant role in the status of the peripheral T lymphocyte compartment in old age. Age-related involution of the thymus was originally described by Hammar (cf. Bodey et al., 1997) as a basic histogenetical rule, and it was assumed to represent "Alternsinvolution". The questions of whether the aged thymus continues to generate lymphocytes and whether the newly generated cells are functional as in the young, have been investigated in many laboratories. Studies over the last decade have made it clear that the T lymphocyte pool in the peripheral blood and lymphoid tissues of aged humans and experimental animals does not show any significant reduction in the total cell number. However, the profile of T lymphocytes is shifted towards increased values of memory/activated vs. naive cell types (Miller, 1991; Globerson, 1995) and cytokine production is shifted from Th1 to Th2 types (Weigle, 1993; Segal et al., 1997). Additional phenotypical and functional changes were observed in decreased expression of HLA-DR and CD7 (Ginaldi et al., 2000), decreased percentage of memory cells exhibiting CD50, and in expression of CD62L among the naïve cells of the old (DeMartinis et al., 2000).

The shift in profile from naïve to memory/activated cell types was originally related to post-thymic processes, namely, accumulation of cells that have been exposed to antigens during the lifespan within the peripheral lymphoid tissues (Livak and Schatz, 1996; Goldrath et al., 2000), However, recent data point to more complex developmental mechanisms, within the framework of thymocytopoiesis (Timm and Thoman, 1999) as well as to possible increased apoptosis of naïve (CD45RA+) cells upon activation (Mountz et al., 1997). Accordingly, the thymus continues to function in aging, yet with an altered output pattern.

Generation of naïve T lymphocytes continues through very old age, at a reduced rate, as indicated from a series of studies. Firstly, substantial levels of naïve T cells were observed in centenarians (Franceschi et al., 1995; Bagnara et al., 2000; Globerson, 2000). Secondly, TdT expression in the thymus was shown in aging persons (70-year-old), despite the subtotal physiological involution (Steinmann and Muller-Hermelink, 1984). Further studies were based on the fact that excision circles created during T cell receptor (TCR) rearrangements (TRECs) are present exclusively in naïve T cells that have not yet undergone any cell division. Relatively high levels of TRECs were observed in elderly subjects, suggesting the occurrence of newly generated T cells (Douek and Koup, 2000). The fact that TRECs are not found following thymectomy conforms to the idea that they are thymus-derived, rather than originating in extrathymic sites (Douek and Koup, 2000).

The belief that occurrence of TRECs represents recent thymic emigrants gains support also from studies on avians (Kong et al., 1998). The chicken TI+ cells are evenly distributed among all of the peripheral T lymphocyte compartments, and their levels in the periphery gradually decline in parallel with age-related thymic involution. These cells disappear following early thymectomy. Measurement of recent thymic emigrants in the periphery thus provides an accurate indication of thymic function (Kong et al., 1998).

The recent human thymic emigrants respond to costimulatory signals, indicating that the thymus retains ability to give rise to functional T lymphopoiesis even late in life (Jamieson et al., 1999). However, not all of the newly generated T cells are fully functional in aging, and the naïve T cell compartment is actually a mosaic of potentially active, as well as anergic cells (Miller, 1991, 1996). The mechanisms underlying anergy are not yet adequately elucidated. Anergy is related, at least in part, to altered patterns of signal transduction, as demonstrated in the lymphocytes of aged mice (Utsuyama et al., 1997b; Fulop et al., 1999; Tamir et al., 2000). The inactive cells may have not completed their differentiation to resume function, or they may have developed from stem cells that were close to a stage of replicative senescence (Globerson, 1999; Effros and Globerson, 2002).

Taken together, the bulk of evidence points to continuous function of the thymus throughout age, yet at a reduced rate and altered patterns. The different aspects of T cell development in aging were studied in a variety of experimental systems, including in vivo assessment of the aged thymus and transplantation into compromised recipient mice, as well as in vitro, particularly in organ culture models of adult and fetal thymic explants (FTOC) (Globerson and Auerbach, 1965; Jenkinson et al., 1982; Eren et al., 1988). The present article reviews these various developmental aspects of thymic involution and function in aging.

2. The thymic microenvironment in aging

Thymic involution involves both the stroma and lymphoid compartments, manifested in the progressive loss of normal organ architecture and cellular composition, and reduction in the output of mature T lymphocytes.

The thymic micro-environment provides signals to the lymphoid cells as a result of cell–cell interactions, locally produced cytokines, chemokines and hormones. Developing thymocytes, in turn, may influence the thymic stroma to form a supportive micro-environment (Mehr et al., 1995; Van Ewijk et al., 2000). The patterns of cross cell–cell interactions that have been well documented for thymocytopoiesis (Boyd et al., 1993) seem to change with age (Globerson, 1997, 2002). In vivo experimental strategies based on transgenic mouse models and transplantation procedures, as well as in vitro systems using cell cultures or thymic organ culture techniques, have led to considerable advancement in understanding of thymic stroma and lymphoid cells interactions, as described in the following sections.

A decrease in the thymic weight/body weight ratio is a hallmark of aging. However, thymic involution is characterized also by qualitative changes, manifested in the altered architecture of cortex and medulla, and the different cell type components (e.g. epithelial cells, macrophages, dendritic and antigen-presenting cells having different properties, in the cortex and medulla (Takeoka et al., 1996; Nabarra and Andrianarison, 1996; Bertho et al., 1997) as further described.

Takeoka et al. used a well-defined panel of monoclonal antibodies (mAbs) that recognize and characterize the thymic miroenvironment of both epithelial and nonepithelial elements (Takeoka et al., 1996). Their results disclosed significant age-related morphometric changes in the thymic microenvironment in 12-month-old C3H/HeJ, C57BL/6 and BALB/c mice, compared to young 4- to 6-week-old, as well as 6-month-old BALB/c mice. In principle, the thymuses of the old mice had normal and distinctive separation of cortical and medullary epithelium. However, there were obvious changes in these regions, including the fact that the cortex and medulla were diffusely irregular and atrophic and cortico medullary junction was poorly defined. The cortex showed small disrupted epithelial networks, the medulla contained clusters of atrophic cells, and the extracellular matrix was increased and contained large irregularly shaped clusters. The thymus of 6-month-old mice expressed some changes within the medullary epithelium and the extracellular matrix, with no difference in the cortical epithelium, suggesting that changes in the medulla precede those of the cortical region.

Nabarra and Andrianarison noticed cellular damages that progressively affected all the thymic stroma in aged mice (Nabarra and Andrianarison, 1996). At the age of about 18–20 months the organ architecture disappeared and there was a drastic decrease in lymphocyte numbers. The cellular integrity of the microenvironment was lost, and one could detect lysis of cellular membranes and formation of a large and clear cytoplasmic layer engulfing a few remaining lymphocytes.

Studies on rats (Elcuman and Akay, 1998), showed no significant age-dependent variations in the immunolocalization of fibronectin, or in the histological structure of the thymus between males and females of the same age. However, there was an age-related increase in the fibronectin content of the thymic capsule, the connective tissue between the lobules around blood vessels, as well as in the medulla and cortex of the thymus. Age-dependent increases in the thymus of both sexes were also noted in the connective-tissue content between lobules, fat cells, Hassall's corpuscles, the thickness of capsule and the ratio of the medulla to the cortex of the lobules. Immunohistochemical studies on cytokeratin (CK) expression by thymic epithelial cell (TEC) subsets (Masunaga et al., 1997) in thymic specimens from neonatal, infantile, and one adult thymus specimen obtained at autopsy have been performed. Analyses were based on the binding of mAbs specific for CK4, CK8, CK13, CK18 and CK19. Simultaneous expression of CK4, CK8, CK13, CK18 and CK19 was observed in the cortex, medulla and subcapsular area. Expression of CK8 and CK19 was overlapped, suggesting the formation of complexes in the cytoplasm of TEC.

Expression of CK4, CK13 and CK18 was attenuated, or absent in the subcapsular area during the early involution stage.

Interestingly, the 6C3 marker (Wu et al., 1989; Sherwood and Weissman, 1990) expressed in the cortical components of the thymus was not detectable in aged thymuses, compared to the young (Globerson, unpublished). Whether cells expressing this marker develop intrinsically in the thymus, or whether they derive from an extrinsic source of stem cells, is as yet unknown.

Analyses of thymic biopsy specimens obtained from 105 patients during cardiac surgery with no underlying immunological abnormalities (Nakahama et al., 1990), showed a gradual decrease in the proportion of medullary dendritic cells and the relative volume of cortical thymocytes, up to the age of 40 years, with no major change in the number of medullary epithelial cells. These findings indicate that in addition to the decreased size, thymic involution has distinct qualitative changes in the morphology of different anatomic regions, cellular components and expression of molecular markers.

Attempts at expansion and maintenance of thymic stromal cells have been based on using a semi-organ culture method, in which the bulk of all thymic stromal cells were cultivated in a mixed monolayer cell culture (Small and Weissman, 1996). A key element in this method was that it depended on the prevention of fibroblast over-growth. These studies revealed that the mixed cell culture could act as an inductive thymic microenvironment in support of T cell development from stem cells. In addition, they indicated the capacity of the stromal cells to expand under suitable in vitro conditions. However, thymic stroma that was obtained from aged mice failed to grow under such conditions (Globerson, unpublished). Cultivation of thymic stroma from aged mice without elimination of fibroblast growth, resulted in a limited appearance of a variety of thymic stroma cells, including epithelial and dendritic cells. It thus appears that the potential for regeneration of the thymic stroma is retained to a certain extent in old age, although it is not readily manifested, and it may be dependent on fibroblast-derived factors. In contrast to the case of the young thymus, growth of fibroblasts from the aged thymus was a priori limited, but it was sufficient to enable stromal cell growth. This observation could be explained on the basis of a failure to replicate, possibly due to replicative senescence of the cells, an absence of critical growth factors, or occurence of inhibitory effects.

Interestingly, vitamin E enhanced T cell differentiation in the thymus through the increase of stromal TEC (Moriguchi, 1998). The effect was related to increased binding of immature T cells to the epithelial cells, via increased expression of ICAM-1. A variety of mechanisms may thus be responsible to the limited growth of stromal elements in the aged thymus. The basis of these changes and their functional relevance still need to be elucidated.

3. The lymphocyte compartment in the aging thymus

The early phases of lymphocyte development in the thymus require cell–cell inter-actions, and is mediated by soluble factors provided by stromal cells within the thymic microenvironment. A review outlining genes that may be involved in these processes has been published recently (Sen, 2001).

Developmental processes in the aging thymus have been shown to be affected at the various stages from the initial homing to the microenviroment, through cell replication and apoptosis, transition from CD4/CD8 double negative (DN) to double positive (DP) subsets, reduced TCR gene rearrangements and the establishment of the ultimate proportion of single positive CD4/CD8 (CD4+CD8− and CD4−CD8+) lineages. These different age-related changes in developmental processes can be divided into those based on intrinsic changes in the stem cells, and those determined by the thymic microenviroment, as discussed in detail in the following sections.

Patterns of changes in cellular subsets and levels of expression of molecular markers seem to differ in different mouse strains, as well as within the same strain, pointing to an age-related impact of stochatic events, in addition to the genetic differences between strains (Dubiski and Cinader, 1992). The large variability observed between mouse strains and intrastrains suggest a wide range of individuality that could occur in the human.

Whereas the total proportion of CD45+ thymocytes remains unchanged in advanced age, thymocyte-membrane densities of CD45 undergo age-related increases, and the proportion of cells expressing this marker at high density is similar in both SJL and BALB/c (Dubiski and Cinader, 1992). Individual variations in relative size of subpopulations in SJL mice of the same age are greater in old than in young mice, and not in BALB/c or C57BL/6 mice.

A shift in CD4/CD8 thymocyte subsets has been observed in the thymus of aged mice of various strains (Dubiski and Cinader, 1992; Thoman, 1995; Yu et al., 1997; Aspinall and Andrews, 2001a; Andrews and Aspinall, 2002), manifested mainly in a decrease in DP and increased values of DN cell types. By the age of 24–27 months the percentage of the total DN population increases two- to threefold over that of the young, while the fraction of DP cells is significantly reduced. Age-related changes were noted in thymus cell subpopulations of BALB/c and SJL mice, in the proportion of cells identified by various markers, and by the membrane density of these markers. In both strains there was an age-related increase in the proportion of single positive CD4+CD8− cells and decrease in DP cells. By 24–27 months of age, the percentage of the total DN population in C57BL/6 mice approached two- to threefold over that of young (2–3 months), while the fraction of DP is significantly reduced (Thoman, 1995; Yu et al., 1997). The studies by Thoman on C57BL/6 mice of various ages showed a similar threefold increase in the percentage of DN cells which express CD3 (from 16.6 to 45.5%) between 4 and 14 months of age (Thoman, 1995).

The DN cells were signified by increased proportions of CD44+CD25−DN, the earliest thymic progenitors, and decreased CD44−CD25+DN cells (Thoman, 1995; Yu et al., 1997; Aspinall and Andrews, 2001a). Differences were noted between different mouse strains, as well as within aged mice of the same strain (Dubiski and Cinader, 1992; Yu et al., 1997). Thus, an increase in CD44+ cells was noted in SJL and C57BL/6, but not in BALB/c thymuses. Cells with high density CD44 were noted in later life in SJL and BALB/c, yet it was more marked in SJL (Dubiski and Cinader, 1992).

These changes are consistent with the possibility that an alternative T cell differentiation pathway plays an increasingly significant role with advancing age. The shifts in thymocyte subsets may thus represent a general "universal" phenomenon, although the quantitative aspects and kinetics of appearance as related to age may differ. Since the transition from CD44+CD25− to CD44−CD25+ involves interleukin-7 (IL-7) effects, attempts have been made to retard thymic involution by treatment with IL-7. Indeed, such effects have been observed (Andrews and Aspinall, 2002; Thoman, 1997), yet this approach seems to be effective only in the thymus of young, and not old mice (Phillips et al., 2002).

3.1. The CD4/CD8 double negative stages

Increased values of myeloid cells were noted when bone marrow (BM) from aged mice was co-cultured with fetal thymus lobes (Globerson, unpublished), or a BM derived stroma cell line (Sharp et al., 1989), consistent with observations on elderly human subjects (Resnitzky et al., 1987). More recent studies, using isolated hematopoietic stem cells from aged mice showed a similar tendency for increased myelopoiesis (Morrison et al., 1996). The aged BM-derived cells in the FTOC showed also increased levels of cells expressing the stem cell marker c-kit (Globerson, 2002) and the adhesion molecule CD44 (Yu et al., 1997), suggesting a limit in downregulation of these markers.

Interleukin-1 (IL-1) is involved in thymocyte development, at the immature DN stage. Although normal thymocytes barely express the interleukin-1 receptor (IL-1R), expression of IL-1R (type I) substantially increased at days 12–15 of FTOC and the CD4/CD8 profile of the IL-1R (type I)+ cells were mostly restricted in the DN and DP subsets. Interestingly, in vitro culture of the thymocytes from aged mice, but not those from young adult or newborn mice, revealed similar results to those of FTOC (Oh and Kim, 1999). In addition, half of the IL-1R+ cells that increased in the later period of FTOC were gammadelta thymocytes. Accordingly, IL-1R is expressed on thymocytes during ex vivo culture, and possibly during normal thymocyte differentiation.

The transition of CD4/CD8 DN cells from the phase of CD44+CD25− to CD44−CD25+, is of particular interest. This stage is also characterized by the onset of TCRb and expression of recombinase activating gene (RAG), and it is subject to the efects of IL-7 (Zlotnik and Moore, 1995). The aging thymus is characterized by a decreased capacity to downregulate CD44 and to express CD25 (Thoman, 1995), as well as RAG-1 and RAG-2 (Ben-Yehuda et al., 1998; Aspinall and Andrews, 2000). A decline in these critical processes is fundamental in leading to thymic involution. Indeed, studies on RAG-transgenic mice have lent support to this hypothesis (Aspinall and Andrews, 2000).

The question of whether the decrease in RAG expression in the aged thymus is based on stem cells properties or on the thymic microenvironment was examined under in vitro experimental conditions (Ben-Yehuda et al., 1998). Fetal thymus explants were co-cultured with either BM cells depleted of mature T cells, or DN CD4/CD8 thymocytes from young, as well as old donors, incubated under conditions enabling T cell development, and then examined at different time intervals for the expression of RAG. The results showed RAG expression in cells that derived from the BM of both young and old donors, and from the thymocytes of the young – but not those of the old. Accordingly, the decrease in RAG expression is not due to aging effects on the BM cells, but rather on the thymic microenvironment. Furthermore, the mere residence of DN cells in the aged thymus rendered them incapable of RAG expression. Whereas the basis of this negative control is still not clear, the results point to manifestation of aging effects on the thymic stroma.

3.2. The T cell receptor repertoire and autoreactive cells

Does the aging thymus retain the capacity for the elimination of anti-self clones? This question was approached by Crizi et al., using Vβ3 and Vβ17a antibodies to determine the presence and functionality of normally deleted T cells bearing potentially self-reactive TCR in peripheral lymphoid tissue and blood from aged (SJL/J × BALB/c) F1, LAF1 and BALB/c mice (Crisi et al., 1996). Although an occasional 20- to 24-month-old mouse exhibited Vβ3+ or Vβ17a+ T cells in their lymph nodes or peripheral blood lymphocytes (PBL) slightly above the range for normal young mice of these I−E+ strains, there was no striking 'escape' from the normal thymic deletion process. However, responsiveness to anti-Vβ3 and -Vβ17a was to a certain degree higher in aged, and particularly in aged thymectomized, than in young animals. This was in contrast to proliferative responses to stimulation with antibody to the normally expressed Vβ8, which were lower in the lymph nodes from aged than from young mice. Increased numbers of 'forbidden' Vβ bearing T cells were seen more frequently in mice 30–36 months old. In spite of the age-related decrease in overall CD4/CD8 T cell ratios in all organs, the mice with relatively high Vβ17a+ T cells exhibited proportionally more CD4+ cells in that Vβ population. The 'forbidden' T cells in the 20- to 24-month-old mice that responded to anti-Vβ stimulation seemed to derive from an extrathymic origin, since they were more readily detectable in the thymectomized mice. Potentially self-reactive CD4 (and CD8) T cells were detectable in the blood only in very aged (30–36 months old) euthymic mice.

Doria et al. transferred syngeneic BM cells from young or old donors to lethally irradiated young and old mice, to investigate whether self reactivity in old mice results from age-related damage of the radioresistant stromal cells and/or of the BM hematopoietic cells (Doria et al., 1997). Three months after irradiation, thymus and spleen cell repopulations and mitotic responses were lower in the old recipients. In contrast, BM cells from young and old donors repopulated the thymus and spleen of recipients of the same age equally well. Interestingly, serum auto-antibodies and glomerular lesions at 3 and 9 months after irradiation were more pronounced in the old than in young recipients, regardless of the age of the BM cell donors. These findings support the possibility that age-related damage of stromal cells induces dysregulation of the immune system leading to the appearance of autoantibodies.

Taken together, these studies suggest both thymic and extrathymic effects which influence the manifestation of autoreactivity in aging.

3.3. Transition to CD4/CD8 single positive populations

This distinct production of CD4+CD8− and CD4−CD8+ thymocytes is regulated by the developmental age of the thymic stroma. The differential expression of Notch receptors and their ligands (especially Jagged1) throughout thymus development (Borowski et al., 2002) plays a key role in the generation of the different CD4 vs. CD8 cell ratios. Jimerez et al. showed that this cell ratio in the thymus changes sharply from fetal to adult values around birth. Differences in the proliferation and emigration rates of the mature thymocyte subsets also contribute to that change (Jimenez et al., 2001).

3.4. Cell replication and programmed cell death

Reduced output of newly generated T cells in aging could be causally related to a decline in cell replication and/or to an increased rate of programmed cell death in the aging thymus. The developmental stage of major cell replications in the thymus is at the DN phase, and most of programmed cell death is related to the DP stage, as part of the negative selection processes.

Studies of Bar-Dayan et al. on the pattern of proliferation and apoptosis of the thymus in 1-month-old, compared to 7 months old female mice showed that the proliferation index of the peripheral cortex of the 1-month-old mice was significantly higher than that of the deep cortex and that of the medulla (×3.6 and ×5.8, respectively) and there was a 45% reduction in the peripheral cortex of the older mice compared to the young (Bar-Dayan et al., 1999). The apoptotic index of the cortico-medullary junction of the young was sixfold higher than that of the cortex, and 18-fold higher than the medulla, and the thymic cortex of the older mice showed values that were 66% higher in the older mice compared to the young. It was thus concluded that these age-related changes might account for the reduction of thymic cortical cellularity during thymic involution.

However, other studies suggest that levels of both mitoses and apoptoses are decreased in the aging thymus. The difference in results obtained from different laboratories may be due to different mouse strains used (e.g. ICR mice in the experiments of Bar Dayan, and C57BL/6 mice in most of the other studies), as well as the age of "old" mice. In addition, major gender differences have been shown in patterns of thymic involution (Bar-Dayan et al., 1999; Aspinall and Andrews, 2001b).

Studies focusing specifically on cell replication in the thymus were performed both in vivo and in vitro. In vivo experiments were based on injection of 3H-thymidine and subsequent analysis of thymocytes by autoradiography (Hirokawa et al., 1988). In addition, in vitro studies examining mitogen (concanavalin A) stimulated cells in thymic organ cultures, and subsequent flowcytometry (Mehr et al., 1996), reinforced the conclusion that the rate of cell replication in the thymus is decreased in advanced age, also following arbitrary triggering of the cells to divide. The results of all of these studies provided further support to the idea that aging is associated with reduced cell replication in the thymus.

Studies monitoring cells in apoptosis in the aged thymus revealed decreased, rather than increased levels. Reduced levels of apoptosis were shown by flow cytometric analyses of thymic cells, indicating a decreased sub-G0 fraction in thymocytes of old, compared to young adult mice. Spontaneous apoptosis following 24 h incubation of thymocytes in vitro was also decreased in thymocytes of the aged mice (Globerson, unpublished).

Interestingly, although the frequency of BM cells that colonized the thymic lobes was reduced with donor age, it had no impact on the total cell number by the first week of culture. This was related to the fact that BM cells of old donors replicated upon seeding in the thymus, while cells of the young started replication only 24 h later (Sharp et al., 1991). However, cells of the old showed a decreased capacity for subsequent sequential cell replications (Sharp et al., 1990a). The data suggest that the stem cells in the aged are a priori in cycle, yet their potential for subsequent replications is limited.

Recent studies on human cells, based on the FTOC experimental model, showed that BM CD34 cells from old donors generate less T cells in vitro than those of the young (Offner and Plum, 1998; Offner et al., 1999). However, the study led to the conclusion that loss of T cell generation capacity was not dependent on thymic involution and it correlated with chemotherapy treatment.

A variety of mechanisms may account for these age-related phenomena. Studies on p53 deficient mice suggested that accelerated development and aging of the immune system may be related to loss of cell cycle regulation, DNA repair and apoptosis (Ohkusu-Tsukada et al., 1999).

4. Stem cell migration and homing to the thymus

The ability of hematopoietic tissues of aged mice to provide cells that are capable of repopulating the thymus of an irradiated young animal persist throughout the lifetime of the donor, as shown in studies of radiation chimeras, and genetically anemic mice (Harrison, 1983), and from in vitro studies in which the BM cells of young and aged mice were applied directly onto fetal thymic explants (Eren et al., 1988). Competitive colonization assays revealed quantitative and qualitative differences between cells of old, compared to young donors, both in vivo (Harrison et al., 1993) and in vitro (Eren et al., 1988). Furthermore, inferiority of the old donor derived BM cells was also noticed in competition with resident cells within intact thymus explants (Fridkis-Hareli et al., 1992). In vitro competitive colonization enabled critical quantitative and qualitative analyses of these age-related changes at the BM level (Sharp et al., 1990a).

To determine the hematopoietic potential of the stem cell pool, BM cells were transplanted to hematologically compromised mice (irradiated, or genetically anemic mice). The results indicated efficient hematological reconstitution, as exhibited by long-term erythropoiesis (Harrison, 1983). Interestingly, T lymphocyte values decreased eventually in such hematologically reconstituted mice (Gozes et al., 1982;

Globerson, 1986), suggesting a selective decline in the capacity to generate T lymphocytes. Accordingly, it is necessary to distinguish between normal hematopoiesis in vivo as related to the chronological age of the individual, the intrinsic potential of stem cells for self-renewal, and the capacity to generate the different blood cell types.

McCormick and Haar used an in vitro model to investigate the BM-thymus axis in aged mice. Erythroid-depleted BM cells from 3-month- and 24-month-old CBA (Thy 1.2) mice were placed in the upper half of a blind-well chamber with thymus supernatant in the lower half (McCormick and Haar, 1991). Experimental cells were treated with thymus supernatant for 1 h prior to migration. Their data showed that pre-thymic stem cells in the aged BM were deficient in their ability to migrate under such conditions to the thymus supernatant. They also demonstrated that treatment of the old BM with supernatant from neonatal thymus cultures improved the thymus migrating ability of the aged BM stem cells. When TEC from newborn CBA/J mice were transplanted into aged syngeneic recipients, this led to enhanced in vitro migration of the host BM cells to supernatants prepared from neonatal thymus cultures, compared to control aged mice (Haar et al., 1989). It thus appears that the decrease in the capacity for migration is related, at least in part, to microenvironmental effects rather than purely intrinsic changes in the migrating BM cells.

To determine whether a reduction in the migratory readiness or the number of precursor cells might account for age-related thymic involution, Perez-Mera et al. studied the effect of mouse age on the in vitro migration of BM-derived hemopoietic precursor cells to thymic supernatant from newborn mice (Perez-Mera et al., 1991). The proportion of BM cells migrating to the thymic supernatants increased with age up to 7 weeks, and then decreased progressively, and this correlated with the changes in thymic weight. The decline of the T cell-precursors capacity for migration in adult mice may thus play a role in thymic involution.

The chemokine stromal derived factor-1 (SDF-1) and its receptor CXCR4 were found to be critical for the engraftment of human stem cells in the BM of SCID mice (Peled et al., 1999, 2000). The possibility that aged cells are unable to down-regulate CXCR4 expression in the BM thus deserves attention. High levels of CD44+ cells in the BM and thymus of aged mice (Yu et al., 1997), would suggest that the cells remain adherent to stroma elements and fail to proceed in subsequent steps of differentiation. The increased levels of fibronectin in the aged thymus (Elcuman and Akay, 1998) is in line with this hypothesis.

Mathematical modeling of the dynamics of cell adherence and detachment from the stromal elements (Mehr et al., 1994, 1995) raise the possibility that such mechanisms account for the generation of T cells in the aged thymus.

It may be speculated that hormonal effects on mobilization of stem cells from the BM to the thymus as described above is based on down-regulation of CD44, and/or CXCR4. Other chemokines (e.g. TECK) may also play a role in the adherence of thymocytes (Norment and Bevan, 2000; Bleul and Boehm, 2000). The effects of aging on expression of CXCR4 receptor and its SDF-1 ligand, and other possible chemokines in the BM still need to be elucidated.

The question of whether stem cells undergo senescence has been approached by both in vivo and in vitro methodologies. The different experimental strategies led to similar conclusions, as described next.

The approach to examine whether aging of haematopoietic stem cells may have any impact on a selective decrease in generation of T lymphocytes is of particular interest. Offner et al. examined the influence of age on the intrinsic capacity of stem cells to generate T cells (Offner et al., 1999). When BM derived CD34+ cells were sorted from healthy donors for transplantation, and introduced into murine fetal thymus lobes, three patterns of T cell precursor development were observed. Hence, either no human cells could be recovered from the cultures ("pattern A"), or maturation stopped at the CD4+CD8−CD3− pre-T cell stage ("pattern B"), or where maturation to DP thymocytes was achieved ("pattern C"). The data showed that in healthy donors over the age of 40, pattern C was limited, suggesting an intrinsic loss of T cell generation capacity from adult BM stem cells. However, this loss correlated weakly with age, irrespective of thymic involution, and data on cells following chemotherapy suggested further supperimposed effects of the treatment.

We examined the effects of aging of the BM separately from the issue of thymic involution, by seeding BM cells from young or old mice onto FTOC under conditions favoring T cell development (Eren et al., 1988). The number of lymphocytes harvested from the cultures under these conditions did not reveal any obvious difference between young and old donor-derived cells. However, a difference became apparent under competitive reconstitution, when identifiable cells of the young and the old donors were seeded onto the same individual thymic lobe (Eren et al., 1988; Sharp et al., 1990a). Identification of T cells that derived from both donors was possible by using congenic mice that express different Thy.1 allotypes (Thy1.1 and Thy1.2). Analysis of a 2-D PAGE of the membrane proteins in FTOC with young or old BM-derived cells indicated two distinct membrane polypeptides that are down-regulated in the young soon after thymic colonization (3 days), and this process is retarded in case of the old BM-derived cells (Francz et al., 1992). Accordingly, age-related effects are manifested in the BM cells at an early phase of interaction with the thymic stroma.

5. Peripheral lymphocyte traffic into the thymus

Mature lymphocytes exit from the thymus to the peripheral blood and lymphoid tissues, and this had long been considered as a one-way traffic of mature cells. However, re-entry of peripheral lymphocytes into the thymus has been noted under several conditions. These include chronic viral infection or antigen stimulation (Fink et al., 1984; Michie and Rouse, 1989; Agus et al., 1991; Haynes and Hale, 1998; Hardy et al., 2001; Chau et al., 2002). The question is then whether re-entry into the thymus has any role in regulation of thymocytopoiesis. Indeed, effects of mature cells were shown both on the developing lymphocytes and stroma cells (Fridkis-Hareli et al., 1994; Mehr et al., 1997; Globerson et al., 1999).

In vitro studies using co-cultures of isolated mature and immature T cell populations revealed effects of the mature lymphocytes on T cell development

(Fridkis-Hareli et al., 1994). Regulatory effects of mature CD4 T cells were also shown in T cell development from human umbilical cord blood cells, under similar in vitro experimental conditions (Globerson et al., 1999). Theoretical modeling of regulation of thymic cellularity suggested that the balance of cell numbers cannot be based solely on intrinsic processes and that one needs to consider also the role of cells entering the thymus from extrinsic sources (Mehr and Perelson, 1997; Van Ewijk et al., 2000).

Mathematical simulation of cell interactions in the thymus led to the hypothesis that T cell development in the thymus is regulated by the more mature thymocyte subsets (Mehr et al., 1997). Interestingly, cross-interactions of cells in the thymus were shown between thymocyte subsets and the thymic microenvironment, and induction of thymic cortex may be regulated by subpopulations of pro-thymocytes (Hollander et al., 1995). Peripheral lymphoid cells of aged mice had different regulatory effects on thymocytopoiesis compared to peripheral cells of the young (Fridkis-Hareli et al., 1994). The presence of CD4 single positive lymphocytes in cultures of CD4/CD8 DN thymocytes originating in different Thy1 congenic donors, respectively, led to a decrease in development of CD4/CD8 DP and CD8 single positive in the old-donor derived cells, as compared with the young ones. In contrast, CD8 single positive cells had no such effect.

The signal(s) leading mature T cells to exit from the thymic medulla represent an additional important aspect of T cell development. Our own studies pointed to T cell exit from the thymus to the peripheral lymphoid tissues upon in vivo antigen stimulation (Bernstein and Globerson, 1974). This was demonstrated in the in vitro experimental model for induction of an antibody response to a hapten conjugated to a protein carrier molecule. When splenocytes from mice immunized to the carrier protein were challenged in vitro with the hapten conjugated to the specific carrier, they produced anti-hapten specific antibodies. However, thymectomy prior to the in vivo immunization interfered with the subsequent splenocyte response in vitro to the hapten. It was thus concluded that in vivo immunization triggers thymic cells to migrate to the peripheral lymphoid tissues. It is tempting to assume that the cells enter cycle before emigration (Scollay and Godfrey, 1995). Since cells in the aging thymus show lower levels of divisions, this may be an additional cause of reduced output of newly generated cells from the aged thymus. Reduced output of the newly generated cells may thus account, at least in part, for the limited generation of T cells.

Altered patterns of T cell development, and the decreased output of naïve cells in aging, may thus be affected by a variety of mechanisms. These include effects of cells in progressive stages of differentiation on the less mature phenotypes, and on stromal elements, as well as incoming mature T cells, that may have been subject to long term exposure to antigens in vivo.

6. Occurrence of B lymphocytes in the aging thymus

Lineages of T cells are the predominant intrathymic population, but small numbers of B cells are also present in the aging thymus, as reported from several laboratories (Andreu-Sanchez et al., 1990; Peled and Haran-Ghera, 1991; Akashi et al., 2000; Flores et al., 2001).

Flores et al. demonstrated the presence of B cells in epithelial and perivascular compartments of human adult thymus (Flores et al., 2001). The phenotypically distinguishable B cells in the thymus are located within both the medulla and perivascular space, and trafficking occurs between these compartments, including B cells trafficking from the periphery. There is an age-related increase in B cells in the thymus, and their numbers correlate with the increase in lymphocyte-rich regions of the perivascular space, that are prominent between the ages 10 and 50. The B cells contain mutated Ig VH sequences, characteristic of post-germinal center B cells.

Recent studies describing the contribution of thymic B cells to the shaping of the T cell repertoire have raised the question as to whether they are produced intrathymically and play a role in this process. Montecino-Rodriguez et al. investigated the B cell developmental potential of CD4−CD8−CD3/TCR− triple-negative thymocytes present in a newly developed, stromal cell-dependent, long term culture system (Montecino-Rodriguez et al., 1996). They showed the ability of selected thymic stromal cells to support distinct stages of B cell development. Some thymic stromal cell lines maintained cells with B cell developmental potential but could not support their differentiation, while other selected lines efficiently supported the development of surface IgM-expressing cells from triple-negative precursors. Thymic surface IgM+ B cells were generated most efficiently from thymocytes harvested from relatively young mice. Hence, these studies demonstrate that the thymic microenvironment exhibits a functional heterogeneity in its ability to support different stages of B cell development.

Studies on the murine AKR leukemia model showed B cell leukemia development in the thymus. This has led to the hypothesis that lack of favorable microenvironment for potential lymphoma cells development in the T cell pathway enables a B cell developmental route (Peled and Haran-Ghera, 1991). The basis of B cells occurrence in the aging thymus may be similar.

Co-cultures of aged BM cells seeded onto fetal thymus explants under conditions favoring T cell development did not show abundance of B cells (Eren et al., 1988). The reconstituted thymus explants showed limited B cell function, as measured by the response to lipopolysaccharide (LPS) stimulation, in contrast to the increased response observed in the intact aged thymus. Development of myeloid cells in the thymus was observed in cultures of adult irradiated thymus in the presence of BM explants (Globerson, 1966). Increased levels of macrophages were observed in the aged thymus in vivo (Globerson, 2001), as well as following seeding of BM cells from aged mice onto lymphoid-depleted fetal thymus explants.

Taken together, it appears that the presence of B cells in the aging thymus may play a role in processes of thymocytopoiesis. Whereas the intact thymic stroma of the fetal thymus does not support B cells, the re-structured stroma of the aging thymus may have a predominance of cells that favor the B cell axis.

7. Regeneration capacity of the aging thymus

The thymus is unique in its plasticity, manifested in the capacity to regenerate following severe lymphoid depletion, as illustrated by spontaneous recovery from acute stress (Haeryfar and Berczi, 2001) and sublethal doses of irradiation (Hirokawa et al., 1988; Basch, 1990). It is also observed in the seasonal changes in reptiles, where periodical thymic involution is followed by regeneration (El Masri et al., 1995; Alvarez et al., 1998; Hareramadas and Rai, 2001). Spontaneous radiation recovery was also shown in vitro, in the FTOC system (Fridkis-Hareli et al., 1991). The fact that this plasticity is not manifested in the age-related involution suggests different, and/or additional mechanisms in aging. The capacity of the thymus to regenerate following severe depletion decreases in advanced age, as revealed from several experimental models. It is noteworthy that irradiation of splenectomized mice resulted in a failure of the thymus to recover (Globerson et al., 1989). Contribution of the spleen to processes of thymic recovery could be related, at least in part, to essential cytokines that are provided by the spleen, and/or effects of chemoattractants.

Timm and Thoman studied the capacity of the thymus to regenerate following irradiation from a radio-resistant precursor population in the thymus (Timm and Thoman, 1999). Analysing the differentiation of this "wave" of thymocytes, it was determined that aging most severely affects the earliest developmental transitions. While the overall rate of differentiation does not appear to be affected in older mice, fewer thymic progenitors initiate differentiation. The reduced expansion of late pre-T cells in the middle-aged is due to the smaller pool size of these cells. Radiation recovery was exhibited in organ cultures of thymic explants, and analysis of the kinetics of this process suggested that the capacity for regeneration depends on availability of radioresitant cells, as well as competence of niches within the thymic stroma (Mehr et al., 1994).

Studies on rats showed that the vascular structures were destroyed on day 3 after irradiation (6 Gy), but recovered by day 7 (Wang et al., 1999). Thymic recovery following irradiation was also examined in vivo by transplantation of identifiable hematopoietic stem cells. The results showed that the thymuses of aged mice retain the ability to attract progenitor cells. Whereas the cortical epithelial cells contributed to this recovery, the medullary epithelial tissue remained inactive for a relatively long period. A defect was also observed after transplantation of the cells directly into the thymus (Basch, 1990), suggesting a failure of the aged thymic microenvironment to support T cell development from progenitor cells.

The capacity of the thymus to regenerate in aging was investigated by Rice and Bucy by applying a single injection of an anti-CD4 (GK1.5) mAb to young and old mice (Rice and Bucy, 1995). They found that in young mice the CD4+ population of T cells was completely depleted from the peripheral blood and subsequently also the lymph nodes and spleen, yet they recovered eventually (80% of CD4+ T cells within 100 days). Although the aged mice had less CD4+ T cells, the depletion was by only 60% after mAb injection. This finding was associated with prolonged circulation of antibody-coated T cells. Moreover, recovery was fivefold less than in young mice, between days 14 and 100 post-injection. Interestingly, the CD4+ T cells that were depleted in both young and old mice by injection of anti-CD4 mAb were preferentially CD45RB (high), suggesting a selective depletion of immature cells. The fact that regeneration was a thymic dependent event, rather than extrathymic process, was supported by the finding that thymectomized young mice failed to recover CD3+CD4+ cells.

Other studies have looked into the effects of glucocorticoids (GC) on thymic transient atrophy and subsequent regeneration. Elevation of blood GC level after stress resulted in increased apoptosis in the thymus, mainly in the DP cells. Interestingly, transgenic mice in which the expression of the GC receptor was down-modulated specifically in thymocytes, showed abnormal thymocyte differentiation, suggesting a significant role of the hormone in T cell development (Ashwell et al., 2000). Studies on T cell development in organ cultures of fetal thymus lobes from transgenic mice, following inhibition of GC biosynthesis, led to the conclusion that the hormonal effect in the thymus may be related to its local synthesis in the thymic epithelium (Lechner et al., 2000).

Studies on reversibility of stress effects in the aging thymus are still required to determine whether this capacity for regeneration is affected by aging. Whereas the thymic stroma may have the capacity for regeneration, extrathymic origin of cells may also contribute to the regeneration of stroma elements (Li et al., 2000).

8. Neuroendocrine control of thymic aging

Thymic involution is characterized by the gradual decrease in size and cellularity, and re-organized morphology. These changes have been viewed as inevitable, irreversible processes of deterioration and were considered as a major cause of immunosenescence, manifested particularly in the peripheral T lymphocytes. However, data gained during the past two decades have revealed that thymic involution may be prevented or reversed, and that changes in the aging thymus actually represent altered developmental processes rather than unilateral deterioration (Globerson, 1997; Globerson and Effros, 1999). Understanding the mechanisms underlying these changes is thus important for the establishment of strategies designed to augment immunological vigor in old age.

The neuroendocrinological aspects of thymic involution have been discussed at length in many publications. Fabris et al. summarized the different endocrinological manipulations that can lead to thymic regrowth even in old animals (Fabris et al., 1997). Examples of such manipulations include: (a) intrathymic transplantation of pineal gland or treatment with melatonin; (b) implantation of a growth hormone (GH) secreting tumor cell line or treatment with exogenous GH; (c) castration or treatment with exogenous luteinizing hormone–releasing hormone (LH–RH); (d) treatment with exogenous thyroxine or triiodothyronine. The effect of GH, thyroid hormones, and LH–RH may be mediated by specific hormone receptors on TEC. Melatonin or other pineal factors may also act through specific receptors, but relevant experimental evidence is still lacking. It is now clear that the thymus is subject to neuroendocrine control, and that certain neuropeptides may be actually produced within the thymus (Rinner et al., 1994, 1998, 1999).

Therefore, collectively, thymic plasticity may be retained through old age, although not in its full capacity.

8.1. Gonadectomy effects on thymic involution

The observations that castration can interfer with thymic involution (Utsuyama and Hirokawa, 1989) has led to several studies on the underlying mechanisms and on the effects of different hormones on the aging thymus. Studies on the effects of gonadectomy on thymic regeneration following irradiation (Utsuyama et al., 1995) are of particular interest in that respect. Gonadectomized (Gx) mice were grafted with thymus lobes from irradiated (0.5 Gy) newborn, 6- or 26-weeks-old mice. One month later, grafted thymuses were recovered and examined for thymocyte numbers, subpopulations and proliferative responses to mitogen (Concananavlin A). The growth of the irradiated thymus was significantly higher in Gx than in sham-operated control mice, and the magnitude of thymic growth was apparently age-dependent, as it was greater for newborns than for older mice. The mitogen-induced proliferative response was also significantly higher in thymocytes of the gonadecomized mice, and the magnitude of the response declined with advancing age of the thymus donors. Although the number of thymocytes was comparable in thymus grafts from 6- and 26-week-old mice, the proliferative response was more pronounced in the young, and it correlated with a significant increase in the percentage of CD4+CD8− cells in the thymus grafts. Noteworthy, thymus grafts from newborn could grow equally well in both Gx and sham-operated recipients, whereas the grafts from 6- and 26-week-old mice could grow well only in the gonadecomized, but not in the intact controls. The effect of gonadectomy on the thymus graft was thus dependent on the age of the thymus donor. Gonadectomy appeared to promote immigration of thymocyte precursors into the thymus and to enhance proliferation and differentiation of thymocytes towards CD4+CD8− T cells, in an age-related manner.

Recently, Leposavic et al. studied rats orchidectomized at the age of 1 month and examined 3–9 months later, in order to elucidate a putative role of male gonadal hormones in the shaping of thymus size and intrathymic T cell maturation (Leposavic et al., 2002). Analyses included thymus weight, thymocyte yield and relative proportions of thymocyte subsets (based on the expression of CD4/CD8 molecules and TCR-alpha/beta). In 4-month-old control rats, the thymus size and cellularity returned to the corresponding levels observed originally in 1-month-old rats, and were retained during the subsequent 6-months period. However, the distribution of the main thymocyte subsets in these rats was changed significantly, probably due to onset of involution. Enlargement of the thymus size and enrichment of thymic lymphoid content were of a limited duration, and changes in the relative proportion of thymocytes were more pronounced as a result of duration of the gonadal deprivation. Changes in gonadal hormones may thus be, at least partly, responsible for the age-related reshaping of the thymus, and as a consequence, T cell development.

Mase and Oishi investigated the effects of castration on the development of lymphoid organs (bursa of Fabricius, thymus, and spleen) in the Japanese quail, during 4–8 weeks of age under a long photoperiod (16L:8D) and the effect of testosterone implantation on the involution of the lymphoid organs under such long, as well as short (8L:16D) photoperiods (Mase and Oishi, 1991). The thymus in intact quails grew rapidly under the long photoperiod, reaching a peak at 6 weeks of age and regressing thereafter. In contrast, development of the lymphoid organs in castrated quail was well correlated with body growth. Testosterone treatment induced a significant reduction in relative thymus weight at 6 weeks of age under both photoperiod conditions. These observations lend further support to the notion that gonadal hormones have an important role in the modulation of thymic growth and involution.

Mechanisms of castration effects on thymic involution were studied by Leposavic et al., focusing on the principal ovarian steroids, 17 beta-estradiol (E2) and progesterone (Leposavic et al., 2001). Adult female rats ovariectomized and treated for 14 days with physiological doses of either E2 or progesterone, were examined in parallel with controls receiving no hormones. Ovariectomy produced a marked increase in thymus weight, which was associated with an increase in the volume and cellularity of both the medulla and cortex. Treatment of the ovary-deprived rats with E2 reduced the thymic weight and cortex volume, and reversed the effects on the volume of medulla. Progesterone treatment following ovariectomy, only prevented the increase in thymus weight and cortical volume, and had no effect on the medullary volume. Ovariectomy also affected the thymocyte profile, increasing the proportion of CD4+8+TCRαβ− cells and producing a corresponding decrease in the relative proportions of all TCRαβ (high) cell subsets. E2 reduced the relative proportion of CD4−8+TCRαβ−, CD4−8+TCRαβ (low) and CD4+8+TCRαβ− cells, while progesterone increased the percentage of CD4−8+TCRαβ− cells. It appeared that these hormones affect both the lymphoid and nonlymphoid compartments of the thymus, yet in different manners. Whereas progesterone increases the volume of the nonlymphoid component of the medulla, E2 has the opposite effect.

Studies attempting to elucidate the role of these hormones were performed by focusing on the receptors to estrogen in the aging thymus, as further described in the following section.

8.2. Expression of receptors to sex-steroidal hormones

Receptors to sex-steroidal hormones are expressed in the thymus, in reticuloendothelial cells (Marchetti et al., 1984; Nilsson et al., 1986; Staples et al., 1999), as well as in early immature lymphoid cells (Amir-Zaltsman et al., 1993; Kohen et al., 1998). We have demonstrated the expression of estrogen receptors (ER) in murine thymocytes of both females and males, and there was no indication of an age-related decrease in levels of ER expressing cells. However, only thymocytes from females showed an increased creatine kinase activity following in vivo treatment with E2, while the males reacted to testosterone (Kohen et al., 1998). Interestingly, cell replication was augmented by in vitro treatment with E2 to FTOC colonized with BM cells or thymocytes from young adults (2 months) and not aged (24 months) mice. Hence, whereas the ER is functional in the aged mice, its triggering does not result in cell replication.

More recent studies showed that ERalpha is necessary in thymic development (Staples et al., 1999). The study was carried out on ERalpha knockout (KO) mice, that have significantly smaller thymuses compared to the wild type (WT) littermates. By establishing BM radiation chimeras between the KO and the WT mice it was found that the small thymus was due to lack of ER in the radiation-resistant tissues, rather than in the hematopoietic cells. However, the KO mice did respond to treatment with E2, and showed thymic atrophy as a result of that treatment. It was thus implied that E2-induced thymic atrophy may be mediated by another receptor pathway. These observations are in line with our findings of ER function in thymic stromal cells, as well as lymphocytes (Kohen et al., 1998) lending further support to the hypothesis that hormonal effects leading to increased cellularity in the thymus cause stem cell migration and their expansion, rather than affecting directly thymocyte cell divisions.

Thymic involution following puberty in males depends on the increasing levels of testosterone, and changes in the thymus are manifested mainly in the decrease of thymic lymphoid-cell elements. Tamoxifen treatment reverses thymic involution in intact adult male rats in a dose-dependent manner (Fitzpatrick et al., 1985). On the other hand, tamoxifen administration at pharmacological doses to adult castrated rats results in thymic regression. Tamoxifen may thus reverse thymic involution by reducing testosterone levels, whereas in the absence of testosterone it has thymolytic effects.

Erlansson et al. used ER KO mice to determine the effects of estrogen on thymic atrophy (Erlandsson et al., 2001). The study was based on male ER-alpha, ER-beta, or ER-alphabeta-double KO mice. Deletion of ER-alpha led to hypoplasia in both the thymus and spleen. ER-alpha(−) mice had a higher frequency of immature DP thymocytes, compared to their control mice. Female oophorectomized ER-beta KO mice treated with E2 had a similar degree of thymic atrophy compared with the WT strain, but showed only limited involution of the thymic cortex and no alteration of thymic CD4/CD8 phenotype expression. Accordingly, expression of ER-alpha, but not ER-beta, seems mandatory for development of full-size thymus and spleen in males, whereas expression of ER-beta is required for E2-mediated thymic cortex atrophy and thymocyte subset shift in females. This may be based on down-regulated activity in the GH/IGF-1 axis in males lacking ER-alpha, and suppressed sensitivity of females lacking ER-beta to E2-mediated suppression of IGF-1.

Although both androgens and estrogens are thymolytic, a significantly decreased percentage of DP thymocytes was observed in mice following treatment with the androgen methyltestosterone, but not with the estrogen ethinylestradiol (Dulos and Bagchus, 2001). To investigate whether the observed thymolytic effects were due to the expression of hormone receptors, thymocytes were incubated with androgens or estrogens to measure apoptosis. There was no direct in vitro effect of androgens on thymocytes, suggesting that cells reacting in vivo to androgens by increased apoptosis are other than thymocytes. Using androgen receptor mutant (Tfm/Y) mice that were treated with androgens, showed no change in thymocyte subpopulations, indicating that the androgen effect on DP cells was based on the function of specific receptors. Thus, androgens indirectly accelerate thymocyte apoptosis in vivo.

8.3. Growth hormone effects

Both GH and IGF-1 have been of particular interest for "rejuvenation" of the involuted thymus, because of their thymopoietic effects and the fact that their serum concentrations decline during aging. Kelley et al. studied in detail the effects of GH on thymic involution, using either transplants of cells secreting the hormone, or injections of the hormone (Kelley et al., 1992, 1998). Indications that GH modulates thymic size by affecting migration of thymocyte progenitors from the BM to the thymus were gained from a variety of in vivo and in vitro studies, as well as a combination of such strategies (Knyszynski et al., 1992). Hematopoietic progenitor cells are thus targets for IGF-1, and this is likely to be important in understanding thymic aging.

GH prevents thymic aging, it reverses the accumulation of BM adipocytes and restores the number of BM erythrocytic and granulocytic lineages (French et al., 2002). Histological evaluation of aged rats treated with either rat, or human GH, displayed clear morphologic evidence of thymic regeneration, reconstitution of hematopoietic cells in the BM, and multi-organ extramedullary hematopoiesis. Aged rats have at least a 50% reduction in the number hematopoietic BM cells, compared with the young. Age-associated decline in leukocytes and increase in adipocytes in the BM, were significantly reversed by in vivo treatment with GH. Restoration of BM cellularity was related primarily to erythrocytic and granulocytic cells, but all cell lineages were represented, and their proportions were similar to those in aged control rats. Enhanced extramedullary hematopoiesis was observed in the spleen, liver, and adrenal glands from animals treated with GH. Effects of GH treatment in vivo were also manifested in increased values of myeloid colony forming units measured in vitro. These data indicated that GH can reverse the accumulation of BM adipocytes and restore the number of BM myeloid cells of both the erythrocytic and granulocytic lineages.

Although treatment of aging rodents with either GH or IGF-1 increased thymic cellularity, it did not restore it to levels observed in young animals (Knyszynski et al., 1992; Montecino-Rodriguez et al., 1998), pointing to additional limiting defects. However, a combination of IGF-1 and young BM cells led to increased thymic cellularity in 18 months old mice, suggesting that optimal therapies for restoring thymus cellularity must address both endocrine and hematopoietic defects that accumulate with age. In vitro studies using the FTOC model show that GH treatment in vivo mobilizes cells from the BM to the thymus (Knyszynski et al., 1992). Similarly, IGF-1 treatment indicated that it potentiates thymic colonization by BM T cell precursors and/or that the hormone affects some other event soon after thymus colonization (Montecino-Rodriguez et al., 1996).

In view of the observed effects of GH and IGF-1 treatment, it seemed that different mechanisms account for the mere increased cellularity in the thymus. These may include mobilization of stem cells from the BM, cell replication and apoptosis. There is no effect of GH on the expression of RAG and differentiation into distinct cell phenotypes, and the ultimate single positive CD4/CD8 T cells expressing the T cell receptor repertoire. The key question is still whether thymic involution is based on one critical, central cause, that leads to a cascade of events, or whether a combination of different independent mechanisms leads to the same final result.

8.4. Neuropeptides in the aging thymus

The scope of neuroendocrine circuits that are networked with the thymus and neuroendocrine-thymic bidirectional interactions is broad (Moll, 1997), including aspects related to aging that will be addressed here. Attention will focus on the various secreted neuropeptides and their specific receptors that transmit communicating signals in this network, as well as the substrates of these neuropeptides in the thymus. It should be noted that thymocytes themselves may also produce some of the agents (e.g. acetylcholine) that function in the dialogue between thymocytes and thymic stroma (Rinner et al., 1994, 1998, 1999).

The data concerning the role of hormones and neuropeptides in thymic involution are equivocal. The structural age-related changes start with a steady decrease of thymocytes, whereas no major variations occur in the number of TEC. The presence of somatostatin (SS) and three different SS receptor (SSR) subtypes were shown in the human thymus (Ferone et al., 2000). Both SS and SSR might play a role in the involution of the human thymus, reinforcing the links between the neuroendocrine and immune systems, at the level of the thymus.

The pituitary-thymus axis was indicated from the effects of hypophysectomy on thymic atrophy with the disturbed immune responses (Utsuyama et al., 1997a). Further studies revealed specific receptors in the thymus. Hence, Kinoshita and Hato showed that binding of pituitary acidophilic cell hormones to their receptors on TECs, augmented the release of thymic hormonal peptides in vitro (Kinoshita and Hato, 2001). Morpho-molecular alterations of cytoplasm preceded nuclear damage in the apoptotic thymocytes.

Studies on in vitro migration capacity of murine lymphocytes in response to three neuropeptides in the thymus showed inhibition of chemotaxis, yet differences were observed in relation to the various neuropeptides, as well as to age, concentrations and locations (Medina et al., 1998). The stimulatory effects that sulfated cholecystokinin octapeptide (CCK-8s), gastrin-releasing peptide (GRP) and neuropeptide Y (NPY) exerted in young and adult mice were not observed in old animals. On the other hand, CCK-8s inhibitory effects on chemotaxis were more striking in old mice, and were observed in every organ studied. It thus appears that stimulatory effects of the neuropeptides disappear or become inhibitory with aging.

Direct effects of neuropeptides and neurotransmitters on TEC function were shown by treating rat TEC in culture with various agents (Head et al., 1998). Neuropeptide control of thymocytopoiesis was inferred from studies on apoptosis in thymic organ cultures (Rinner et al., 1994), showing that the effect was due to TEC. Fetal thymus explants responded to cholinergic stimuli in increased apoptosis, when co-cultured with cortical TEC lines. This was not detectable when the epithelial cell lines were from the thymic medulla. Restructuring of the thymic microenvironment in aging may thus involve changes in such stromal cell components, and/or in the availability of the relevant neuropeptides. Cholinergic stimulation of TEC in vivo could be provided by innervation (Head et al., 1998; Bulloch et al., 1998) and/or by thymic lymphocytes (Rinner et al., 1999).

Although these experiments indicate a response to cholinergic stimulation, the possibility of adrenergic effects under physiological conditions in vivo need also be considered (Rinner et al., 1998). It should be noted in that respect that the density of noradrenergic sympathetic nerves and the concentration of noradrenaline (NA) increase dramatically in the thymus of old mice and rats (Madden and Felten, 2001).

The hypothesis that maturational processes within the hypothalamo-pituitary-gonadal axis and thymus are reciprocally regulated via neural pathways was tested in the thymus of adult orchidectomized rats (Leposavic et al., 2000). The thymus weight was significantly increased in the orchidectomized, compared to sham-operated controls. In 1-week old rats, the increase in thymus weight was accompanied by a proportional increase in the content of both catecholamines and serotonin (5-HT); that subsequently remained unaltered. In these animals, the density of both thymus nerve fibers and cells stained with sucrose phosphate glyoxylic acid (SPG) also remained unchanged. In 7-week old rats the rise in the thymus weight was followed by a proportional increase in the content of all monoamines, except for NA that was reduced. The reduction in both NA content and concentration reflected a diminished density of SPG-positive nerve profiles. In older orchidectomized rats, the increase in thymus weight was neither paralleled by a proportional increase in dopamine content nor in 5-HT, while the content of NA was decreased. In all orchidectomized rats, the pattern of intrathymic distribution of SPG-positive fibers and cells remained unchanged. Orchidectomy had no effect on acetylcholinesterase (AchE) activity and on the density of AchE-positive nerves and cells in the thymus.

Studies on adrenergic nerve fibers in juvenile, adult and old rats showed that chemical sympathectomy with neurotoxin 6-OH dopamine led to disappearance of the greater part of the fibers (Cavallotti et al., 1999). The results suggest that: (a) total innervation of the thymus increases with age; (b) adrenergic nerve fibers do not change with age; (c) the content of NA in the thymus increases with age; and (d) NPY-like immunoreactive structures in the thymus decrease with age.

Further indication on age-related increased density of sympathetic noradrenergic innervation and concentration of norepinephrine was documented in relation to thymic involution (Madden and Felten, 2001). In old mice (18 months BALB/c) compared to the young (2 months), thymocyte CD4/CD8 co-expression was altered by beta-adrenoceptor blockade. In nadolol-treated old mice, the frequency of the immature DN population was increased, and the proportion of the DP population was reduced. A corresponding increase was observed in the frequency of mature single positive CD4−8+, but not CD4+8− cells, with no nadolol effect on CD3 (high) expression in the DP population. The percentage of CD8+44 (low) naïve cells in peripheral blood increased in nadolol-treated mice. Hence, the age-associated increase in sympathetic noradrenergic innervation of the thymus seems to modulate thymocyte maturation. It was thus proposed that pharmacological manipulation of noradrenergic innervation may provide a novel means of increasing naïve T cell output and improving T cell reactivity to incognate antigens with age.

9. Effects of dietary restriction

Dietary restriction is known to retard the aging processes. The question was therefore raised as to whether it retards thymic involution. Indeed, dietary restricted mice showed higher levels of T cell frequencies in their peripheral tissues, compared to controls (Miller and Harrison, 1985), and decreased thymic size (Weindruch and Suffin, 1980). However, our own studies using long-lived transgenic mice (alpha-MUPA) that are genetically dietary-restricted, had involuted thymuses in old age similarly to their age-matched controls (Miskin et al., 1999). Hence, the improved thymic size observed in the case of enforced dietary restriction may not be related to the dietary restriction per se, but rather to other associated mechanisms.

10. Extrathymic T cell development

Studies using a variety of murine experimental models indicate that T cell maturation can occur outside the thymus. Extrathymic T cell development has been shown in several tissues, including the BM, intestine and liver. Accordingly, observations on naïve T cells in the peripheral blood of elderly subjects (Franceschi et al., 1995; Bagnara et al., 2000) may represent thymic function, but an additional compensatory contribution of extrathymic T cell development cannot be excluded.

10.1. The bone marrow

The BM in aged mice contains increased numbers (twofold) of mature T lymphocytes, possibly representing an effector function that this tissues resumed in old age (Garcia-Ojeda et al., 1998; Sharp et al., 1990b). These cells could derive from the peripheral lymphoid tissues, yet, local differentiation from stem cells has also been considered. Hence, T cell development in the BM has been clearly indicated (Garcia-Ojeda et al., 1998; Antica and Scollay, 1999). Although these studies did not examine BM of aged mice, they do point to the possibility that the BM can serve as an alternative site for T cell development. The increased values of mature T cells in the BM of aged mice (Sharp et al., 1990b) may thus be due, at least in part, to locally differentiated cells.

10.2. The intestine

The intestine is a major site of extrathymic T cell maturation in mice, and there is abundant evidence that the gut mucosa is an immune effector organ postnatally. However, there has been little critical evidence that extrathymic T cell maturation occurs to any significant extent in human gut.

A large population of T lymphocytes was noted in human fetal intestinal mucosa, proliferating in the absence of foreign antigens and expressing mRNA transcripts for pre-TCR-alpha (Howie et al., 1998). There is also a substantial proportion of CD7+ lamina propria lymphocytes that do not express CD3 in the fetal gut that may be differentiating into CD3+ cells. Rearranged TCR-beta transcripts of fetal lamina propria and blood lymphocytes, were cloned and sequenced, and virtually no overlap of clonality was observed between blood and intestine, suggesting that gut T cells may not be derived from the blood. In addition, 30 days after transplantation of intestinal CD3−7+ cells to SCID mice, proliferating T cells and pre-TCR-alpha transcripts were abundant. A population of precursor T cells thus exists in the human intestine before birth, and it may be differentiating into mature T cells in situ. The question whether the gut generates T cells in aging humans is still open.

10.3. The liver

The liver is one of the important hematopoietic organs even after birth, and T cell development in the liver was recently described (Abo et al., 2000). The adult liver still comprises c-kit+ stem cells and it gives rise to extrathymic T cells, natural killer (NK) cells, as well as granulocytes. The T cells generated in the liver of mice express intermediate levels of TCR, including both NK1.1+ (NKT cells) and NK1.1 (neg) subsets. The number of extrathymic T cells is limited in the young, and it increases with age. It is of importance to note that the number and function of extrathymic T cells are elevated even in the young under conditions when the mainstream of T cell differentiation in the thymus is rather suppressed (e.g. stress, infection, malignancy, pregnancy, autoimmune disease, chronic graft-versus-host diseases and others). Theoretically, extrathymic T cells may comprise anti-self-reactive forbidden clones and mediate cytotoxicity against abnormal self-cells (e.g. malignant tumor cells, microbially infected hepatocytes, and regenerating hepatocytes). In that respect, overactivation or continuous activation of extrathymic T cells may have harmful outcomes. Abo et al. proposed that mechanisms underlying a shift of T cell develop-ment from the thymus to the liver might involve regulation by the autonomic nervous system as well as by cytokines.

11. Concluding remarks

Recent advancement in understanding of thymocytopoiesis and thymic involution points to a series of cell properties and processes that change with age, suggesting multi-factorial mechanisms, and they may vary in patterns of manifestation in different individuals. The following conclusions can be drawn from the data gained so far.

- Age-related changes have been observed along the stem cell/thymus axis, particularly in the dynamics of cellular and molecular events in the thymus.
- A comprehensive consideration of thymocytopoiesis in aging requires critical evaluation of the intrinsic properties and developmental potential of cells both in the stromal and lymphoid components, as well as microenvironmental effects on the cells. In vitro experimental models enable elucidation of such mechanisms.
- Microenvironmental structural and molecular parameters include cytokine profiles, chemo-attractants, growth factors, hormones and neuropeptides.

- Mature T cells may affect the differentiation of less mature cell types, either directly or via effects on microenvironmental elements.
- Compensation for aging effects at the tissue, or cell population level, have been demonstrated. Understanding the basis of compensatory mechanisms is a key element in designing strategies to prevent, or minimize possible hazards of aging effects.
- Developmental pathways that are unconventional in the young adult may become dominant as alternative pathways in aging.
- Further molecular genetic approaches will be needed to establish mechanisms underlying the changes and differential decrease in the potential for T cell development in aging, to pave the way for relevant intervention strategies.

References

Abo, T., Kawamura, T., Watanabe, H., 2000. Physiological responses of extrathymic T cells in the liver. Immunol. Rev. 174, 135–149.

Agus, D.B., Surh, C.D., Sprent, J., 1991. Reentry of T cells to the adult thymus is restricted to activated T cells. J. Exp. Med. 173, 1039–1046.

Akashi, K., Richie, L.I., Miyamoto, T., Carr, W.H., Weissman, I.L., 2000. B lymphopoiesis in the thymus. J. Immunol. 164, 5221–5226.

Alvarez, F., Razquin, B.E., Villena, A.J., Zapata, A.G., 1998. Seasonal changes in the lymphoid organs of wild brown trout, Salmo trutta L: a morphometrical study. Vet. Immunol. Immunopathol. 64, 267–278.

Amir-Zaltsman, Y., Mor, G., Globerson, A., Thole, H., Kohn, F., 1993. Expression of estrogen receptors in thymocytes. Endocr. J. 1, 211–217.

Andreu-Sanchez, J.L., Faro, J., Alonso, J.M., Paige, C.J., Martinez, C., Marcos, M.A., 1990. Ontogenic characterization of thymic B lymphocytes. Analysis in different mouse strains. Eur. J. Immunol. 20, 1767–1773.

Andrews, D., Aspinall, R., 2002. Age-associated thymic atrophy is linked to a decline in IL-7 production. Exp. Gerontol. 37, 455–463.

Antica, M., Scollay, R., 1999. Development of T lymphocytes at extrathymic sites. J. Immunol. 163, 206–211.

Ashwell, J.D., Vacchio, M.S., Galon, J., 2000. Do glucocorticoids participate in thymocyte development? Immunol. Today 21, 644–646.

Aspinall, R., Andrews, D., 2000. Thymic atrophy in the mouse is a soluble problem of the thymic environment. Vaccine 18, 1629–1637.

Aspinall, R., Andrews, D., 2001a. Age-associated thymic atrophy is not associated with a deficiency in the CD44(+)CD25(−)CD3(−)CD4(−)CD8(−) thymocyte population. Cell. Immunol. 212, 150–157.

Aspinall, R., Andrews, D., 2001b. Gender-related differences in the rates of age associated thymic atrophy. Dev. Immunol. 8, 95–106.

Bagnara, G.P., Bonsi, L., Strippoli, P., Bonifazi, F., Tonelli, R., DwAddato, S., Paganelli, R., Scala, E., Fagiolo, U., Monti, D., Cossarizza, A., Bonafe, M., Franceschi, C., 2000. Hemopoiesis in healthy old people and centenarians: well-maintained responsiveness of CD34+ cells to hemopoietic growth factors and remodeling of cytokine network. J. Gerontol. Biol. Sci. 55A, B61–B66.

Bar-Dayan, Y., Afek, A., Bar-Dayan, Y., Goldberg, I., Kopolovic, J., 1999. Proliferation, apoptosis and thymic involution. Tissue Cell 31, 391–396.

Basch, R.S., 1990. Thymic repopulation after irradiation in aged mice. Aging Immunol. Infect. Dis. 2, 229–235.

Ben-Yehuda, A., Friedman, G., Wirtheim, E., Abel, L., Globerson, A., 1998. Checkpoints in thymocyto-poiesis in aging: expression of the recombination activating genes RAG-1 and RAG-2. Mech. Ageing Dev. 102, 239–247.

Bernstein, A., Globerson, A., 1974. Thymus cell traffic induced by antigen stimulation. Cell. Immunol. 14, 171–181.

Bertho, J.M., Demarquay, C., Moulian, N., Vandermeeren, A., Berrihaknin, S., Gourmelon, P., 1997. Phenotypic and immunohistological analyses of the human adult thymus: evidence for an active thymus during adult life. Cell Immunol. 179, 30–40.

Bleul, C.C., Boehm, T., 2000. Chemokines define distinct microenvironments in the developing thymus. Eur. J. Immunol. 30, 3371–3379.

Bodey, B., Bodey, B., Jr., Siegel, S.E., Kaiser, H.E., 1997. Involution of the mammalian thymus, one of the leading regulators of aging. In Vivo 11, 421–440.

Borowski, C., Martin, C., Gounari, F., Haughn, L., Aifantis, I., Grassi, F., von Boehmer, H., 2002. On the brink of becoming a T cell. Curr. Opin. Immunol. 14, 200–206.

Boyd, R.L., Tucek, C.L., Godfrey, D.I., Izon, D.J., Wilson, T.J., Davidson, N.J., Bean, A.G.D., Ladyman, H.M., Ritter, M.A., Hugo, P., 1993. The thymic microenvironment. Immunol. Today 14, 445–459.

Bulloch, K., McEwen, B.S., Nordberg, J., Diwa, A., Baird, S., 1998. Selective regulation of T-cell development and function by calcitonin gene-related peptide in thymus and spleen. An example of differential regional regulation of immunity by the neuroendocrine system. Ann. N.Y. Acad. Sci. 840, 551–562.

Cavallotti, C., Artico, M., Cavallotti, D., 1999. Occurrence of adrenergic nerve fibers and of noradrenaline in thymus gland of juvenile and aged rats. Immunol. Lett. 70, 53–62.

Chau, L.A., Rohekar, S., Wang, J.J., Lian, D., Chakrabarti, S., Zhang, L., Zhong, R., Madrenas, J., 2002. Thymic re-entry of mature activated T cells and increased negative selection in vascularized allograft recipients. Clin. Exp. Immunol. 127, 43–52.

Crisi, G.M., Tsiagbe, V.K., Russo, C., Basch, R.S., Thorbecke, G.J., 1996. Evaluation of presence and functional activity of potentially self-reactive T cells in aged mice. Int. Immunol. 8, 387–395.

DeMartinis, M., Modesti, M., Profeta, V.F., Tullio, M., Loreto, M.F., Ginaldi, L., Quaglino, D., 2000. CD50 and CD62L adhesion receptor expression on naive (CD45RA+) and memory (CD45RO+) T lymphocytes in the elderly. Pathobiology 68, 245–250.

Doria, G., Mancini, C., Utsuyama, M., Frasca, D., Hirokawa, K., 1997. Aging of the recipients but not of the bone marrow donors enhances autoimmunity in syngeneic radiation chimeras. Mech. Ageing Dev. 95, 131–142.

Douek, D.C., Koup, R.A., 2000. Evidence for thymic function in the elderly. Vaccine 18, 1638–1641.

Dubiski, S., Cinader, B., 1992. Age-related polymorphism of thymus subpopulations in inbred mice. Thymus 20, 183–193.

Dulos, G.J., Bagchus, W.M., 2001. Androgens indirectly accelerate thymocyte apoptosis. Int. Immunopharmacol. 1, 321–328.

Effros, R.B., Globerson, A., 2002. Hematopoietic cells and replicative senescence. Exp. Gerontol. 37, 191–196.

El Masri, M., Saad, A.H., Mansour, M.H., Badir, N., 1995. Seasonal distribution and hormonal modulation of reptilian T cells. Immunobiology 193, 15–41.

Elcuman, E.A., Akay, M.T., 1998. Age-dependent immunolocalization of fibronectin and histological changes in the thymus of rats. Vet. Res. Commun. 22, 525–532.

Eren, R., Zharhary, D., Abel, L., Globerson, A., 1988. Age-related changes in the capacity of bone marrow cells to differentiate in thymic organ cultures. Cell. Immunol. 112, 449–455.

Erlandsson, M.C., Ohlsson, C., Gustafsson, J.A., Carlsten, H., 2001. Role of oestrogen receptors alpha and beta in immune organ development and in oestrogen-mediated effects on thymus. Immunology 103, 17–25.

Fabris, N., Mocchegiani, E., Provinciali, M., 1997. Plasticity of neuroendocrine-thymus interactions during aging. Exp. Gerontol. 32, 415–429.

Ferone, D., Pivonello, R., Van Hagen, P.M., Waaijers, M., Zuijderwijk, J., Colao, A., Lombardi, G., Bogers, A.J., Lamberts, S.W., Hofland, L.J., 2000. Age-related decrease of somatostatin receptor number in the normal human thymus. Am. J. Physiol. Endocrinol. Metab. 279, E791–E798.

Fink, P.J., Bevan, M.J., Weissman, I.L., 1984. Thymic cytotoxic T lymphocytes are primed in vivo to minor histocompatibility antigens. J. Exp. Med. 159, 436–451.

Fitzpatrick, F.T., Kendall, M.D., Wheeler, M.J., Adcock, I.M., Greenstein, B.D., 1985. Reappearance of thymus of ageing rats after orchidectomy. J. Endocrinol. 106, R17–R19.

Flores, K.G., Li, J., Hale, L.P., 2001. B cells in epithelial and perivascular compartments of human adult thymus. Human Pathol. 32, 926–934.

Franceschi, C., Monti, D., Samsoni, P., Cossarizza, A., 1995. The immunology of exceptional individuals: the lesson of centenarians. Immunol. Today 16, 12–16.

Francz, P.I., Fridkis-Hareli, M., Abel, L., Bayreuther, K., Globerson, A., 1992. Differential expression of membrane polypeptides on fetal thymic stroma co-cultured with bone marrow cells from young and old mice. Mech. Ageing Dev. 64, 99–109.

French, R.A., Broussard, S.R., Meier, W.A., Minshall, C., Arkins, S., Zachary, J.F., Dantzer, R., Kelley, K.W., 2002. Age-associated loss of bone marrow hematopoietic cells is reversed by GH and accompanies thymic reconstitution. Endocrinology 143, 690–699.

Fridkis-Hareli, M., Abel, L., Globerson, A., 1992. Patterns of dual lymphocyte development in co-cultures of foetal thymus and lymphohaemopoietic cells from young and old mice. Immunology 77, 185–188.

Fridkis-Hareli, M., Mehr, R., Abel, L., Globerson, A., 1994. Developmental interactions of CD4 T cells and thymocytes: age-related differential effects. Mech. Ageing Dev. 73, 169–178.

Fridkis-Hareli, M., Sharp, A., Abel, L., Globerson, A., 1991. Thymocyte development in an in vitro constructed chimera of irradiated fetal thymus and lymphohemopoietic cells. Thymus 18, 225–235.

Fulop, T., Jr., Gagne, D., Goulet, A.C., Desgeorges, S., Lacombe, G., Arcand, M., Dupuis, G., 1999. Age-related impairment of p56lck and ZAP-70 activities in human T lymphocytes activated through the TcR/CD3 complex. Exp. Gerontol. 34, 197–216.

Garcia-Ojeda, M.E., Dejbakhsh-Jones, S., Weissman, I.L., Strober, S., 1998. An alternative pathway for T cell development supported by the bone marrow microenvironment: recapitulation of thymic maturation. J. Exp. Med. 187, 1813–1823.

Ginaldi, L., De Martinis, M., Modesti, M., Loreto, F., Corsi, M.P., Quaglino, D., 2000. Immunopheno typical changes of T lymphocytes in the elderly. Gerontology 46, 242–248.

Globerson, A., 1966. In vitro studies on radiation lymphoid recovery of mouse spleen. J. Exp. Med. 123, 25–32.

Globerson, A., 1986. Immunoregulatory cells in aging. In: Facchini, A., Haaijman, J.J., Labo', G. (Eds.), Topics in Ageing Research in Europe, vol. 9. pp. 9–15. Eurage, Rijswijk, The Netherlands.

Globerson, A., 1995. T lymphocytes and aging. Int. Arch. Allergy Immunol. 107, 491–497.

Globerson, A., 1997. Thymocytopoiesis in aging: the bone marrow-thymus axis. Arch. Gerontol. Geriat. 24, 141–155.

Globerson, A., 1999. Hematopoietic stem cells and aging. Exp. Gerontol. 34, 137–146.

Globerson, A., 2000. Commentary: hemopoiesis in healthy old people and centenarians: well maintained responsiveness of CD34+ cells to hemopoietic growth factors and remodeling of cytokine network. J. Gerontol. Biol. Sci. 55A, B69–B70.

Globerson, A., 2001. Haematopoietic stem cell ageing, In: Bock, G., Goode, J.A. (Eds.), Aging Vulnerability: Causes and Interventions. Novartis Foundation Symposium 235, Wiley, Chichester, pp. 85–100.

Globerson, A., 2002. Haematopoietic stem cells and the thymus in ageing. In: Mattson, M.P., Van Zant, G. (Eds.), Stem Cells: A Cellular Fountain of Youth? Amsterdam, Elsevier Science. Chapter 3, pp. 43–72.

Globerson, A., Auerbach, R., 1965. In vitro studies on thymus and lung differentiation following urethan treatment. Wistar Inst. Monogr. 4, 3–19.

Globerson, A., Effros, R.B., 1999. Ageing of lymphocytes and lymphocytes in the aged. Immunol. Today 21, 515–521.

Globerson, A., Eren, R., Abel, L., Ben-Menahem, D., 1989. Developmental aspects of T lymphocytes in aging. In: Goldstein, A.L. (Ed.), Biomedical Advances in Aging. Plenum Press, New York and London. pp. 363–373. Chapter 34.

Globerson, A., Kollet, O., Abel, L., Fajerman, I., Ballin, A., Nagler, A., Slavin, S., Ben-Hur, H., Hagay, Z., Sharp, A., Lapidot, T., 1999. Differential effects of CD4+ and CD8+ cells on lymphocyte development from human cord blood cells in murine fetal thymus explants. Exp. Hematol. 27, 282–292.

Goldrath, A.W., Bogatzki, Y., Bevan, M.J., 2000. Naive T cells transiently acquire a memory-like phenotype during homeostasis-driven proliferation. J. Exp. Med. 192, 557–564.

Gozes, Y., Umiel, T., Trainin, N., 1982. Selective decline in differentiating capacity of immunohemopoietic stem cells with aging. Mech. Ageing Dev. 18, 251–259.

Haar, J.L., Taubenberger, J.K., Doane, L., Kenyon, N., 1989. Enhanced in vitro bone marrow cell migration and T-lymphocyte responses in aged mice given subcutaneous thymic epithelial cell grafts. Mech. Ageing Dev. 47, 207–219.

Haeryfar, S.M., Berczi, I., 2001. The thymus and the acute phase response. Cell. Mol. Biol. (Noisy-le-grand) 47, 145–156.

Hardy, C.L., Godfrey, D.I., Scollay, R., 2001. The effect of antigen stimulation on the migration of mature T cells from the peripheral lymphoid tissues to the thymus. Dev. Immunol. 8, 123–131.

Hareramadas, B., Rai, U., 2001. Thymic structural changes in relation to seasonal cycle and testosterone administration in wall lizard Hemidactylus flaviviridis (Ruppell). Indian J. Exp. Biol. 39, 629–635.

Harrison, D.E., 1983. Long-term erythropoietic repopulating ability of old, young and fetal stem cells. J. Exp. Med. 157, 1496–1504.

Harrison, D.E., Jordan, C.T., Zhong, R.K., Astle, C.M., 1993. Primitive hemopoietic stem cells: direct assay of most productive populations by competitive repopulation with simple binomial, correlation and covariance calculations. Exp. Hematol. 21, 206–219.

Haynes, B.F., Hale, L.P., 1998. The human thymus. A chimeric organ comprised of central and peripheral lymphoid components. Immunol. Res. 18, 61–78.

Head, G.M., Mentlein, R., VonPatay, B., Downing, J.E., Kendall, M.D., 1998. Neuropeptides exert direct effects on rat thymic epithelial cells in culture. Dev. Immunol. 6, 95–104.

Hirokawa, K., Utsuyama, M., Katsura, Y., Sado, T., 1988. Influence of age on the proliferation and peripheralization of thymic T cells. Arch. Pathol. Lab. Med. 112, 13–21.

Hollander, G.A., Wang, B., Nichogiannopoulou, A., Platenburg, P.P., Van Ewijk, W., Burakoff, S.J., Gutierrez-Ramos, J.C., Terhorst, C., 1995. Developmental control point in induction of thymic cortex regulated by a subpopulation of prothymocytes. Nature 373, 350–353.

Howie, D., Spencer, J., DeLord, D., Pitzalis, C., Wathen, N.C., Dogan, A., Akbar, A., MacDonald, T.T., 1998. Extrathymic T cell differentiation in the human intestine early in life. J. Immunol. 161, 5862–5872.

Jamieson, B.D., Douek, D.C., Killian, S., Hultin, L.E., Scripture-Adams, D.D., Giorgi, J.V., Marelli, D., Koup, R.A., Zack, J.A., 1999. Generation of functional thymocytes in the human adult. Immunity 10, 569–575.

Jenkinson, E.J., Franchi, L.L., Kingston, R., Owen, J.J.T., 1982. Effect of deoxyguanosine on lymphopoiesis in the developing thymus rudiment in vitro: application in the production of chimeric thymus rudiments. Eur. J. Immunol. 12, 583–587.

Jimenez, E., Vicente, A., Sacedon, R., Munoz, J.J., Weinmaster, G., Zapata, A.G., Varas, A., 2001. Distinct mechanisms contribute to generate and change the CD4:CD8 cell ratio during thymus development: a role for the Notch ligand, Jagged1. J. Immunol. 166, 5898–5908.

Kelley, K.W., Arkins, S., Li, Y.M., 1992. Growth hormone, prolactin, and insulin-like growth factors: new jobs for old players. Brain Behav. Immun. 6, 317–326.

Kelley, K.W., Meier, W.A., Minshall, C., Schacher, D.H., Liu, Q., VanHoy, R., Burgess, W., Dantzer, R., 1998. Insulin growth factor-I inhibits apoptosis in hematopoietic progenitor cells. Implications in thymic aging. Ann. N.Y. Acad. Sci. 840, 518–524.

Kinoshita, Y., Hato, F., 2001. Cellular and molecular interactions of thymus with endocrine organs and nervous system. Cell. Mol. Biol. (Noisy-le-grand) 47, 103–117.

Knyszynski, A., Adler-Kunin, S., Globerson, A., 1992. Effects of growth hormone on thymocyte development from progenitor cells in the bone marrow. Brain Behav. Immun. 6, 327–340.

Kohen, F., Abel, L., Sharp, A., Amir-Zaltsman, Y., Somjen, D., Luria, S., Mor, G., Thole, H., Globerson, A., 1998. Estrogen-receptor expression and function in thymocytes in relation to gender and age. Dev. Immunol. 5, 277–285.

Kong, F., Chen, C.H., Cooper, M.D., 1998. Thymic function can be accurately monitored by the level of recent T cell emigrants in the circulation. Immunity 8, 97–104.

Lechner, O., Wiegers, G.J., Oliveira-Dos-Santos, A.J., Dietrich, H., Recheis, H., Waterman, M., Boyd, R., Wick, G., 2000. Glucocorticoid production in the murine thymus. Eur. J. Immunol. 30, 337–346.

Leposavic, G., Obradovic, S., Kosec, D., Pejcic-Karapetrovic, B., Vidic-Dankovic, B., 2001. In vivo modulation of the distribution of thymocyte subsets by female sex steroid hormones. Int. Immunopharmacol. 1, 1–12.

Leposavic, G., Pejcic-Karapetrovic, B., Kosec, D., 2002. Alterations in thymopoiesis in intact and peripubertally orchidectomized adult rats of different age. Mech. Ageing Dev. 123, 401–411.

Leposavic, G., Ugresic, N., Pejcic-Karapetrovic, B., Micic, M., 2000. Castration of sexually immature rats affects sympathetic innervation of the adult thymus. Neuroimmunomodulation 7, 59–67.

Li, Y., Hisha, H., Inaba, M., Lian, Z., Yu, C., Kawamura, M., Yamamoto, Y., Nishio, N., Toki, J., Fan, H., Ikehara, S., 2000. Evidence for migration of donor bone marrow stromal cells into recipient thymus after bone marrow transplantation plus bone grafts: a role of stromal cells in positive selection. Exp. Hematol. 28, 950–960.

Livak, F., Schatz, D.G., 1996. T-cell receptor alpha locus V(D)J recombination by-products are abundant in thymocytes and mature T cells. Mol. Cell. Biol. 16, 609–618.

Madden, K.S., Felten, D.L., 2001. Beta-adrenoceptor blockade alters thymocyte differentiation in aged mice. Cell. Mol. Biol. (Noisy-le-grand) 47, 189–196.

Marchetti, P., Scambia, G., Reiss, N., Kaye, A.M., Cocchia, D., Iacobelli, S., 1984. Estrogen-responsive creatine-kinase in the reticulo-epithelial cells of rat thymus. J. Ster. Biochem. 20, 835–839.

Mase, Y., Oishi, T., 1991. Effects of castration and testosterone treatment on the development and involution of the bursa of fabricius and the thymus in the Japanese quail. Gen. Comp. Endocrinol. 84, 426–433.

Masunaga, A., Sugawara, I., Nakamura, H., Yoshitake, T., Itoyama, S., 1997. Cytokeratin expression in normal human thymus at different ages. Pathol. Int. 47, 842–847.

McCormick, K.R., Haar, J.L., 1991. Bone marrow-thymus axis in senescence. Am. J. Anat. 191, 321–324.

Medina, S., Del Rio, M., Manuel Victor, V., Hernanz, A., De la Fuente, M., 1998. Changes with ageing in the modulation of murine lymphocyte chemotaxis by CCK-8S, GRP and NPY. Mech. Ageing Dev. 102, 249–261.

Mehr, R., Fridkis-Hareli, M., Abel, L., Segel, L., Globerson, A., 1995. Lymphocyte development in irradiated thymuses: dynamics of colonization by progenitor cells and regeneration of resident cells. J. Theor. Biol. 177, 181–192.

Mehr, R., Perelson, A.S., 1997. Blind T-cell homeostasis and the CD4/CD8 ratio in the thymus and peripheral blood. Acquir. Immune Defic. Syndr. Hum. Retrovirol. 14, 387–398.

Mehr, R., Perelson, A.S., Fridkis-Hareli, M., Globerson, A., 1997. Regulatory feedback pathways in the thymus. Immunol. Today 18, 581–585.

Mehr, R., Segel, L., Sharp, A., Globerson, A., 1994. Colonization of the thymus by T cell progenitors: models for cell–cell interactions. J. Theor. Biol. 170, 247–257.

Mehr, R., Ubezio, P., Kukulansky, T., Globerson, A., 1996. Cell cycle in the aging thymus. Aging: Immunol. Infec. Dis. 6, 133–140.

Michie, S.A., Rouse, R.V., 1989. Traffic of mature lymphocytes into the mouse thymus. Thymus 13, 141–148.

Miller, R.A., 1991. Aging and immune function. Int. Rev. Cytol. 124, 187–215.

Miller, R.A., Harrison, D.E., 1985. Delayed reduction in T cell precursor frequencies accompanies diet-induced lifespan extension. J. Immunol. 134, 1426–1429.

Miller, R.A., 1996. The aging immune system: primer and prospectus. Science 273, 70–74.

Miskin, R., Masos, T., Yahav, S., Shinder, D., Globerson, A., 1999. AlphaMUPA mice: a transgenic model for increased life span. Neurobiol. Aging 20, 555–564.

Moll, U.M., 1997. Functional histology of the neuroendocrine thymus. Microsc. Res. Tech. 38, 300–310.

Montecino-Rodriguez, E., Clark, R., Dorshkind, K., 1998. Effects of insulin-like growth factor administration and bone marrow transplantation on thymopoiesis in aged mice. Endocrinology 139, 4120–4126.

Montecino-Rodriguez, E., Johnson, A., Dorshkind, K., 1996. Thymic stromal cells can support B cell differentiation from intrathymic precursors. J. Immunol. 156, 963–967.

Moriguchi, S., 1998. The role of vitamin E in T-cell differentiation and the decrease of cellular immunity with aging. Biofactors 7, 77–86.

Morrison, S.J., Wandycz, A.M., Akashi, K., Globerson, A., Weissman, I.L., 1996. The aging of hematopoietic stem cells. Nat. Med. 2, 1011–1016.

Mountz, J.D., Wu, J., Zhou, T., Hsu, H.C., 1997. Cell death and longevity: implications of Fas-mediated apoptosis in T-cell senescence. Immunol. Rev. 160, 19–30.

Nabarra, B., Andrianarison, I., 1996. Ultrastructural study of thymic microenvironment involution in aging mice. Exp. Gerontol. 31, 489–506.

Nakahama, M., Mohri, N., Mori, S., Shindo, G., Yokoi, Y., Machinami, R., 1990. Immunohistochemical and histometrical studies of the human thymus with special emphasis on age-related changes in medullary epithelial and dendritic cells. Virchows Arch. B Cell Pathol. Incl. Mol. Pathol. 58, 245–251.

Nilsson, B., Bergqvist, A., Lindblom, D., Ljungberg, O., Sodergard, R., von Schoultz, B., 1986. Characterization and localization of specific oestrogen binding in the human thymus. Gynecol. Obstet. Invest. 21, 150–157.

Norment, A.M., Bevan, M.J., 2000. Role of chemokines in thymocyte development. Semin. Immunol. 12, 445–455.

Offner, F., Kerre, T., Smedt, M., Plum, J., 1999. Bone marrow CD34 cells generate fewer T cells in vitro with increasing age and following chemotherapy. Br. J. Haematol. 104, 801–808.

Offner, F., Plum, J., 1998. The role of interleukin-7 in early T-cell development. Leuk. Lymphoma 30, 87–99.

Oh, S.H., Kim, K., 1999. Expression of interleukin-1 receptors in the later period of foetal thymic organ culture and during suspension culture of thymocytes from aged mice. Immunol. Cell Biol. 77, 491–498.

Ohkusu-Tsukada, K., Tsukada, T., Isobe, K., 1999. Accelerated development and aging of the immune system in p53-deficient mice. J. Immunol. 163, 1966–1972.

Peled, A., Haran-Ghera, N., 1991. Prevention of spontaneous AKR T cell lymphomagenesis by 24-666, a virus isolated from an AKR B cell lymphoma. Virology 181, 528–535.

Peled, A., Kollet, O., Ponomoyov, T., Petit, I., Franitza, S., Grabovsky, V., Slav, M.M., Nagler, A., Lider, O., Alon, R., Zipori, D., Lapidot, T., 2000. The chemokine SDF-1 activates the integrins LFA-1, VLA-4, and VLA-5 on immature human CD34(+) cells: role in transendothelial/stromal migration and engraftment of NOD/SCID mice. Blood 95, 3289–3296.

Peled, A., Petit, I., Kollet, O., Magid, M., Pnomaryov, T., Byk, T., Nagler, A., Ben-Hur, H., Many, A., Shultz, L., Lider, O., Alon, R., Zipori, D., Lapidot, T., 1999. Dependence of human stem cell engrafment and repopulation of NOD/SCID mice on CXCR4. Science 283, 845–848.

Perez-Mera, M.L., Carnero, D.G., Rey-Mendez, M., 1991. Effect of age on the proportion of mouse bone marrow cells migrating in response to newborn thymus supernatant. Cell migration and thymus evolution in mice. Thymus 18, 237–241.

Phillips, J.A., Brondstetter, T.I., English, C.A., Thoman, M.L., 2002. International Conference on Immunology and Aging, June 2002 (abstract).

Resnitzky, P., Segal, M., Barak, Y., Dassa, C., 1987. Granulopoiesis in aged people: inverse correlation between bone marrow cellularity and myeloid progenitor cell numbers. Gerontology 33, 109–114.

Rice, J.C., Bucy, R.P., 1995. Differences in the degree of depletion, rate of recovery, and the preferential elimination of naive CD4+ T cells by anti-CD4 monoclonal antibody (GK1.5) in young and aged mice. J. Immunol. 154, 6644–6654.

Rinner, I., Felsner, P., Liebmann, P.M., Hofer, D., Wolfler, A., Globerson, A., Schauenstein, K., 1998. Adrenergic/cholinergic immunomodulation in the rat model–in vivo veritas? Dev. Immunol. 6, 245–252.

Rinner, I., Globerson, A., Kawashima, K., Korstako, W., Schauenstein, K., 1999. A possible role for acetylcholine in the dialogue between thymocytes and thymic stroma. Neuroimmunomodulation 6, 51–55.

Rinner, I., Kukulansky, T., Felsner, P., Skreiner, E., Globerson, A., Kasai, M., Hirokawa, K., Korsatko, W., Schauenstein, K., 1994. Cholinergic stimulation modulates apoptosis and differentiation of murine thymocytes via a nicotinic effect on thymic epithelium. Biochem. Biophys. Res. Commun. 203, 1057–1062.

Scollay, R., Godfrey, D.I., 1995. Thymic emigration: conveyor belts or lucky dips? Immunol. Today 16, 268–273.

Segal, R., Dayan, M., Globerson, A., Habot, B., Shearer, G., Mozes, E., 1997. Effect of aging on cytokine production in normal and experimental systemic lupus erythematosus afflicted mice. Mech. Ageing Dev. 96, 47–58.

Sen, J., 2001. Signal transduction in thymus development. Cell. Mol. Biol. (Noisy-le-grand) 47, 197–215.

Sharp, A., Brill, S., Kukulansky, T., Globerson, A., 1991. Developmental changes in bone marrow thymocyte progenitors and Thy1+ cells in aging. Ann. N.Y. Acad. Sci. 621, 229–238.

Sharp, A., Kukulansky, T., Globerson, A., 1990a. In vitro analysis of age-related changes in the developmental potential of bone marrow thymocyte progenitors. Eur. J. Immunol. 20, 2541–2546.

Sharp, A., Kukulansky, T., Malkinson, Y., Globerson, A., 1990b. The bone marrow as an effector T cell organ in aging. Mech. Ageing Dev. 52, 219–233.

Sharp, A., Zipori, D., Toledo, J., Tal, S., Resnitzky, P., Globerson, A., 1989. Age related changes in hemopoietic capacity of bone marrow cells. Mech. Ageing Dev. 48, 91–99.

Sherwood, P.J., Weissman, I.L., 1990. The growth factor IL-7 induces expression of a transformation-associated antigen in normal pre-B cells. Int. Immunol. 2, 399–406.

Small, M., Weissman, I.L., 1996. CD3−4−8− thymocyte precursors with interleukin-2 receptors differentiate phenotypically in coculture with thymic stromal cells. Scand. J. Immunol. 44, 115–121.

Staples, J.E., Gasiewicz, T.A., Fiore, N.C., Lubahn, D.B., Korach, K.S., Silverstone, A.E., 1999. Estrogen receptor alpha is necessary in thymic development and estradiol-induced thymic alterations. J. Immunol. 163, 4168–4174.

Steinmann, G.G., Muller-Hermelink, H.K., 1984. Immunohistological demonstration of terminal transferase (TdT) in the age-involuted human thymus. Immunobiology 166, 45–52.

Takeoka, Y., Chen, S.Y., Yago, H., Boyd, R., Suehiro, S., Shultz, L.D., Ansari, A.A., Gershwin, M.E., 1996. The murine thymic microenvironment: changes with age. Int. Arch. Allergy Immunol. 111, 5–12.

Tamir, A., Eisenbraun, M.D., Garcia, G.G., Miller, R.A., 2000. Age-dependent alterations in the assembly of signal transduction complexes at the site of T cell/APC interaction. J. Immunol. 165, 1243–1251.

Thoman, M.L., 1995. The pattern of T lymphocyte differentiation is altered during thymic involution. Mech. Ageing Dev. 82, 155–170.

Thoman, M.L., 1997. Early steps in T cell development are affected by aging. Cell. Immunol. 178, 117–123.

Timm, J.A., Thoman, M.L., 1999. Maturation of CD4+ lymphocytes in the aged microenvironment results in a memory-enriched population. J. Immunol. 162, 711–717.

Utsuyama, M., Hirokawa, K., 1989. Hypertrophy of the thymus and restoration of immune functions in mice and rats by gonadectomy. Mech. Ageing Dev. 47, 175–185.

Utsuyama, M., Hirokawa, K., Mancini, C., Brunelli, R., Leter, G., Doria, G., 1995. Differential effects of gonadectomy on thymic stromal cells in promoting T cell differentiation in mice. Mech. Ageing Dev. 81, 107–117.

Utsuyama, M., Kobayashi, S., Hirokawa, K., 1997a. Induction of thymic hyperplasia and suppression of splenic T cells by lesioning of the anterior hypothalamus in aging Wistar rats. J. Neuroimmunol. 77, 174–180.

Utsuyama, M., Wakikawa, A., Tamura, T., Nariuchi, H., Hirokawa, K., 1997b. Impairment of signal transduction in T cells from old mice. Mech. Ageing Dev. 93, 131–144.

Van Ewijk, W., Hollander, G., Terhorst, C., Wang, B., 2000. Stepwise development of thymic microenvironments in vivo is regulated by thymocyte subsets. Development 127, 1583–1591.

Wang, Y., Tokuda, N., Tamechika, M., Hashimoto, N., Yamaguchi, M., Kawamura, H., Irifune, T., Choi, M., Awaya, A., Sawada, T., Fukumoto, T., 1999. Vascular and stromal changes in irradiated and recovering rat thymus. Histol. Histopathol. 14, 791–796.

Weigle, W.O., 1993. The effect of aging on cytokine release and associated immunologic functions. Prob. Aged 13, 551–569.

Weindruch, R.H., Suffin, S.C., 1980. Quantitative histologic effects on mouse thymus of controlled dietary restriction. J. Gerontol. 35, 525–531.

Wu, Q., Tidmarsh, G.F., Welch, P.A., Pierce, J.H., Weissman, I.L., Cooper, M.D., 1989. The early B lineage antigen BP-1 and the transformation-associated antigen 6C3 are on the same molecule. J. Immunol. 143, 3303–3308.

Yu, S., Abel, L., Globerson, A., 1997. Thymocyte progenitors and T cell development in aging. Mech. Ageing Dev. 94, 103–111.

Zlotnik, A., Moore, T.A., 1995. Cytokine production and requirements during T-cell development. Curr. Opin. Immunol. 7, 206–213.

Advances in
Cell Aging and
Gerontology

Effective immunity during late life, a possible role for the thymus

Jeffrey Pido-Lopez and Richard Aspinall

*Department of Immunology, Faculty of Medicine, Imperial College of Science Technology and
Medicine, Chelsea and Westminster Hospital, 369 Fulham Road, London SW10 9NH, UK*
Correspondence address: Tel.: +44-208-746-5973; fax: +44-208-746-5997. E-mail: jpido@ic.ac.uk (J.P.-L.)

Contents

1. Compromised immunity in the aged; a potential role for the thymus

Over the next 40 years it is projected that approximately 25% of the total UK population will be above the age of 65. UK mortality figures reveal that at the turn of the last century over 95% of deaths were of individuals above the age of 45 with 84% being accounted for by over 65 year olds. Diseases including bacterial meningitis, tuberculosis, herpes zoster, infectious endocarditis and respiratory tract, intra-abdominal, urinary tract, skin and soft tissue infections are observed to be at a higher prevalence in the aged population with mortality rates for these diseases being 2–3 times higher in elderly compared to younger patients (Yoshikawa, 2000). In addition to increased susceptibility to infections, increased incidence of certain cancers and specific autoimmune diseases and reduced responsiveness to vaccination are also more commonly observed in the elderly population which is suggestive of an age-associated deterioration of the immune system (Gardner, 1980; Bernstein et al., 1986;

Advances in Cell Aging and Gerontology, vol. 13, 79–91

Beverly, 2000). With an older population anticipated within the next half century, identifying the causes of immune system deterioration and elucidating methods of alleviating this process is important if we are to prevent a possible pressure on future health resources.

T lymphocytes, generated by the thymus, play a key role in an efficiently functioning immune system and deficiencies in immunity with increasing age have been suggested to be a consequence of deterioration in the T cell response. This decline in T cell responses may in part arise from the on-going process of age-associated thymic atrophy (Steinmann, 1986; Kampinga et al., 1997). Due to thymic atrophy, an age-related decrease in the contribution of thymic-derived naïve T cells to the peripheral T cell pool is observed, leading to reductions in the numbers of naïve T cells with age (Kurashima et al., 1995). Despite a declining thymic output, existing peripheral T cells with an established T cell receptor (TCR) repertoire by puberty are able to undergo cellular expansion in order to maintain T cell numbers at a generally constant level throughout life (Rocha et al., 1989; Hannet et al., 1992). However, these existing T cells may develop defects during their replicative lifespan, furthermore such cells like other cells in the body have a limited replicative capacity (Effros and Pawelec, 1997). The theoretical consequence of this on-going reduction in thymic T cell output is an age-related decline in the efficiency of newly-generated T cells to replace those existing defective, senescent ones. This in turn leads to the gradual accumulation of faulty T cells which ultimately results in compromised immune responses during old age (Murasko et al., 1987; Bernstein et al., 1998; Aspinall, 2000; Haynes et al., 2000). The efficient replacement of defective, senescent T cells with those possessing a high replicative capacity may therefore be a valid means of maintaining a competently functioning immune system during later life. One possible route of achieving this goal is by generating either T lymphocytes that do not have a replicative limit or alternatively T cells that have significantly extended replicative capacities in the periphery. A second alternative route is through the reversal of thymic atrophy and the maintenance of T cell generation by the thymus at sufficient levels to permit the efficient replacement of defective T cells. In this review we will consider the possible methods of achieving the latter approach; first, in order to do this, the elements that bring about the process of thymic atrophy require definition.

2. Current understanding of age-related thymic atrophy

In many species, age-associated thymic atrophy is visually illustrated by the obvious gradual size reduction of the true thymic tissue during increasing age, that is the consequence of significant losses in the number of thymocytes. These developing thymocytes can be divided into specific subpopulations based upon their expression of the classic T cell surface markers; CD3, CD4 and CD8. The most immature thymocyte subpopulation are termed double-negatives (DN), these cells possess low levels of CD4 but do not express CD3 or CD8 (Rahemtulla et al., 1991; Wu et al., 1991). The DN cells mature to become the most abundant thymocyte subpopulation in the thymus known as double-positive (DP) thymocytes, which express CD3, CD8 and CD4. Finally, DP cells after successfully undergoing the thymic selection process, develop into the most mature thymocyte subpopulation known as single-positives (SP) which express CD3 and either CD4 or CD8. In the mouse, the earliest DN T cell precursors can be further subdivided according to CD44 and CD25 expression. The most immature of DN cells express CD44 but lack CD25 and these cells have the capacity to become B lymphocytes, NK cells or dendritic cells, as well as T cells (Shortman and Wu, 1996). CD44+CD25− thymocytes subsequently develop into CD44+CD25+DN, which are also capable of giving rise to dendritic cells and which then go on to mature into T cell lineage-committed CD44−CD25+DN cells and finally into CD44−CD25−DN (Petrie et al., 1992; Wu et al., 1996) (Fig. 1). Early studies in both mice and humans were undertaken to identify the location of the lesion/s responsible for thymic atrophy along the T cell developmental pathway and which thymocyte subpopulations were specifically involved. These studies involved the quantification of the major thymocyte subpopulations and worked on the notion that an age-associated decline in the contribution of a specific subpopulation would cause a proportional reduction in the numbers of thymocyte subpopulations downstream to it, so that consequently a bottleneck effect is observed. In the human studies, a significant decrease in the total number of DN, DP and SP cells was observed with increasing age. However, the relative contribution that each thymocyte subpopulation made to the total thymocyte population in the thymus was found to remain relatively constant at different ages, indicating the absence of a lesion in the thymic T cell developmental pathway between the late DN to the SP stages (Bertho et al., 1997; Flores et al., 1999). Furthermore, a study in mice revealed similar results, with the DP thymocyte subpopulation showing the most dramatic decline in numbers but with the proportions of each thymocyte subpopulation remaining unchanged with increasing age (Aspinall, 1997). Although these studies failed to locate the cause of any lesions in the T cell developmental pathway, they provided us with a clue as to where a possible lesion may occur, namely at the very early stage of development.

3. Pathophysiology of thymic atrophy

Having deduced that the lesion provoking atrophy of the thymus most likely occurs at the earliest stage of T cell development, a number of theories accounting for age-related losses in thymocyte numbers were postulated. One of the initial hypotheses suggested thymic atrophy to be the result of a progressive reduction in the number of T cell progenitors emigrating from the bone marrow to the thymus during increasing age. An experiment analysing the number of progenitor cells in the bone marrow of old and young mice found significantly lower precursor cell numbers in the older mice providing some support for this hypothesis (Eren et al., 1990). This is further sustained by studies showing the lower capacity of bone marrow progenitors from old mice in comparison to bone marrow progenitors from younger mice to repopulate the thymocyte-depleted thymi of young mice (Tyan, 1976, 1977). These findings subsequently led to a strategy for reversing thymic atrophy involving the use of bone marrow obtained from juvenile mice to supply T cell progenitors to the thymus of aged mice. Experiments involving thymic reconstitution of irradiated old mice using bone marrow grafts from young mice, however, failed to produce any significant improvement, with old graft recipients being observed to have lower thymocyte numbers than younger recipients (Mackall et al., 1998). A second theory proposes that developing thymocytes were less capable of undergoing differentiation with age. Studies to test this hypothesis, however, have found little difference in the ability of progenitors either from old or young bone marrow to develop and repopulate thymic tissues in vitro (Aspinall and Andrew, 2001). These results therefore cumulatively suggest that the defect at the root of thymic atrophy is unlikely to be located within the developing thymocytes but rather in their environment.

The initial findings of experiments involving the quantification of thymocyte subpopulations prompted our own analysis of the early DN thymocyte subpopulations (i.e. CD44+CD25−; CD45−CD25−; CD44−CD25+ and CD44−CD25− cells). This study revealed that the earliest CD44+, CD25−DN cells did not decline in numbers significantly with age. However, significant reductions in the three downstream DN subpopulations were noted, suggesting the existence of a lesion located between the CD44+CD25− to the CD44+CD25+ phase of development or within the CD44+CD25+DN stage itself (Aspinall and Andrew, 2001). Because gene rearrangement of the β TCR locus initiates at around the CD45+CD25+DN stage, problems with TCR gene rearrangement/production may be accountable for the thymic atrophy process during ageing. Results of a study involving F5 and F5 RAG$^{-/-}$ mice and other transgenic mice possessing and expressing productively rearranged αβ TCR transgenes provide evidence for this theory. In such manipulated mice, thymic atrophy with increasing age was not observed (Aspinall, 1997).

4. Reversal of thymic atrophy

With the notion that there is no intrinsic age-related defect in T cell progenitors and that the lesion is located around the CD44+CD25+DN stage of development, we can assume that environmental factors affecting the DN stage of T cell development are likely to be at the core of thymic atrophy. Several factors have been shown to be capable of positively affecting the early stages of thymocyte development and can potentially be employed to correct the lesion in the developmental pathway, augment thymopoiesis and reverse thymic atrophy.

4.1. Interleukin-7

Several experiments indicate the importance of the cytokine interleukin-7 (IL-7) during the development of T cells within the thymus. Although IL-7 is observed to play a role in the expansion of mature SP thymocytes, its stimulatory effect is particularly evident and probably of greater importance in the DN thymocyte subpopulation (Hare et al., 2000). In vivo studies in mice involving the administration of IL-7 neutralising antibodies resulted in the severe involution of the thymus. However, upon termination of the antibody treatment a subsequent reversion of thymic atrophy was observed, suggesting a possible role for thymic IL-7 in the age-associated atrophy of the thymus (Bhatia et al., 1995). Experiments involving the use of IL-7 signalling-deficient mice (i.e. IL-7, IL-7 receptor α and common γ chain transgenics) further illustrated the importance of IL-7 with the gene-deficient mice showing severely hypoplastic thymi and extreme T cell lymphopenia (Peschon et al., 1994; von Freeden-Jeffry et al., 1995; He and Malek, 1996). Analysis of the thymocyte subpopulations in IL-7-deficient mice revealed a significant block in thymocyte development at the transition from the CD44+CD25−DN stage of development to the CD44+CD25+DN and CD44−CD25+DN stages (Moore et al., 1996). Because the affected DN subpopulations are at a particular stage of development where rearrangement of the β TCR genes and generation of the β TCR chain takes place, a key role for IL-7 during TCR gene rearrangement was suggested (Crompton et al., 1997). A number of studies provide some supporting evidence for this proposal (Muegge et al., 1993; Tsuda et al., 1996). Additionally, other studies have indicated an ability of IL-7 to promote the survival of early thymocytes through the prevention of apoptosis, partially via the upregulation of the Bcl2 anti-apoptotic protein (Murray et al., 1989). IL-7 has also been observed to directly stimulate early thymocytes to proliferate in vitro as well as SP thymocytes in vivo (Hitoaki et al., 1989; Murray et al., 1989; Mackall et al., 1998; Hare et al., 2000). These several qualities make IL-7 an ideal candidate for reversing thymic atrophy and increasing thymic T cell production. An initial experiment involving the injection of IL-7 into old mice yielded some promising results, with the study finding immediate, rapid increases in the numbers of viable DN thymocytes and increases in DP and SP thymocytes several days after the administration of IL-7 (Andrew and Aspinall, 2001). Increases in naïve CD4 T cell numbers were also noted several weeks after the final IL-7 administration, which may be a consequence of increased thymic output (Aspinall and Andrew, 2000). However, the observed ability of IL-7 to induce naïve T cell proliferation cannot be dismissed as a possible alternative cause for the observed increase in naïve T cell numbers (Soares et al., 1998). This uncertainty emphasises the need for a more accurate method of analysing thymic output. These several studies strongly indicate IL-7's capacity to increase thymopoiesis, but whether these IL-7-induced increases in thymic function subsequently lead to the replacement of residing peripheral T cells with new ones and the rejuvenation of the T cell pool remains to be determined. Clearly, in order to elucidate the effect of IL-7-induced thymopoiesis on the periphery a reliable means of identifying peripheral T cells that have just recently emigrated from the thymus is essential.

4.2. Stem cell factor

The receptor for stem cell factor (SCF), c-kit (CD117) can be found on the surface of CD44+CD25−DN and CD44+CD25+DN thymocytes and ligation of this receptor by its ligand promotes the expansion of these cells (Rodewald et al., 1995). Studies involving the use of SCF signalling-deficient mice (i.e. SCF and c-kit knockouts) have produced results similar to those observed in the IL-7 signalling-deficient mice (Rodewald et al., 1990; Rodewald, 2001). Interestingly, the absence of both IL-7 and SCF signalling was found to result in the complete abrogation of thymocyte development, suggesting a compensatory effect of IL-7 or SCF in the absence of either of these thymic factors (Di Santo et al., 1999). A study involving the treatment of old mice with SCF alone or with IL-7, however, failed to display any effect of SCF or any supplementary affect of SCF on IL-7-induced increases in viable DN thymocyte numbers (Andrew and Aspinall, 2001).

4.3. Growth hormone

Studies involving the augmentation of growth hormone levels in animals have resulted in notable increases in thymic cellularity. The implantation of growth hormone-producing cell lines (GH3) into rats was found to increase the numbers of DN and DP thymocyte subpopulations, while administration of agents promoting growth hormone release resulted in increases in the DN subpopulation in old mice (Kelley et al., 1986; Koo et al., 2001). Furthermore, a recent study in humans reported notable increases in naïve T cell numbers after the administration of growth hormone (personal communication, Antonio Pires) suggesting this agent's potential for augmenting thymic function and rejuvenating the T cell pool. The mechanism by which growth hormone exerts its stimulatory effect on the thymus, however, remains to be defined.

4.4. Thymosin α1

Thymosin α1 (Tα1) is a 28 amino acid peptide produced by thymic epithelial cells located in the outer cortex and medulla of the thymus (Hirokawa et al., 1982). In vitro studies involving the use of bone marrow and progenitor cell cocultures have revealed Tα1's capacity to stimulate thymocyte maturation (Knutsen et al., 1999). In addition, the administration of Tα1 in mice has resulted in notable increases in murine thymic size and thymocyte numbers, further demonstrating Tα1's ability to enhance thymocyte growth (Beuth et al., 2000). Tα1 is likely to exert its stimulatory effect on developing thymocytes partly by acting as an anti-apoptotic agent (Baumann et al., 1996). Interestingly, the addition of Tα1 to thymocyte and thymic epithelial cell cocultures was found to increase the production of IL-7 by epithelial cells, suggesting that the enhancing effect of Tα1 on thymopoiesis may be through the induction of IL-7 (Knutsen et al., 1999).

4.5. Thymic factors

Several other factors found in the thymus may also play possible roles in the age-dependent atrophy of the thymus. Levels of thymic inhibitory factors such as leukemia inhibitory factor (LIF), oncostatin M (OSM) and SCF have been observed to increase with age (Sempowski et al., 2000), while thymic mRNA expression of a number of cytokines including IL-2, IL-9, IL-10, IL-13 and IL-14 has been found to decrease with age (Sempowski et al., 2000). The mechanisms by which these factors affect the thymus require further investigation but their understanding may lead to alternative strategies for stimulating thymic T cell generation.

5. Assessment of thymic output

The observed role of IL-7 on thymopoiesis suggests its genuine potential as a means of reversing age-associated thymic atrophy and increasing the output of new T cells from the thymus to the periphery. The recent study showing IL-7's efficacy in stimulating increased thymopoiesis suggests that a subsequent increase in thymic output may contribute to the replacement of existing T cells in the peripheral pool with "younger" T cells that possess higher replicative capacities. Clearly, in order to perform such a study a means of identifying and quantifying recent thymic emigrant (RTE) T cells in the periphery is vital.

The ability to distinguish between T cells that have just recently entered the periphery from the thymus and those pre-existing cells in the peripheral T cell pool poses a major problem. Initially, this had been undertaken through the detection of T cell surface markers that allow phenotypic discrimination of circulating human T cells into distinct "naïve" or "memory" populations. The naïve T cell subpopulation in which RTEs are included, was found to be associated with the expression of CD45RA and CD62L whilst memory T cells express the shorter CD45RO cell surface protein instead of CD45RA, and varying amounts of CD62L. However, phenotypic measures have been shown to be imprecise in their ability to differentiate between RTEs and the CD45RA-expressing T cells that have reverted back from the memory, CD45RO status (Bell and Sparshott, 1990). Indeed, despite ongoing age-related thymic atrophy, the numbers of circulating naïve CD45RA+ T cells have been found to remain relatively constant between mid- to late-life in humans (Erkeller-Yuksel et al., 1992). Additionally, studies in old mice have shown that RTEs rapidly express memory cell surface markers, further putting the reliability and applicability of phenotypic analysis into question (Timm and Thoman, 1999). An additional cell surface marker, CD103 expressed by a fraction of phenotypically naïve CD8 T cells has been proposed to allow the better identification of CD8+ RTEs in humans (McFarland et al., 2000). Nonetheless, in some animals, the problems with phenotypic assays can be completely overcome. For example, in the chicken a definitive cell surface marker known as chT1 is expressed by avian RTEs which decline in levels with age and after partial thymectomy, while in rats high expression of RT6 and CD45RC have been observed in RTEs (Hosseinzadeh and Goldschneider, 1993; Kong et al., 1998). In mice, peripheral RTEs can be detected and kinetically followed in the peripheral circulation after direct intrathymic injections e.g. with FITC (Scollay et al., 1980). Alternatively, lethally-irradiated mice reconstituted with bone marrow grafts can be used to study thymic output in a murine model. These bone marrow grafts generate T cells expressing specific congenic markers that allow their identification from the recipients' T cells in the periphery (Mackall et al., 1998). A study involving the use of this latter technique to assess the effect of IL-7 upon the murine thymus indicated IL-7's ability to increase thymic output in mice (Mackall et al., 2001). The interventionist nature of intrathymic injections and of bone marrow transplantation, however, poses a problem with their use and makes them inapplicable for human studies.

A PCR-based method utilizing a key feature of the α TCR rearrangement process has become a commonly used technique of assessing thymic output in humans (Douek et al., 1998; Markert et al., 1999; Pido-Lopez et al., 2001). The assay relies upon the detection of extrachromosomal DNA molecules known as TCR rearrangement excision circles (TRECs). TRECs are generated in the majority of developing thymocytes as a result of TCR δ locus deletion, which occurs prior to the rearrangement of TCRα genes in the thymus. The generation of TRECs is initiated by the specific cleavage of the δREC and ψJα deleting elements located upstream of the most 5' Dδ and upstream of the most 5' Jα gene, respectively. The excised δREC and ψJα portions are then ligated to one another to form the circular extrachromosomal TREC molecule (Fig. 2). Due to the precision of TCR δ locus deletion all TRECs possess a region, spanning between the δREC and ψJα gene portions consisting of a conserved nucleotide sequence, thus allowing the detection and quantification of RTE T cells that contain them via PCR. Due to the lack of allelic exclusion during TCRα gene rearrangement, two TRECs molecules are generated post-gene rearrangement. Additionally, TRECs are incapable of replicating during cell division, so that they become diluted over cell divisions (Douek et al., 1998). Human studies involving the measurement of TRECs revealed a gradual reduction in TRECs with increasing age, echoing the decline in thymic function resulting from age-associated atrophy of the thymus (Douek et al., 1998; Zhang et al., 1999; Aspinall et al., 2000). Our analysis of TRECs continued to reveal a reduction of these DNA molecules during ageing but also a significantly higher level of TRECs in females compared to age-matched males. This result suggests a more effective thymus in females and indicates thymic output as a possible factor for the higher life expectancy in females compared to males (Pido-Lopez et al., 2001). The TREC assay has also been applied for the assessment of thymic output in animals including primates (sooty mangabey, chimpanzee, rhesus macaques) (Wykrzykowska et al., 1998; Chakrabarti et al., 2000) and chickens (Kong et al., 1998). However, in mice certain features of murine TCRδ locus excision prevent the utilisation of the TREC assay (Shutter et al., 1995). Although the TREC assay generally provides a simple and relatively reliable means of detecting increases in human thymic output (i.e. increased TRECs levels indicating increased RTE numbers), interpretation of TREC reductions are less straightforward. Because cell replication has a diluting effect on TRECs, any decreases in the levels of TRECs may not necessarily be solely due to a reduction in thymic output (Hazenberg et al., 2000; Douek et al., 2001). Nevertheless, the non-invasive nature of the TREC assay makes it an ideal method for determining the effect of thymic stimulatory agents on thymic output in humans and potentially as an ideal alternative to the interventionist assays currently employed for assessing thymic output in animal models.

6. Conclusion

With an anticipated increase in the world's elderly population, means of improving the quality of life in the aged, particularly in the reduction of disease incidence, are essential to prevent possible strains on health resources. With increasing age, a reduction in immune competence is observed which results in compromised immune responses and increased susceptibility to disease. T cells play a central role in immune responses and deficiencies in this arm of the immune system may be the major factor involved in the deterioration of immunity in old age. The inability of existing T cells to replicate indefinitely and the gradual reduction in T cell production by the thymus will inevitably lead to an accumulation of defective, senescent T cells during later life. This outcome is likely to account for the reductions in the efficacy of the T cell responses seen in the aged. The reversal of the age associated thymic atrophy process provides us with a credible solution to the problem of compromised T cell immunity during ageing. The replacement of senescent T cells in the periphery with new ones from the thymus could generate a rejuvenated T cell pool capable of providing protective immunity to the aged. A number of agents have the potential capacity to reverse thymic atrophy and permit the rejuvenation of the peripheral T cell pool. A number of tools are available to allow the elucidation of these thymic stimulatory agents' effect on thymic output and to determine their value. The potential benefits of increasing thymic output during old age has been reiterated throughout this review; however a number of questions need to be acknowledged. Will the stimulation of thymopoiesis trigger defective negative selection in the thymus so that repopulating cells are more likely to be self-reactive clones that increase the risk of autoimmunity? Would complete rejuvenation of the immune system be necessary or would partial replacement of existing T cells be sufficient in achieving effective immunity? Would the thymus require to function at pre-pubescent levels continuously or for only a limited time period to permit adequate protection? It is not difficult to envisage the reversion of age-dependent thymic atrophy in the future; however the above questions must be sufficiently addressed if a thymus-based solution to the problem of immune system deficiency during old age is to be advocated. The effects of other factors such as nutrition and stress, along with the contribution of extra-thymic T cell production and the effects of thymic inhibitory agents on this mechanism will also require consideration in order to achieve a maximal effect from therapeutic regimens.

Acknowledgements

This work was funded by grants from the Wellcome Trust (grant number 051541), Research into Ageing and the Luard Foundation (J. P-L is the Luard Scholar).

References

Aspinall, R., 1997. Age-associated thymic atrophy in the mouse is due to a deficiency affecting rearrangement of the TCR during intrathymic T cell development. J. Immunol. 158, 3037–3045.
Aspinall, R., 2000. Longevity and the immune response. Biogerontology 1, 273–278.
Aspinall, R., Andrew, D., 2000. Thymic atrophy in the mouse is a soluble problem of the thymic microenvironment. Vaccine 18, 1629–1634.

Aspinall, R., Pido, J., Andrew, D., 2000. A simple method for the measurement of sjTREC levels in blood. Mech. Ageing Dev. 21, 59–67.

Andrew, D., Aspinall, R., 2001. IL-7 and not stem cell factor reverses both the increase in apoptosis and the decline in thymopoiesis seen in aged mice. J. Immunol. 166, 1524–1530.

Aspinall, R., Andrew, D., 2001. Age-associated thymic atrophy is not associated with a deficiency in the CD44(+)CD25(−)CD3(−)CD4(−)CD8(−) thymocyte population. Cell. Immunol. 212, 150–157.

Baumann, C., Badamchian, M., Goldstein, A.L., 1996. Thymosin α1 antagonism of induced apoptosis of developing thymocytes requires activation for multiple signal transduction pathways. FASEB J. 10, A441.

Bell, E.B., Sparshott, S.M., 1990. Interconversion of CD45R subsets of CD4 T cells in vivo. Nature 348, 163–166.

Bernstein, E.D., Gardner, E.M., Abrutyn, E., Gross, P., Murasko, D.M., 1998. Cytokine production after influenza vaccination in a healthy elderly population. Vaccine 16, 1722–1731.

Bertho, J.M., Demarquay, C., Moulian, N., Van Der Meeren, A., Berrih-Aknin, S., Gourmelon, P., 1997. Phenotypic immunohistological analysis of the human adult thymus: evidence for an active thymus during adult life. Cell. Immunol. 179, 30–37.

Beuth, J., Schierholz, J.M., Meyer, G., 2000. Thymosin α1 application augments immune response and down-regulates tumor weight and organ colonization in BALB/c mice. Cancer Lett. 159, 9–13.

Beverly, P., 2000. Is immune senescence reversible? Vaccine 18, 1721–1725.

Bhatia, S.K., Tygrett, L.T., Grabstein, K.H., Waldschmidt, T.J., 1995. The effect of in vivo IL-7 deprivation on T cell maturation. J. Exp. Med. 181, 1399–1409.

Chakrabarti, L.A., Lewin, S.R., Zhang, L., Gettie, A., Luckay, A., Martin, L.N., Skulsky, E., Ho, D.D., Cheng-Mayer, C., Marx, P.A., 2000. Age-dependent changes in T cell homeostasis and SIV load in sooty mangabeys. J. Med. Primatol. 29, 158–165.

Crompton, T., Outram, S.V., Buckland, J., Owen, M.J., 1997. A transgenic T cell receptor restores thymocyte differentiation in interleukin-7 receptor α chain deficient mice. Eur. J. Immunol. 27, 100–106.

Di Santo, J.P., Aifantis, I., Rosmaraki, E., Garcia, C., Feinberg, J., Fehling, H.J., Fischer, A., von Boehmer, H., Rocha, B., 1999. The common cytokine receptor gamma chain and the pre-T cell receptor provide independent but critically overlapping signals in early alpha/beta T cell development. J. Exp. Med. 189, 563–568.

Douek, D.C., McFarland, R.D., Keiser, P.H., Gage, E.A., Massey, J.M., Haynes, B.F., Polis, M.A., Haase, A.T., Feinberg, M.B., Sullivan, J.L., Jamieson, B.D., Zack, J.A., Picker, L.J., Koup, R.A., 1998. Changes in thymic function with age and during treatment of HIV infection. Nature 369, 690–695.

Douek, D.C., Betts, M.R., Hill, B.J., Little, S.J., Lempicki, R., Metcalf, J.A., Casazza, J., Yoder, C., Adelsberger, J.W., Stevens, R.A., Baseler, M.W., Keiser, P., Richman, D.D., Davey, R.T., Koup, R.A., 2001. Evidence for increased T cell turnover and decreased thymic output in HIV infection. J. Immunol. 167, 6663–6668.

Effros, R.B., Pawelec, G., 1997. Replicative senescence of T cells: does the Hayflick limit lead to immune exhaustion? Immunol. Today 18, 450–455.

Eren, R., Globerson, A., Abel, L., Zharhary, D., 1990. Quantitative analysis of bone marrow thymic progenitors in young and aged mice. Cell. Immunol. 127, 238–244.

Erkeller-Yuksel, F.M., Deneys, V., Yuksel, B., Hannet, I., Hulstaert, F., Hamilton, C., Mackinnon, H., Stokes, L.T., Munhyeshuli, V., Vanlangendonck, F., 1992. Age-related changes in human blood lymphocyte subpopulations. J. Pediatr. 120, 216–222.

Flores, K.G., Li, J., Sempowski, G.D., Haynes, B.F., Hale, L.P., 1999. Analysis of the human thymic perivascular space during aging. J. Clin. Invest. 104, 1031–1036.

Gardner, I.D., 1980. The effect of aging on susceptibility to infection. Rev. Infect. Dis. 2, 801–810.

Hannet, I., Erkeller-Yuksel, F., Lydyard, P., Deneys, V., DeBruyere, M., 1992. Developmental and maturational changes in human blood lymphocyte subpopulations. Immunol. Today 13, 215–218.

Hare, K.J., Hare, K.J., Jenkinson, E.J., Anderson, G., 2000. An essential role for the IL-7 receptor during intrathymic expansion of the positively selected neonatal T cell repertoire. J. Immunol. 165, 2410–2415.

Haynes, B.F., Markert, M.L., Sempowski, G.D., Patel, D.D., Hale, L.P., 2000. The role of the thymus in immune reconstitution in aging, bone marrow transplantation, and HIV-1 infection. Annu. Rev. Immunol. 18, 529–560.

Hazenberg, M.D., Otto, S.A., Cohen Stuart, J.W., Verschuren, M.C., Borleffs, J.C., Boucher, C.A., Coutinho, R.A., Lange, J.M., Rinke de Wit, T.F., Tsegaye, A., van Dongen, J.J., Hamann, D., de Boer, R.J., Miedema, F., 2000. Increased cell division but not thymic disfunction rapidly affects the T cell receptor excision circle content of the naïve T cell population in HIV-1 infection. Nat. Med. 6, 1036–1043.

He, Y.W., Malek, T.R., 1996. Interleukin-7 receptor alpha is essential for the development of gamma delta+T cells, but not natural killer cells. J. Exp. Med. 184, 289–293.

Hirokawa, K., McClure, J.E., Goldstein, A.L., 1982. Age-related changes in localization of thymosin in human thymus. Thymus 4, 19–29.

Hitoaki, O., Ito, M., Sudo, T., Hattori, M., Kano, S., Katsura, Y., Minato, N., 1989. IL-7 promotes thymocyte proliferation and maintains immunocompetent thymocytes bearing alpha beta or gamma delta T-cell receptors in vitro: synergism with IL-2. J. Immunol. 143, 2917–2922.

Hosseinzadeh, H., Goldschneider, I., 1993. Recent thymic emigrants in the rat express a unique antigen phenotype and undergo post thymic maturation in peripheral lymphoid tissue. J. Immunol. 150, 1670–1679.

Kampinga, J., Groen, H., Klatter, F.A., Pater, J.M., van Petersen, A.S., Roser, B., Nieuwenhuis, P., Aspinall, R., 1997. Post thymic T cell development in the rat. Thymus 24, 173–178.

Kelley, K.W., Brief, S., Westly, H.J., Novakofski, J., Bechtel, P.J., Simon, J., Walker, E.B., 1986. GH3 pituitary adenoma cells can reverse thymic aging in rats. Proc. Natl. Acad. Sci. USA 83, 5663–5668.

Kong, F., Chen, C.H., Cooper, M.D., 1998. Thymic function can be accurately monitored by the level of recent T cell emigrants in the circulation. Immunity 8, 97–104.

Knutsen, A.P., Freeman, J.J., Mueller, K.R., Roodman, S.T., Bouhasin, J.D., 1999. Thymosin α1 stimulates maturation of CD34+ stem cells into CD3+4+ cells in an *in vitro* thymic epithelia organ coculture model. J. Immunopharmacol. 21, 15–26.

Koo, G.C., Huang, C., Camacho, R., Trainor, C., Blake, J.T., Sirotina-Meisher, A., Schleim, K.D., Wu, T.J., Cheng, K., Nargund, R., McKissick, G., 2001. Immune enhancing effect of a growth hormone sercretogue. J. Immunol. 166, 4195–4200.

Kurashima, C., Utsuyama, M., Kasai, M., Ishijima, S.A., Konno, A., Hirokawa, K., 1995. The role of thymus in the aging of Th cell subpopulations and age-associated alteration of cytokine production by these cells. Int. Immunol. 7, 97–104.

Mackall, C.L., Plunt, J.A., Morgan, P., Farr, A.G., Gress, R.E., 1998. Thymic function in young/old chimeras: substantial thymic T cell regenerative capacity despite irreversible age-associated thymic involution. Eur. J. Immunol. 28, 1886–1893.

Mackall, C.L., Fry, T.J., Bare, C., Morgan, P., Galbraith, A., Gress, R.E., 2001. IL-7 increases both thymic-dependent and thymic-independent T cell regeneration after bone marrow transplantation. Blood 97, 1491–1497.

Markert, M.L., Boeck, A., Hale, L.P., Kloster, A.L., McLaughlin, T.M., Batchvarova, M.N., Douek, D.C., Koup, R.A., Kostyu, D.D., Ward, F.E., Rice, H.E., Mahaffey, S.M., Schiff, S.E., Buckley, R.H., Haynes, B.F., 1999. Transplantation of thymus tissue in complete Di George syndrome. N. Engl. J. Med. 341, 1180–1189.

McFarland, R.D., Douek, D.C., Koup, R.A., Picker, L.J., 2000. Identification of human recent thymic emigrant phenotype. PNAS 97, 4215–4420.

Moore, T.A., von Freeden-Jeffry, U., Murray, R., Zlotnik, A., 1996. Inhibition of gamma delta T cell development and early thymocyte maturation in IL-7−/− mice. J. Immunol. 157, 2366–2373.

Muegge, K., Vila, M.P., Durum, S.K., 1993. Interleukin-7: a cofactor for V(D)J rearrangement of the T cell receptor beta gene. Science 261, 293–298.

Murasko, D.M., Weiner, P., Kaye, D., 1987. Decline in mitogen induced proliferation of lymphocytes with increasing age. Clin. Exp. Immunol. 70, 440–448.

Murray, R., Suda, T., Wrighton, N., Lee, F., Zlotnik, A., 1989. IL-7 is a growth and maintenance factor for mature and immature thymocyte subsets. Int. Immunol. 1, 526–531.

Peschon, J.J., Morrissey, P.J., Grabstein, K.H., Ramsdell, F.J., Maraskovsky, E., Gliniak, B.C., Park, L. S., Ziegler, S.F., Williams, D.E., Ware, C.B., 1994. Early lymphocyte expansion is severely impaired in interleukin 7 receptor-deficient mice. J. Exp. Med. 1980, 1955–1960.

Petrie, H.T., Scollay, R., Shortman, K., 1992. Commitment to the T cell receptor alpha beta or gamma delta lineages can occur just prior to the onset of CD4 and CD8 expression among immature thymocytes. Eur. J. Immunol. 22, 2185–2190.

Pido-Lopez, J., Imami, N., Aspinal, R., 2001. Both age and gender affect thymic output: more recent thymic migrants in females than males as they age. Clin. Exp. Immunol. 125, 409–413.

Rahemtulla, A., Fung-Leung, W.P., Schilham, M.W., Kundig, T.M., Sambhara, S.R., Narendran, A., Arabian, A., Wakeham, A., Paige, C.J., Zinkernagel, R.M., 1991. Normal development and function of CD8+ cells but markedly decreased helper activity in mice lacking CD4. Nature 353, 180–185.

Rocha, B., Dautigny, N., Pereira, P., 1989. Peripheral T lymphocytes: expansion potential and homeostatic regulation of pool sizes and CD4/CD8 ratios in vivo. Eur. J. Immunol. 19, 905–911.

Rodewald, H.R., Kretzschmar, K., Swat, W., Takeda, S., 1995. Intrathymically expressed c-kit ligand (stem cell factor) is a major factor driving expansion of very immature thymocytes in vivo. Immunity 3, 313–318.

Rodewald, H.R., 2001. Essential requirement for c-kit and common gamma chain in thymocyte development cannot be overruled by enforced expression of Bcl2. J. Exp. Med. 193, 1431–1436.

Scollay, R.G., Butcher, E.C., Weissman, I.L., 1980. Thymus cell migration. Thymus cell migration. Quantitative aspects of cellular traffic from the thymus to the periphery in mice. Eur. J. Immunol. 10, 210–218.

Sempowski, G.D., Hale, L.P., Sundy, J.S., Massey, J.M., Koup, R.A., Douek, D.C., Patel, D.D., Haynes, B.F., 2000. Leukemia inhibitory factor, oncostatin M, IL-6 and stem cell factor mRNA expression in human thymus increases with age and is associated with thymic atrophy. J. Immunol. 164, 2180–2187.

Shortman, K., Wu, L., 1996. Early T lymphocyte progenitors. Annu. Rev. Immunol. 14, 29.

Shutter, J., Cain, J.A., Ledbetter, S., Rogers, M.D., Hockett, R.D. Jr., 1995. A delta T-cell receptor deleting element transgenic reporter construct is rearranged in alpha beta but not gamma delta T-cell lineages. Mol. Cell. Biol. 15, 7022–7031.

Steinmann, G.G., 1986. Changes in the human thymus during aging. Curr. Top. Pathol. 75, 43–88.

Soares, M.V., Borthwick, N.J., Maini, M.K., Janossy, G., Salmon, M., Akbar, A.N., 1998. IL-7 dependent extra thymic expansion of CD45RA+ T cells enables preservation of a naïve repertoire. J. Immunol. 161, 5909–5914.

Timm, J.A., Thoman, M.L., 1999. Maturation of CD4+ lymphocytes in the aged microenvironment results in a memory-enriched population. J. Immunol. 162, 711–716.

Tsuda, S., Rieke, S., Hashimoto, Y., Nakauchi, H., Takahama, Y., 1996. IL-7 supports D-J but not V-DJ rearrangement of the TCR-beta gene in fetal liver progenitor cells. J. Immunol. 156, 3233–3238.

Tyan, M.L., 1976. Impaired thymic regeneration in lethally irradiated mice given bone marrow from aged donors. Proc. Soc. Exp. Biol. Med. 152, 33–38.

Tyan, M.L., 1977. Age-related decrease in mouse progenitors. J. Immunol. 118, 846–851.

von Freeden-Jeffry, U., Vieira, P., Lucian, L.A., McNeil, T., Burdach, S.E., Murray, R., 1995. Lymphopenia in interleukin (IL)-7 gene-deleted mice identifies IL-7 as a nonredundant cytokine. J. Exp. Med. 181, 1519–1526.

Wu, L., Scollay, R., Egerton, M., Pearse, M., Spangrude, G.J., Shortman, K., 1991. CD4 expressed on earliest T lineage precursor cells in the adult murine thymus. Nature 349, 71–76.

Wu, L., Shortman, K., 1996. Thymic dendritic cell precursors: relationship to the T lymphocyte lineage and phenotype of the dendritic cell progeny. J. Exp. Med. 184, 903–908.

Wykrzykowska, J.J., Rosenzweig, M., Veazey, R.S., Simon, M.A., Halvorsen, K., Desrosiers, R.C., Johnson, R.P., Lackner, A.A., 1998. Early regeneration of thymic progenitors in rhesus macaques infected with simian immunodeficiency virus. J. Exp. Med. 187, 1767–1778.

Yoshikawa, A., 2000. Epidemology and unique aspects of aging and infectious diseases. Clin. Infect. Dis. 36, 931–936.

Zhang, L., Lewin, S.R., Markowitz, M., Lin, H.H., Skulsky, E., Karanicolas, R., He, Y., Jin, X., Tuttleton, S., Vesanen, M., Spiegel, H., Kost, R., van Lunzen, J., Stellbrink, H.J., Wolinsky, S., Borkowsky, W., Palumbo, P., Kostrikis, L.G., Ho, D.D., 1999. Measuring recent thymic emigrants in blood of normal and HIV-1-infected individuals before and after effective therapy. J. Exp. Med. 190, 725–732.

Advances in
Cell Aging and
Gerontology

Alterations in signal transduction in T lymphocytes and neutrophils with ageing

Tamas Fülöp[a], Katsuiku Hirokawa[b], Gilles Dupuis[c],
Anis Larbi[a] and Graham Pawelec[d]

[a]Institut Universitaire de Gériatrie, Centre de Recherche sur le Vieillissement, Service de Gériatrie,
Département de Médecine, Faculté de Médecine, Université de Sherbrooke, Sherbrooke, Québec, Canada
Correspondence address: 1036 rue Belvedere sud, Sherbrooke, Québec, Canada, J1H 4C4.
Tel.: + 1-819-829-7131; fax: + 1-819-829-7141. E-mail: tfulop@courrier.usherb.ca (T.F.)
[b]Department of Pathology and Immunology Ageing and Developmental Sciences, Tokyo-Medical and Dental
University Graduate School, 1-5-45 Yushima, Bunkyo-ku, Tokyo 113-8519, Japan
[c]Département de Biochimie, Faculté de Médecine, Université de Sherbrooke, Sherbrooke, Québec, Canada
[d]Tübingen Ageing and Tumour Immunology Group, Section for Transplantation Immunology and
Immunohaematology, Second Department of Internal Medicine, University of Tubingen Medical School,
Zentrum für Medizinsche Forschung, Waldhörnlestr. 22, D-72072 Tubingen, Germany

Contents

1. Introduction

It is widely accepted that cell-mediated immune functions (cytotoxicity, delayed-type hypersensitivity, etc.) decline with age (Makinodan and Kay, 1980; Hirokawa et al., 1988). These age-associated immunological changes render an individual more susceptible to infection and possibly cancer, as well as to age-associated autoimmune phenomena and may also contribute to atherosclerosis and Alzheimer's disease (Ben-Yehuda and Weksler, 1992; Ershler, 1993; Fülöp et al., 1986; Wick and Grubeck-Loebenstein, 1997).

There is still no clear consensus as to why cell-mediated immunity declines with age. It is generally believed that age-related immune deficiency develops coincident with the gradual involution of the thymus gland and consequently that thymic-related (T cells) functions are the most profoundly affected. However, thymic atrophy alone is unlikely to account for immunosenescence, as thymic weight and volume change little between the ages of 5 and 84 years (Steimann, 1986). Recently, it has been suggested that the alterations observed with ageing are a reflection of an accumulation of relatively inert memory T cells and a consequent reduction of reactive naïve T cells (Effros, 2000; Bruunsgaard et al., 2000). Nevertheless, none of these findings alone can explain satisfactorily the decline of cell-mediated immunity with age.

The proliferation and clonal expansion of T lymphocytes is mostly controlled by interactions between the cytokine interleukin-2 (IL-2) and its cellular receptor (Taniguchi and Minami, 1993; Liparoto et al., 2002) following T cell receptor (TCR) ligation (Hombach et al., 2001). It is now well-accepted that the induction of IL-2 secretion decreases with age in mice (Effros and Walford, 1983; Miller, 1991a), rats (Gilman et al., 1982; Iwai and Fernandes, 1989) and humans (Gillis et al., 1981; Nagel et al., 1989). It has been previously reported that the age-related decline in IL-2 production and activity in spleen lymphocytes isolated from rats was paralleled by a decline in IL-2 mRNA levels (Wu et al., 1986). Subsequent studies in humans (Nagel et al., 1988, 1989) and in mice (Grossmann et al., 1990) also showed that the age-related decline in the induction of IL-2 secretion was correlated to a decline in the induction of IL-2 mRNA levels. These results suggest then, that the age-related T cell proliferative abnormality is due to defects in the transcription of IL-2 mRNA and in subsequent IL-2 secretion. It seems that such defects occur in the transduction of mitogenic signals following TCR stimulation. Indeed, recent studies suggest that alterations in, for example, tyrosine kinase activity, intracellular free calcium, inositol phosphates, protein kinase C (PKC) etc. may all contribute to changes in signal transduction with ageing (Grossmann et al., 1990; Varga et al., 1990a,b; Fülöp, 1994; Fülöp et al., 1993, 1995, 1999; Ghosh and Miller, 1995; Garcia and Miller, 1998; Utsuyama et al., 1997; Miller, 2000; Pawelec et al., 2001). Although the signalling machinery in T cells is extremely complicated and many steps remain to be clarified, the age-related change in T cell signal transduction (Fulop, 1994; Miller, 2000; Zeng and Hirokawa, 1994) may be one of the most important causes of the cell-mediated immune response decline with ageing.

Until recently, the majority consensus remained that phagocytic cell functions (of neutrophils and monocytes/macrophages) would not change very much with ageing (Hirokawa, 1998). Currently, however, much experimental data are emerging which show that many effector functions of phagocytic cells mediated by receptors are altered with ageing (Fülöp et al., 1985a,b; Biasi et al., 1996). Obviously, one explanation put forward for this decrease is the alteration in signal transduction under specific stimulation, such as by GM-CSF, LPS or FMLP (Fülöp, 1994; Seres et al.,

1993; Fülöp et al., 2001). These alterations in the effector function might have serious consequences for infections and inflammatory processes.

Thus, considering the decrease of cell-mediated immune response with ageing and the possibility that alterations in signal transduction may be one of the causes, we will review the signal transduction changes in T lymphocytes from healthy elderly subjects that would translate into a modification of the activation of transcription factors involved in IL-2 gene expression leading to decreased IL-2 production, as well as in neutrophils (PMN).

2. Signal transduction and age-related changes in T lymphocytes

2.1. Immunological changes with ageing in T lymphocytes

A wide variety of age-related abnormalities have been reviewed by Thoman and Weigle (1989). In vitro culture of T cells from older humans and also experimental animals reveals an impaired proliferative response induced by antigens, mitogenic lectins, and antibody directed to the CD3 components of the T cell receptor complex (Douziech et al., 2002). This defect in T cell proliferative capacity leads to decreases in helper activity, delayed type hypersensitivity and cytotoxicity (Makinodan and Kay, 1980; Zatz and Goldstein, 1985; Linton and Thoman, 2001). There are also changes in the ratio of T cell subsets (Effros, 2000; Bruunsgaard et al., 2000; Miller, 1991b).

It has been demonstrated that both IL-2 and the IL-2 receptor are necessary for antigen-driven T cell proliferation and that IL-2 interacting with its receptor drives the cell cycle from G1 to S phase (Serfling et al., 1995). Age-related changes in the production of cytokines have been reported, most notably for IL-2 (Chang et al., 1983; Erschler et al., 1985; Thoman and Weigle, 1988), but also for others, including IL-1, IL-4, IL-6, IL-10 and tumour necrosis factor (TNF) (Bruunsgaard et al., 2000; Rink et al., 1998). Thus, age-related functional changes can be identified in the T cell compartment and the activation of cell cycle entry is required for the execution of these effector functions by T cells.

2.2. T cell receptor complex and co-receptor signal transduction changes with ageing

2.2.1. General
Signal transduction events can be arbitrarily divided into early and late stages. The early events include changes in plasma membrane potential, tyrosine phosphorylation of CD3 components and cytosolic proteins, intracellular mobilization and flux of calcium, and the activation and redistribution of PKC (Peterson and Koretzky, 1999; Kane et al., 2000; Hermiston et al., 2002). The late events include DNA, RNA, and protein synthesis, production of lymphokines, expression of lymphokine receptors and clonal expansion (Peterson and Koretzky, 1999; Kane et al., 2000; Hermiston et al., 2002).

Occupation of the TCR by antigen in the context of the major histocompatibility complex or by antibodies directed against the clonotypic portion of the TCR/CD3 complex as well as the CD3 subunits, or by lectins (collectively, "Signal 1") results in a rapid early cascade of intracellular signal events that leads to activation of cyto-plasmic and nuclear factors that are necessary, but not sufficient, for IL-2 gene transcription transcription (Kane et al., 2000; Peterson and Koretzky, 1999; Hermiston et al., 2002). Indeed, CD28 and other co-stimulatory pathways (collec-tively, "Signal 2") must be activated to fully activate T cells (Holdorf et al., 2000; Bernard et al., 2002). The most remarkable progress in T cell signal transduction in recent years has been the establishment of the central role of protein tyrosine kinases (PTKs). There are three distinct PTK families, namely, src, syk and Tec (Hubbard and Till, 2000). Engagement of the TCR stimulates the kinase activity of tyrosine kinases, such as p59fyn, p56lck and ZAP-70 (Peterson and Koretzky, 1999; Kane et al., 2000; Hermiston et al., 2002), which have been demonstrated to be tightly associated with different components of the CD3 complex and other T cell antigens (CD4, CD8). The activation of the src-like cytoplasmic tyrosine kinases (lck, fyn) leads to phosphorylation of a number of ITAM motif-containing proteins (Isakov, 1997) such as the zeta chain (June et al., 1990) and this promotes the recruitment of the syk family of protein kinases, ZAP-70 and Syk and induces their activation. Lck and Fyn functions are also regulated by the tyrosine phosphatase CD45 (Weiss and Littman, 1994), as well as by some recently-identified phosphatase complexes (Chow et al., 1993). The assembly of active PTK and a number of adaptor proteins leads to activation of a number of signalling pathways (Peterson and Koretzky, 1999; Kane et al., 2000; Tomlinson et al. 2000; Cochran et al., 2001; Hermiston et al., 2002).

The linker of activated T cells (LAT) was originally viewed as a predominant 36-kDa Grb2-associated tyrosine phosphoprotein isolated from stimulated T cell lysates (Zhang et al., 1998; Pivniouk and Geha, 2000; Norian and Koretzky, 2000; Zhang and Samelson, 2000; Leo et al., 2002). LAT contains a transmembrane domain and several intracellular tyrosine residues that are phosphorylated following TCR liga-tion (Zhang et al., 1989). A number of signalling molecules, including Grb2, the p85 subunit of phosphatidylinositol 3-kinase (PI3-K), phospholipase C (PLC), SLP-76 and Cbl, are recruited to phosphorylated LAT. LAT is then a strong candidate for recruiting a Grb2-Sos complex to the membrane and initiating Ras activation after TCR engagement (Genot and Cantrell, 2000).

The main pathways activated during TCR complex stimulation include the Ca2+/calcineurin and the PKC pathway. The key regulators are p21ras/c-Raf-1 for the mitogen-activated protein (MAP) kinase pathway (Campbell et al., 1995; Izquierdo et al., 1995; Fanger, 1999) and PLC1 for the Ca2+ (calcineurin, PKC) pathway (Granja et al., 1991). Tyrosine phosphorylation of Grb2-like proteins activates the ras signal transduction pathway, which has been shown to play a major role in cell differentiation and proliferation (Weissman, 1994; Campbell et al., 1995). Grb2 binds to a mammalian guanine nucleotide exchange factor, Sos, which activates p21ras (Ras) by accelerating the exchange of GDP for GTP (Buday and Downward, 1993). The GTP-bound form of p21ras is active and p21ras is known to lie upstream of the MAPK cascade (Franklin et al., 1994). The GTP-bound p21ras interacts with the serine/threonine kinase, Raf-1, leading to the activation of MAP kinases. There are actually three MAP kinase subfamilies: the extracellular signal-regulated kinase (ERK1 and 2), the p38 and the JN kinase (JNK) (Hardy and Chaudhri, 1997;

Arbabi and Maier, 2002). Between ERK1 and 2, ERK 2 seems to play an important role in IL-2 secretion (Pastor et al., 1995; Jabado et al., 1997; Dasilva et al., 1998; Dumont et al., 1998). Another kinase that may regulate Th1 development is the p38 MAP kinase that is activated by the upstream kinases MKK3 and MKK6. Furthermore, p38 that responds primarily to stressful and inflammatory stimuli also has a role in transducing the mitogenic signal (Hunt et al., 1999; Park et al., 2000). JNK1 and JNK2 seem to play a role in T cell activation by regulating the production of various cytokines (Dasilva et al., 1998). The downstream signalling pathways mediated by the family of MAP kinases are considered essential for normal cell growth, proliferation and function. Moreover, the Ras signalling pathway has been implicated in the activation of the NF-AT transcription factor, suggesting that the Ras pathway is important in IL-2 gene expression (Genot et al., 1996).

Phosphorylation of PLC1 increases its activity and causes cleavage of PIP2 resulting in generation of the secondary messengers inositol 1,4,5-trisphosphate and diacylglycerol (DAG) (Berridge and Irvine, 1989; Fraser et al., 1993). The rise in intracellular calcium activates the calcium/calmodulin-dependent serine/threonine kinases and phosphatases, such as calcineurin (Liu et al., 1991; Wang et al., 1995), which allows the cytoplasmic component of nuclear factors to be activated and to move to the nucleus (Crabtree, 2001). For instance, NF-ATp will combine with newly-formed Fra-1 (a member of the fos family) and JunB proteins (induced by the PKC pathway (June et al., 1990)) to create the NF-AT complex (Rao et al., 1997; Masuda et al., 1998). DAG binds directly to and activates PKC in conjunction with Ca2+ in the case of some isozymes like cPKC (I, II (Nishizuka, 1988; Ron and Kazanietz, 1999)), which in turn phosphorylates a series of proteins. PKC activation contributes to the activation of MAPK, most probably via c-Raf-1 activation. Activated MAP kinases can phosphorylate, at least in vitro, a variety of transcription factors including c-jun, c-myc, c-fos, p62TCF (Elk-1). Finally, it has been shown that TCR-mediated signal transduction causes the dissociation of the NFkB transcription factor from the inhibitory factor IkB, probably via PKC-dependent phosphorylation of IkB (Pimentel-Muinos et al., 1995).

Although the pathways are still not fully established, signalling through CD28 involves an association with phosphatidylinositol 3 kinase (PtdIns 3-kinase) that may involve activation of AkT/PkB and other kinases (Harada et al., 2001). PtdIns 3-kinase could be a potent activator of the Ca2+ independent PKC, and isoforms.

Very recently, great progress has been made in our comprehension of how these events are spatially linked to form an optimal signalling complex. The notion of the immune synapse emerged and is now widely accepted. This is an informational synapse that relays information across a quasistable cell–cell junction during TCR interactions with MHC-peptide complex (Dustin, 2002). A redistribution of the signalling components takes place in two major compartments: central supramolecular activation clusters (cSMACS) enriched in TCR and CD28, and peripheral supramolecular activation clusters (pSMACS) containing LFA-1 and talin (Monks et al., 1998). Another major advance in our understanding is the existence of special membrane domains called rafts, small regions of detergent-resistant complexes in the membrane (Brown and London, 1998). Immunofluorescent studies have confirmed the membrane localisation of LAT. LAT is further targeted as with other signalling molecules, including members of the Src PTK family, heterotrimeric G proteins and Ras, to glycolipid-enriched microdomains as a consequence of postranslational palmitoylation (Simons and Ikonen, 1997; Zhang et al., 1998). After TCR ligation, phosphorylated TCR, ZAP-70, Shc and PLC also localise to these microdomains (Montixi et al., 1998). The inducible assembly of signalling complexes within these microdomains is a prerequisite for efficient TCR signalling (Moran and Miceli, 1998; Xavier et al., 1998; Janes et al., 1999, 2000; Schaeffer and Schwartzberg, 2000). The combination of these signalling events results in the formation/activation of transcription factors.

2.2.2. Age-related changes

It is well accepted that the early signalling events during stimulation via the TCR/CD3 complex are altered with ageing in T lymphocytes (Whisler et al., 1991a,b; Shi and Miller, 1992; Pawelec et al., 2001), as is the expression of early-activation surface markers such as CD69 and CD71 (Lio et al., 1996). These changes might arise from alterations to the cell membrane, decrease in numbers of TCR or changed TCR re-expression after stimulation; they might also be due to changes in the TCR signalling pathways or the alterations of co-receptors.

A question that is still not completely settled is whether the TCR receptor number does change with ageing. This could be due to a decrease of expression or re-expression. In T lymphocytes there are experimental data for many changes in TCR receptor number, but the bulk of the experimental evidence suggests that with normal ageing the TCR number does not change. However, its re-expression could be altered as a consequence of the alteration in CD28 and signalling (Schrum et al., 2000).

2.2.2.1. The role of the plasma membrane
Specificity and fidelity of signal transduction are crucial for cells to respond efficiently to changes in their environment. This is achieved in part by the differential localisation of proteins that participate in signalling pathways. The lipid bilayer of the plasma membrane is organized into cholesterol and glycosphingolipid-rich microdomains, also called rafts. Recently, it was shown that TCR ligation induces a redistribution of tyrosine-phosphorylated proteins into lipid rafts (Montixi et al., 1998; Moran and Miceli, 1998; Xavier et al., 1998; Janes et al., 1999, 2000; Kane et al., 2000; Kabouridis et al., 2000). These include p56lck, phospholipase C, GRB2, ZAP-70 and the phosphorylated TCR itself. Collectively these experimental data indicate that the plasma membrane, via its special organization in lipid rafts, plays an important role in signal transduction via the TCR. The rafts seem to be part of the immunological synapse which forms over the course of 5–10 min in response to MHC–TCR signalling and co-stimulation and persists for several hours. Formation of the immunological synapse seems to be correlated with induction of the complete cellular activation (Kane et al., 2000). However, it has been known for some time that there is an alteration in the

plasma membrane physico-chemical status leading to increased rigidity and decreased fluidity with ageing (Zs-Nagy et al., 1986; Fraeyman et al., 1993; Denisova et al., 1998) and this may affect raft formation.

What is our actual knowledge concerning changes to these immune synapses and lipid rafts with ageing? Very limited information exists concerning this issue in the context of TCR signalling. Recently the group of Miller (Tamir et al., 2000; Eisenbraun et al., 2000; Garcia and Miller, 2001) became interested in this aspect of T cell signalling and published very interesting studies in mice. They demonstrated an alteration in several components of this signalling complex with ageing in memory (Eisenbraun et al., 2000) as well as in naïve T cells (Garcia and Miller, 2001). The most important findings concern the reduced activation of several raft-associated or recruited proteins, such as LAT, PKC, and Vav in T cells of aged mice. There was an age-associated decline in the proportion of CD4+ T cells that redistributes LAT and Vav to the T cell-APC (antigen-presenting cell) synapse upon ligation of TCR. The levels of TCR did not change with age. Not only the redistribution of LAT to the lipid rafts decreased, but this was also accompanied by a proportionate decline in tyrosine phosphorylation. Similarly, but at the level of the nucleus, the transcription factor NF-ATc had a decreased redistribution upon stimulation. Most probably the decrease of the recruitment of the signalling molecules to the synapses is responsible for decreased NF-ATc translocation to the nucleus, playing a role in decreased IL-2 gene expression. The reasons for this altered tyrosine phosphorylation-mediated activation is not yet known. Miller et al. found that the composition of the synapses does not seem to change with ageing. However, as the formation of SMACs requires actin polymerization, while F-actin cap formation in turn requires phosphorylation of ITAMs, studying F-actin assembly and relocalisation they also found an alteration with ageing. They conducted similar experiments in separated memory CD4+ T cells (characterised by high levels of Pgp expression) (Eisenbraun et al., 2000) and found similar alterations; in addition, c-Cbl relocalisation occurred in a high proportion of these cells, suggesting that the association of c-Cbl with components of the TCR complex might interfere with recruitment of LAT and probably PKC into the synapses. Their studies on naïve CD4+ T cells showed two age-related alterations: Firstly, twofold diminution in the proportion of T cell/APC conjugates that could relocalise any of the signalling molecules tested (i.e. LAT, Vav, PLC, etc.); and secondly ageing diminished, by a factor of 2, the frequency of cells showing NF-AT migration among those capable of generating immune synapses containing essential signalling molecules. Together, these data show that alterations in the immune synapses and lipid rafts can be found in naïve and memory cells with ageing. How all these affect the distinct functioning of the cells will be the target of future research.

We were interested to investigate in human T cells whether the alteration of the cholesterol content of the cell membrane, and in consequence of lipid rafts, could modulate the activation of various signalling pathways and functions (Fülöp et al., 2001). We used a well-known cholesterol-extracting molecule, methyl cyclodextrin (MCD), which is known to disrupt rafts in T cells and as a consequence to alter signal transduction upon TCR ligation (Ilangumaran and Hoessli, 1998; Kabouridis et al., 2000). Unlike other cholesterol-binding agents that incorporate into membranes, MCD are strictly surface acting and selectively extract plasma membrane cholesterol (Pitha et al., 1988). In agreement with previous findings, it was confirmed that the cholesterol content of the T cell plasma membrane was significantly increased with ageing (Zs-Nagy et al., 1986; Denisova et al., 1998) and this could explain its increasing rigidity with age. However, the effect of MCD was quite different on the cholesterol content and signalling molecules of T lymphocytes of young versus elderly subjects. The extraction of the cholesterol in T cells of young healthy subjects had a very dramatic signal-disrupting effect, while in T cells of healthy elderly subjects, lck and ERK 1 and 2 phosphorylation was enhanced compared to the non-treated T cells; proliferation of these cells was also improved, although without attaining the level of young subjects. These data further support an alteration at the level of the lipid rafts and the idea that membrane cholesterol plays a critical role in the homeostatic regulation of signalling pathways in T cells, which is altered with ageing. Thus, the manipulation of cholesterol content might have an immunomodulating effect. Recent data from our laboratory (Fülöp et al., 2002, in press), obtained in T cells of elderly subjects, showed a decreased phosphorylation of LAT recruited to lipid rafts in accordance with the data found in mice. There is still much to be learned in relation to this new concept of immune signal transduction during TCR ligation with ageing in different T cell sub-populations.

2.2.2.2. Various signalling pathways   As mentioned above, the first step in TCR-mediated signal transduction is the activation of various tyrosine kinases leading to the tyrosine phosphorylation of several downstream proteins (Whisler et al., 1998; Garcia and Miller, 1998). The level of tyrosine phosphorylation of p59fyn and ZAP-70 kinases is impaired in T cells from old mice activated through the TCR/CD3 complex (Shi and Miller, 1992; Zeng and Hirokawa, 1994). In human T cells, an age-related defect is observed in tyrosine-specific protein phosphorylation after activation via TCR/CD3 complexes, CD4 and IL-2 receptors (Quadri et al., 1996). In addition, a reduction in p59fyn activity was found in some elderly subjects without compensation by p56lck activity (Whisler et al., 1997). We have also recently shown a substantial decrease in p56lck activity in T lymphocytes of healthy elderly subjects (Fülöp et al., 1999). Consequently, as in mice, ZAP-70 activity is also decreased in T cells. Not unexpectedly, there are some discrepancies in the PTK activity measurement with ageing, but together these results suggest that the activation and function of the early signalling PTKs induced by TCR ligation are altered with ageing.

It is now well documented that other early events related to protein tyrosine phosphorylation following TCR activation are altered with ageing; such as the generation of myo-inositol 1,4,5-trisphosphate (IP3), intracellular free calcium mobilization and PKC translocation (Miller, 2000; Pawelec et al., 2001). Few studies have been conducted concerning inositol phosphate formation with ageing. However, the overall available data are conflicting (Miller, 1989; Roth, 1989; Viani et al., 1991; Fülöp et al., 1989, 1992). Miller et al. (Miller, 1989), studying IP3 production in T lymphocytes of old mice, could not demonstrate a decrease, but Proust et al. (1987) did report a decrease. In humans, we (Varga et al., 1990a,b) found decreased production of IP3 with ageing in peripheral blood lymphocytes stimulated by anti-CD3 mAb, while the resting level of IP3 was increased, as was also reported by Proust et al. (1987) in mice. Others' studies as well as ours, concerning the changes in cytoplasmic free calcium ion concentration as an index of the very early events in the T cell activation process, have shown that the anti-CD3 mAb-induced mobilization of cytoplasmic free Ca2+ declines with age (Miller et al., 1987; Miller, 1987; Gupta, 1989; Grossmann et al., 1989; Whisler et al., 1991a,b; Fülöp, 1994). There are some studies concerning PKC activity with ageing in immune cells. Proust et al. (1987) have shown defects in translocation of PKC in T lymphocytes of old mice. Gupta (1989) has shown the same variations in human T lymphocytes with ageing using anti-CD3 mAb. In human B cells the same alteration in PKC translocation was described as in T cells (Kawanishi, 1993). Our recent studies demonstrated an altered PKC isozyme distribution and translocation in human T lymphocytes with ageing under anti-CD3 mAb stimulation (Fulop et al., 1995). These alterations in PKC activation might markedly contribute to the observed impairment of T cell activation with ageing (Ohkusu et al., 1997). Moreover, the results suggest that an inability to elevate PKC activity after TCR stimulation may originate from alterations in the early signal transduction events (Kawanishi and Joseph, 1992; Wilkinson and Nixon, 1998; Pascale et al., 1998). These events depend on the activation of PLC1, which is itself activated by tyrosine phosphorylation following TCR ligation. One study measuring PLC activity in murine T lymphocytes failed to demonstrate age-related changes (Utsuyama et al., 1993). Recently, PLC1 tyrosine phosphorylation was found to be decreased in old mice (Grossmann et al., 1995). No definitive data exist in humans, but considering the upstream and downstream alterations in events depending on the PLC1 tyrosine phosphorylation we can suppose that it is very likely to be altered.

Only a few data exist concerning phosphatase activity in T cells with ageing. There is more and more experimental evidence that the balance between tyrosine kinases and phosphatases is essential for the maintenance of the resting status and for activation (Hermiston et al., 2002). CD45 is a receptor-like PTP expressed on all nucleated hematopoietic cells. One key function of CD45 is to serve as a positive regulator of src tyrosine kinases, by opposing Csk function, and dephosphorylating the negative regulatory C-terminal tyrosine of src tyrosine kinases. CD45-protein tyrosine phosphatase activity in old cells after CD3-stimulation is not changed compared to young cells (Whisler et al., 1997). However, it may be necessary to re-assess the behaviour of CD45 in terms of its involvement in the immunological synapse, from which it is usually excluded upon T cell activation (Leupin et al., 2000). Csk is a ubiquitously expressed cytosolic PTK; it plays a negative regulatory role in cells by inhibiting intracellular processes induced by src tyrosine kinases. The Csk SH2 domain interacts specifically with several tyrosine phosphorylated molecules and among them with the recently identified adaptor-Csk-binding protein/phosphoprotein associated with glycosphingolipid enriched microdomains (Cbp/PAG) (Janes et al., 2000). Cbp/PAG has been shown to be palmitoylated and targeted to rafts. In resting human T cells Cbp/PAG is constitutively phosphorylated and this results in recruitment of Csk to the raft. This interaction increases the catalytic activity of Csk on its substrate, thereby inhibiting src tyrosine kinases activity. However, this inter-action is reversible. The dephosphorylation of Cbp/PAG releases Csk and promotes the activation of src kinases upon TCR stimulation. This represents a sort of thresh-old regulator in T cell activation. Thus far, no data exist on changes of activities of these factors with ageing. However, it can be supposed that the interaction between the Cbp/PAG and Csk is altered, and therefore the release of Csk could be also altered. Further studies are needed. Together, these data suggest that early events in human T cell activation are altered with ageing.

Data are starting to accumulate which show that events more distal from tyrosine kinases are also altered with ageing (Kirk et al., 1999; Li et al., 2000a). Data indicate that the Ras-MAPK/ERK pathways are also changed with ageing (Gorgas et al., 1997; Hutter et al., 2000). Whisler et al. (1996, 1997) have shown that 50% of old subjects had a reduction in MAPK activation. Considering these alterations in the signal transduction pathways following specific antigen presentation to the TCR, several approaches have been tested to modulate signal transduction with a view to increasing IL-2 production and ultimately increasing proliferation (clonal expan-sion). Thus, for example, stimulation with phorbol ester in combination with calcium ionophore resulted in greater MAPK activation in T cells of old subjects, but still not to the same extent as in T cells of young subjects, suggesting signalling defects between the TCR and the inducers of MAPK (Whisler et al., 1996). These alterations were also shown in mice (Gorgas et al., 1997). ERK 2 activation was shown to be correlated with the ability of T cells to produce IL-2 and to proliferate. Thus, diminished ERK2 activation may represent the rate-limiting step for IL-2 produc-tion by T cells of old individuals (Whisler et al., 1996). Furthermore, we have recently shown that there is an alteration with ageing in MAPK/ERK as well as in p38 activation in T cells of the elderly compared to young subjects following TCR stimulation (Douziech et al., 2002), as already demonstrated (Whisler et al., 1996; Beiqing et al., 1997).

2.2.2.3. Proto-oncogenes and transcription factors   Two laboratories have studied the effect of ageing on the expression of fos and jun in lymphocytes. Unfortunately, the results of these studies are contradictory. Song et al. (1992) showed that the induction of c-jun mRNA levels by PHA decreased with age in human peripheral blood lymphocytes, although neither c-fos, nor jun B mRNA levels were altered by age. In contrast, Sikora et al. (1992) reported that the induc-tion of c-fos mRNA, but not c-jun mRNA levels, decreased with age in splenocytes isolated from mice. These investigators also showed that the induction of AP-1 binding activity in splenocytes decreased with age. Data on the effect of age on the NFAT complex (NFAT) show an age-related decline in NFAT binding to nuclear extracts of T lymphocytes from rats (Pahlavani et al., 1995). Thus, it is possible that the age-related changes in NFAT binding activity may arise from changes in the induction of either c-fos or c-jun expression (Macian et al., 2001). However, the current studies are not conclusive as to whether age-related changes in fos or jun expression are responsible for the decline in NFAT binding. The age-related decline observed in NFAT binding activity could also arise through changes in its cytoplasmic component i.e. NFAT-p. A decrease in calcium signal generation with age (Miller et al., 1987; Miller, 1986; Gupta, 1989; Grossmann et al., 1989; Whisler et al., 1991a,b; Fülöp, 1994), through calcineurin, could then contribute to the decreased NFAT binding activity as observed in nuclear extracts isolated from splenocytes of old rats. The other important transcription factor for IL-2 secretion is NFkB (Shreck et al., 1992). NFkB protein is constitutively expressed and remains in the cytoplasm, bound to an inhibitory protein IkB, prior to activation. When stimulated, T cells generate reactive oxygen species, changing the redox status and leading to phosphorylation and ubiquitination of IkB. This results in dissociation of IkB from NFkB followed by the degradation of IkB by the proteasome. This in turn results in the translocation of active NFkB to the nucleus. Studies on NFkB showed a decrease in its activation in mice and humans mostly due to a decreased inactivation of IkB by the proteasomes (Ponnappan et al., 1999). The decrease in proteasome activity with ageing was also reported in tissues of old rats and cultured human fibroblasts undergoing replicative senescence. The alteration of proteasome activity with ageing was attributed to oxidative stress. However, Daynes et al. reported results contradictory to this (Spencer et al., 1997). They found a constitutive activation of NFkB in tissues of old animals, which they attributed also to the free radical production increase and which could be normalized by dietary antioxidants or by a stimulator of PPAR. They suggested that this constitutive activation of NFkB might be responsible for the altered cytokine production occurring with ageing. It is in fact difficult to explain these disparate results, but as they seem to be related to oxidative stress, the cellular stress level could have differential effects on cells, tissues and organisms. Based on these results, we conclude that ageing does influence the activation of transcription factors following T cell stimulation, which may result in decreased IL-2 production. However, more studies are needed to elucidate the complete mechanism of this decline under different experimental circumstances.

2.2.2.4. Co-receptors   For an effective and sustained T cell activation, the co-receptors of T cells must be ligated by structures on the antigen-presenting cells. Alteration either in the number or signal transduction capacity of co-receptors leads to a decreased T cell response. One of the most important and well-characterised co-receptors on T cells is the CD28 molecule (Holdorf et al., 2000). CD28 interacts with CD80 and CD86 on APC. The activation of CD28 is very important for the re-expression of TCR, for the recruitment and stabilization of T cell lipid rafts and for the stimulation of the PI3K signalling pathway (Viola et al., 1999). Furthermore, CD28 co-stimulation synergizes with TCR activation and induces IL-2, IL-4, IL-5, TNF and GM-CSF cytokine production (Ghiotto-Ragueneau et al., 1996). Co-stimulatory requirements for T lymphocyte activation are influenced by previous T cell antigen exposure. More co-stimulation is required for the activation of naïve cells than for memory cells (Suresh et al., 2001). Moreover, the activation of CD28 protects T cells from activation-induced cell death (AICD). It is now well-established that CD28 expression is decreased with ageing in both CD8+ and CD4+ lymphocytes (Effros, 2000). This diminution of CD28 signalling can contribute to the decrease of telomerase up-regulation as well as to decreased IL-2 production, leading to decreased proliferation. Thus, changes either in co-receptor number or signal transduction could have far-reaching effects on T cell functions with ageing including decrease in proliferative response (Vallejo et al., 1998). It was also recently shown that AICD is increased in T cells from old mice, as a direct consequence of their decreased levels of CD28-mediated co-stimulation, which otherwise may protect stimulated cells from apoptosis (Engwerda et al., 1996). One of the mechanisms by which CD28 signalling protects against apoptosis is by preventing CD95L up-regulation and by increasing the expression of the anti-apoptotic proteins c-FLIPshort and Bcl-xL (Kirchhoff et al., 2000a,b).

The expression and activity of another co-receptor, CD152 (CTLA-4), which inhibits the activation of T lymphocytes, might be also up regulated with ageing (Chambers and Allison, 1997; Sansom, 2000; Hu et al., 2001). Few data exist concerning the regulation of other co-receptors with ageing such as CD40L (CD154) and ICOS, which is a third member, together with CD152, of the CD28 co-receptor family (Armitage et al., 1992; Weyand et al., 1998; Fernandez Gutierrez et al., 1999; Dong et al., 2001). ICOS may modulate the immune response by increasing the secretion of IL-10 (Buonfiglio et al., 2000). Preliminary data indicate that with ageing there is an increase in ICOS expression on T cell clones (Pawelec et al., 2000). Recently, it has been shown that LFA-4, an integrin receptor, may also act as a co-stimulatory molecule in synergy with CD28 (Geginat et al., 2000). The CD154/CD40 ligand pair also seems to be decreased with ageing (Lio et al., 1998). This area also requires deeper exploration in the near future for a better understanding of immune response changes and the appearance of immune-related diseases with ageing.

2.2.3. Lymphokine receptors

A decrease in IL-2 production (Gillis et al., 1981; Miller and Stutman, 1981; Thoman and Weigle, 1988) has been a consistent finding in alterations of the immune response with ageing. Data concerning changes with ageing to the IL-2 receptor (CD25) on T lymphocytes are very controversial. However, we have shown that the density of IL-2 receptors was not modified with ageing (Fülöp et al., 1991). The signal transduction pathways elicited by IL-2 are well-documented even if the respective role of each signalling pathway is not yet fully determined (Gesbert et al., 1998). One of the main signalling pathways elicited by IL-2 in T lymphocytes is the Jak-Stat pathway (Gilmour et al., 1995; Frank et al., 1995) and the activation of p56lck (Minami et al., 1993). We confirmed the time-dependent increase in tyrosine phosphorylation of JAK3, STAT3 and STAT5 in T cells of young subjects. JAK3 tyrosine phosphorylation by the IL-2R subunit is essential for the activation of the STAT pathway leading to T cell proliferation. In the case of T cells from elderly subjects the level of tyrosine phosphorylation of JAK3 could not be modulated by IL-2 and was already increased in unstimulated cells (Fülöp et al., 2001). This higher level of JAK3 phosphorylation in cells not stimulated in vitro may be explained by an increased persistent activation with ageing in vivo, in agreement with the suggestion that a chronic inflammatory status may be present with ageing (Moulias et al., 1999; Franceschi et al., 2000). However, this increased

JAK3 phosphorylation does not translate into STAT5 activation neither in the resting state, nor during IL-2 stimulation. STAT5 activation is essential for IL-2-induced proliferation of T cells (Arlinger et al., 1999). The lack of activation of STAT5 to transduce signals could be related to age-dependent changes in the targeting of IL-2/JAK3 complexes into specific plasma membrane domains, which may limit accessibility of the activation kinase JAK3 to downstream molecules. A similar situation has been shown to apply in the case of ZAP-70 in CD4+ T cells upon anti-CD3 stimulation in old mice (Garcia and Miller, 1998). Recently, it has been demonstrated that IL-2 signalling also requires the presence of lipid rafts (Mamor and Julius, 2001), namely by the presence in rafts of subunits of the IL-2R. There is controversy over whether exogenously added IL-2 can revert age-related changes in signal transduction and functions, but recent studies demonstrate that this does not seem to be the case (Bruunsgaard et al., 2000). The inefficacy of IL-2 could be related to its altered signalling pathways with ageing (Fülöp et al., 2001), but no data exist concerning lipid rafts in this context. This alteration in the JAK-STAT signalling pathway upon IL-2 stimulation of T cells of elderly subjects further suggests that there are severe alterations in T cell signal transduction through the TCR and IL-2R receptors with ageing.

2.3. T lymphocyte subpopulations

It is well known that the priming status of T lymphocytes determines their response to stimulation as well as their ultimate function. In this context the distinction between naïve and memory cells becomes essential. Furthermore, important recent findings indicate that there is a difference between naïve and memory and effector cells in term of lipid raft distribution and protein content (Kane et al., 2000; Lanzavecchia and Sallusto, 2000). Naïve T cells have fewer rafts in their plasma membrane and require CD28 co-stimulation to amplify TCR signalling by recruiting rafts to the TCR-ligand contact site. By contrast, effector and/or memory T cells have more rafts in their plasma membrane; thus amplification of signalling is able to occur in the absence of CD28 co-stimulation.

It is generally quite well accepted that the number of T cells does not change markedly with ageing. In contrast, there is a consensus that ageing is accompanied by changes in the proportions of T cell sub-populations. There is a higher number of T cells with the CD45RO+ "memory" phenotype and much less with the CD45RA+ "naïve" phenotype in PBMC, although this is of course an oversimplification, albeit still a useful one (Effros, 2000). There are other T cell surface markers that change during ageing in humans, such as CD62L, the adhesion molecule CD49d (DeMartinis et al., 2000a) and ICAM-3 (CD50) (DeMartinis et al., 2000b). As with CD28 expression, these changes affect both the percentage of cells expressing the molecule or the density of the molecule at the cell surface. At this time there is little information on the real functional consequences of these changes. Whether these changes could explain the altered cytokine secretion profile with ageing (Karanfilov et al., 1999) is still a matter of debate. There is also discussion as to which surface marker should be used for defining naïve and memory cells (Wada et al., 1998;

Fagnoni et al., 2000; Ginaldi et al., 2001); nevertheless it appears that CD45 isoform expression is a good overall marker. Finally, no matter which marker we use the numbers of naïve cells dramatically decreases with ageing, mainly among CD8+ cells (Fagnoni et al., 2000). These alterations lead to decreased proliferative responses, decreased response to new antigens, but possibly a better response to antigens already encountered.

CD28 may be considered as a biomarker of ageing in T cells. The proportion of CD28+ T cells decreases in vivo with ageing and in in vitro culture models (Fagnoni et al., 1996, 2000; Pawelec et al., 1997). Effros et al., (2000) have shown a decreased percentage of T lymphocytes that are CD28+ in the CD8 T cell subpopulation. In addition, the average telomere lengths in the CD28-negative T cells are decreased, indicating that these cells have undergone numerous cell divisions. This type of proliferative senescence might be responsible for the accumulation of oligoclonal CD28-negative populations in elderly subjects (Posnett et al., 1994). Very few data exist relating these changes in T cell subpopulations to the signal transduction changes observed in PBLs. Some data seem to suggest that the signal transduction changes demonstrated in the whole population of T cells with ageing are also observed in isolated naïve T cells (Garcia and Miller, 2001). However, several findings indicate that the alterations rather reflect the behaviour of the accumulated memory T cells with ageing. More in-depth analyses are needed to establish the exact contribution of the observed signal transduction changes, in the various T cell subpopulations, with ageing.

Other recently-described T regulatory cells, such as CD4+/CD25+ and NK T cell subsets, could also play an important role in ageing and autoimmune diseases such as diabetes mellitus type 1 (Kukreja et al., 2002). It was shown that these peripheral T cell subsets actively contribute to the maintenance of self-tolerance. What the role of these T cell subsets are in ageing is actually unknown. Future research will certainly bring further information on these cells, in relation to ageing, and help to integrate them into the complicated network of T cell subpopulations.

2.4. The role of oxidative stress in signal transduction

In addition to the increase in cholesterol content, alterations to the membranes of cells associated with ageing could be due to oxidation by free radicals (Harman, 1956). Oxidative stress has been shown to damage cell membranes (Halliwell and Gutteridge, 1989), altering in vitro binding activity of AP-1 (Sen and Packer, 1996) and suppressing in vitro Con-A induced T cell proliferation and IL-2 production (Pahlavani and Harris, 1997). On the other hand, it is also well known that for effective signal transduction the cells need free radicals that modulate AP-1 and NFkB activities (Fagnoni et al., 1996; Pahl and Bauerle, 1994; Fagnoni et al., 1996). The MAPK pathway components are also very redox sensitive (Li et al., 2000b). Aside from the negative action of oxidative stress, in recent years several studies have emphasised the importance of intracellular antioxidant levels for preserving immune function (Meydani et al., 1997). Recent progress in understanding the mechanisms of action of antioxidants on cellular metabolism has shown that antioxidants may modulate signal transduction factors (Palmer and Paulson, 2000), transcription of genes involved in cell-mediated immunity and cytokine production (Pahlavani and Harris, 1997). Vitamin E is widely recognised as a major lipid-soluble chain-breaking antioxidant in the biological membrane. Vitamin C (ascorbate) is one of the main aqueous-phase antioxidants within cells. Ascorbate is the first line of defence in the control of redox state, sparing other endogenous antioxidants from consumption (Stahl and Sies, 1997) such as GSH. These vitamins are supplied by nutrition and as such have been shown to modulate the immune response of elderly (Serafini, 2000).

We recently found that vitamin E inhibited both PHA and anti-CD3 mAb induced proliferation of T cells in vitro independently of the age-groups (Douziech et al., 2002). These were unexpected results in view of the majority of data in the literature stating the T cell proliferation-enhancing effect of vitamin supplementation, either in old animals or humans (Wu et al., 2000). However, our results are consistent with studies where no modulation of T cell proliferation was observed after vitamin supplementation of elderly subjects (De Waart et al., 1997). In earlier studies, there was a difference between the sensitivity of Con A- and PHA-stimulated cells to vitamin E (Serafini, 2000), where Con A- but not PHA-induced responses were enhanced by vitamin E, implying mainly helper T cell targeting, as was the case in our study. We also found that the TCR-stimulated mitogenic response was significantly decreased by vitamin E in both age groups. There are several hypotheses to explain this inhibition of anti-CD3 mAb stimulation. First, the concentration of Vitamin E, which was 2–4 times higher than in the sera, could have been too high, if there is a threshold of vitamin E immuno-enhancing activity, suggested by Meydani et al. (1997). However, in an intervention study where the Vitamin E level increase was threefold, an increased proliferative response was nevertheless found. It could also be supposed that vitamin E in vitro in this concentration decreases so drastically the free radical production that no efficient signal transduction could occur. Similar inhibition was observed by antioxidants when T cells were stimulated by PMA and calcium ionophore (Chaudhri et al., 1988). This is corroborated by the fact that even the decreased ERK and p38 activation in T cells of elderly is completely suppressed by vitamin E. As already mentioned the MAPK signalling pathway is particularly sensitive to the redox status of the cells. In contrast, vitamin C did not affect either PHA- or anti-CD3 mAb-mediated T cell stimulation of any age group. In concordance with this finding no modulation of the ERK or p38 activation could be elicited by vitamin C in any age group compared to anti-CD3 stimulation. This signifies that the two antioxidant vitamins might have different mechanisms of action.

3. Signal transduction and age-related changes in polymorphonuclear leukocytes

Polymorphonuclear leukocytes (PMN) are short-lived cells that play important roles in both host defence and acute inflammation (Yamamoto and Sasada, 1999; Burg and Pillinger, 2001). They represent the first line of defence against an assault. They are committed to die in circulation within 18 h unless activated, which results in the initiation of an inflammatory response leading to chemotaxis (Akgul et al., 2001).

Certain pro-inflammatory cytokines prolong the life span and the functional survival of PMN (Whyte et al., 1999). Among these molecules are GM-CSF, LPS and IL-6. For a long time, a paradigm existed stating that PMN functions would not change with ageing (Hirokawa, 1998). However, the results are controversial and very few groups work on PMN signal transduction and function with ageing (Fülöp et al., 1985a,b; Vlahos and Matter, 1992; Tortorella et al., 1999; Wenisch et al., 2000; Lord et al., 2001). It is well known that ageing results in a predisposition to inflammation as well as to infections (Franceschi et al., 2000). Over the past few years, it has been demonstrated that PMN-specific receptor-driven effector functions are altered with ageing (Fülöp et al., 1987; Varga et al., 1997). An altered signal transduction mechanism was demonstrated with ageing in PMN (Fülöp et al., 1985a,b, 2001; Biasi et al., 1996; Seres et al., 1993). Our group as well as others have also demonstrated that GM-CSF was unable to rescue the PMN of elderly healthy subjects from apoptosis (Fülöp et al., 1997).

3.1. FMLP receptor

The bacterial tetrapeptide product formyl-methionyl-leucyl-phenylalanine (FMLP) receptor is coupled to a Pertussis toxin-sensitive G protein (Varga et al., 1988; Mcleish et al., 1989a,b; Varga et al., 1989). The transduction pathway of the receptor involves the activation of PLC, which induces the production of IP3, leading to an increase of intracellular free calcium on the one hand, and of DAG on the other hand, which stimulates the translocation of PKC to the membrane leading to the phosphorylation of members of the MAPK family (Chang and Wang, 1999).

Stimulation of the cells by FMLP induces, via the production of IP3 and the opening of calcium channels in the membrane, an increase in intracellular free calcium. This increase is normally very rapid and returns to the pre-stimulation level relatively quickly. There is a slight difference between young and elderly subjects in the intracellular free calcium kinetics stimulated by FMLP in PMN (Biasi et al., 1996). The amount of the intracellular free calcium inside the cells is higher under FMLP stimulation in PMN of young than elderly subjects (Fülöp, 1994). The return of the intracellular free calcium must be tightly regulated, because if it remains high this could lead to cell death via the activation of certain intracellular proteases such as calpains or endonucleases. Thus, we associate ageing with a decrease in the early phase of signal transduction in PMN.

The induction of PKC via the ras pathway in turn induces the activation of MAPK family members when the PMN are stimulated by FMLP (Zu et al., 1998; Heuertz et al., 1999; Yagisawa et al., 1999). MAPKs are a family of serine/threonine kinases that are activated by a cascade of protein kinase reactions (Kyriakis and Avrach, 1996), which are not completely elucidated in human neutrophils even after FMLP stimulation (Zu et al., 1998). In rat neutrophils the activation of Lyn is associated with binding to the Shc adaptor protein and allows the G protein-coupled receptors to modulate the activity of the Ras/ERK cascade (Chang and Wang, 1999). In rat neutrophils the FMLP-stimulated p38 pathway remains unclear, even if the role of PLC/Ca2+ and PKC was evoked (Chang and Wang, 2000).

Nevertheless, investigation of human neutrophils has suggested that p38 MAPK is involved in an intracellular cascade that regulates stress-activated signal transduction (Zu et al., 1998). p38 MAPK can phosphorylate transcription factors, thereby regulating gene expression, and can also phosphorylate other proteins to stimulate NADPH oxidase activity, adhesion and chemotaxis (Kyriakis and Avrach, 1996; Zu et al., 1998; Heuertz et al., 1999; Yagisawa et al., 1999; Chang and Wang, 2000). FMLP has been shown to induce the activities of ERK1 and 2, playing a role in neutrophil adherence and respiratory burst activity, as well as inducing p38 and contributing to chemotaxis and superoxide anion production (Zu et al., 1998). We therefore associate ageing with a decrease of ERK and p38 tyrosine phosphorylation (Fülöp et al., unpublished data) indicating decreased activity of these MAPKs. These alterations could explain the decrease found in effector functions of PMN with ageing such as superoxide anion production.

3.2. GM-CSF receptor

GM-CSF is a powerful modulator of granulopoiesis and the priming of mature PMN to a second stimulation such as FMLP. GM-CSF is able to rescue PMN from apoptosis by interacting with its specific receptor on the PMN plasma membrane. GM-CSF has been shown to activate three distinct pathways in various cells: (1) the JAK-STAT pathway; (2) the Ras-Raf-1-MEK-MAP kinase pathway; and (3) PI3-kinase intracellular signalling events (Sato et al., 1993; Watanabe et al., 1997). Recently, the MAPK and PI3K pathways were suggested to be involved with the GM-CSF anti-apoptotic effect in PMN. These signalling pathways induce the executioner phase of apoptosis, mediated by a family of cysteine proteases, the caspases, as well as members of the bcl-2 family, which are key players in the regulation of apoptosis (Kroemer et al., 1998). Our recent studies suggest that ageing is accompanied by a decrease in GM-CSF-signal transduction. The Jak/STAT and MAPK pathways were found to be altered with ageing in PMN upon GM-CSF receptor stimulation. There is no decrease in the GM-CSFR number with ageing, however.

3.3. Effect of signal transduction changes on effector functions

3.3.1. Apoptosis

Apoptosis plays an important role in the regulation of tissue development, differentiation and homeostasis as well as in several pathologies such as cancer, neurodegenerative and immune diseases (Wang, 1997). Mature neutrophils undergo spontaneous apoptosis very rapidly when maintained in vitro; this apoptotic sensitivity regulates both their production and survival. Although neutrophils appear to be committed to apoptotic death in vitro and in vivo, the life span and functional activity of mature PMN can be extended in vitro by incubation with pro-inflammatory cytokines, including GM-CSF.

Bcl-2, Bcl-X (Bcl-XL and Bcl-XS), Bad and Bax are members of the Bcl-2 family of proteins and play important roles in regulating cell survival and apoptosis. Bcl-2 and Bax are homologous proteins that have opposing effects on apoptosis

(Chang and Wang, 1999). Bcl-2 serves to prolong cell survival, but as PMN do not contain Bcl-2, however, other homologous proteins were identified such as A1 and Mcl-1 (novel hematopoietic specific homologues of Bcl-2) (Chang and Wang, 2000). The ratio of Bax to Mcl-1 is important for death or survival of PMN (Chang and Wang, 2000). GM-CSF was suggested to increase Mcl-1 expression. These proteins act as ion channels and adapter proteins regulating the release of mitochondrial proteins into the cytosol in order to activate the caspases that are the terminal effectors of apoptosis.

We have demonstrated that the Jak2-STAT5 signal transduction pathway is altered with ageing and this alteration contributes to the decreased rescue of PMN from apoptosis by GM-CSF with ageing. In the later phase of the apoptotic pathway the bcl-2 family members, Mcl-1 and Bax, play an important role. We found that the expression of Bax was identical in young and elderly PMN undergoing spontaneous apoptosis. After GM-CSF stimulation for 18 h Bax expression was significantly decreased by the GM-CSF and was almost identical to that found in freshly isolated PMN of young subjects. In contrast, in PMN of elderly subjects, GM-CSF was unable to modulate the expression of Bax. The MAPK ERK inhibitor PD98059 and the MAPK p38 inhibitor SB203580 had no effect on Bax expression in any age group. Mcl-1 was significantly increased in PMN of young subjects following 18 h of GM-CSF stimulation compared to spontaneous apoptosis. In contrast, in the PMN of elderly subjects, there was a similar decrease after 18 h of culture either during stimulation or spontaneous apoptosis. The ratio of Bax/Mcl-1 was 0.2 in the PMN of young after 18 h of treatment with GM-CSF, while it was 0.84 in PMN of the elderly.

Our results suggest that the modulation by GM-CSF of the expression of the bcl-2 family members, Mcl-1 and Bax, as well as their relative ratio, is very important for the rescue of PMN from apoptosis. In PMN of young subjects, Mcl-1 predominates after 18 h of GM-CSF treatment relative to Bax content. In contrast, in the PMN of the elderly, there is almost no difference between Mcl-1 and Bax content under GM-CSF stimulation. The relationship between these alterations in the expression of the bcl-2 family members with ageing and the alteration of the JAK2/STAT signalling pathway is not yet fully understood. Nevertheless, as JAK2 is related to the expression of Bcl-2 there might also be a link between JAK2 and Mcl-1. All these alterations leading to the decreased rescue of PMN from apoptosis might contribute to the increased incidence of infection with ageing.

3.3.2. Free radical production

One of the most accepted theories of ageing is the free radical theory of ageing, suggested for the first time by Harman (1956). Since then a huge amount of experimental evidence has been put together for supporting this theory. FMLP receptor-stimulation by FMLP induced the production of superoxide anion as well as hydrogen peroxide by phagocytic cells (monocytes and polymorphonuclear neutrophils) (Fülöp et al., 1986, 1989, 1997). This free radical production participates in the destruction of invading organisms (Varga et al., 1988, 1990a,b) and as such in host defence. Superoxide anion may also combine with NO, which the production is also induced by FMLP, to form peroxinitrite (NOO−), a very reactive free radical.

Moreover, these free radicals may also contribute to the peroxidation of lipoproteins in blood vessel walls. It was demonstrated some time ago that the production of free radicals by PMN of elderly subjects was decreased under FMLP stimulation, while the number of FMLP receptors did not change. Moreover, with ageing, superoxide anion production is also significantly decreased after 24 and 48 h of PMN culture in the presence of GM-CSF.

4. Conclusions and future perspectives

The decline of the immune response with ageing is revealed by a decrease in delayed-type hypersensitivity responses, diminished responses to vaccination, increased susceptibility to serious illness after viral infection and a weakening of protective responses against latent infections. The cause of these defects in age-related immune responses is unknown. A number of possibilities have been suggested, namely an alteration of the intracellular biochemical cascade pathways that transduce signals from surface receptors such as the TCR, cell surface co-stimulatory molecules or cytokine receptors. It is well-documented that there is an alteration in several signalling pathways with ageing (Pawelec et al., 2001; Chakravarti and Abraham, 1999). These alterations include a decrease in the activity of PTK, a decline in inducible ERK function and a reduced activation of PKC leading to the decreased activation of transcription factors such as NF-AT and NFkB. Recent findings conceptualising the interaction between APC and T cells under the form of immunological synapses including specialized plasma membrane domains called lipid rafts help to better understand the mechanism of signal transduction. The sparse data related to ageing in this context indicate an alteration in the composition and the signalling molecule content of lipid rafts and enforce the importance of cholesterol in the signal transduction of TCR through lipid rafts.

Clearly, further, studies are needed to better understand the role of lipid rafts in signal transduction of T cells with ageing under TCR stimulation and whether alternative way(s) exist(s) to abrogate these age-related membrane alterations. Other important aspects for future research in signal transduction should address the question of phosphatase activity and co-receptor signal transduction. Finally, the separation of T cells into various well-defined sub-populations is needed to understand whether these changes are age-related only, or related to the increase of memory cells or other subsets during life, due to encounters of the immune system with pathogens.

References

Akgul, C., Moulding, D.A., Edwards, S.W., 2001. Molecular control of neutrophil apoptosis. FEBS Lett. 487, 318–322.
Arbabi, S., Maier, R.V., 2002. Mitogen-activated protein kinases. Crit. Care Med. 30, S74–S79.
Arlinger, M., Frucht, D.M., OShea, J.J., 1999. Interleukin/interferon signaling A 1999 perspective. The Immunologist 7, 139–146.
Armitage, R.J., Sato, T.A., Macduff, B.M., Clifford, K.N., Alpert, A.R., Smith, C.A., Fanslow, W.C., 1992. Identification of a source of biologically active CD40 ligand. Eur. J. Immunol. 22, 2071–2076.

Beiqing, L., Carle, K.W., Whisler, R.L., 1997. Reductions in the activation of ERK and JNK are associated with decreased IL-2 production in T cells from elderly humans stimulated by the TCR/CD3 complex and costimulatory signals. Cell. Immunol. 182, 79–88.

Ben-Yehuda, A., Weksler, M.E., 1992. Host resistance and the immune system. Clin. Ger. Med. 8, 701–711.

Bernard, A., Lamy, A.L., Alberti, I., 2002. The two-signal model of T-cell activation after 30 years. Transplantation 73, S31–S35.

Berridge, M.J., Irvine, R.F., 1989. Inositol phosphates and cell signalling. Nature 341, 197–205.

Biasi, D., Carletto, A., Dellagnola, C., Caramaschi, P., Montesanti, F., Zavateri, G., Zeminian, S., Bellavite, P., Bambara, L.M., 1996. Neutrophil migration, oxidative metabolism, and adhesion in elderly and young subjects. Inflammation 20, 673–681.

Brown, D.A., London, E., 1998. Functions of lipid rafts in biological membranes. Annu. Rev. Cell Dev. Biol. 14, 111–136.

Bruunsgaard, H., Pedersen, A.N., Schroll, M., Skinhoj, P., Pedersen, B.K., 2000. Proliferative responses of blood mononuclear cells (BMNC) in a cohort of elderly humans: role of lymphocyte phenotype and cytokine production. Clin. Exp. Immunol. 119, 433–440.

Buday, L., Downward, J., 1993. Epidermal growth factor regulates p21ras through the formation of a complex of receptor, Grb2 adapter protein, and Sos nucleotide exchange factor. Cell 73, 611–620.

Buonfiglio, D., Bragardo, M., Redoglia, V., Vaschetto, R., Bottarel, F., Bonissoni, S., Bensi, T., Mezzatesta, C., Janeway, C.A., Jr., Dianzani, U., 2000. The T cell activation molecule H4 and the CD28-like molecule ICOS are identical. Eur. J. Immunol. 30, 3463–3467.

Burg, N.D., Pillinger, M.H., 2001. The neutrophil: function and regulation in innate and humoral immunity. Clin. Immunol. 99, 7–17.

Campbell, J.S., Seger, R., Graves, J.D., Graves, L.M., Jensen, A.M., Krebs, E.G., 1995. The MAP kinase cascade. Rec. Progr. Hormone Res. 50, 131–159.

Chakravarti, B., Abraham, G.N., 1999. Aging and T-cell-mediated immunity. Mech. Ageing Dev. 108, 183–206.

Chambers, C.A., Allison, J.P., 1997. Co-stimulation in T cell responses. Curr. Opin. Immunol. 9, 396–404.

Chang, L.C., Wang, J.P., 1999. Examination of the signal transduction pathways leading to activation of extracellular signal-regulated kinase by formyl-methionyl-leucyl-phenylalanine in rat neutrophils. FEBS Lett. 454, 165–168.

Chang, L.C., Wang, J.P., 2000. Activation of p38 mitogen-activated protein kinase by formyl-methionyl-leucyl-phenylalanine in rat neutrophils. Eur. J. Pharmacol. 390, 61–66.

Chang, M.P., Makinodan, T., Peterson, W.J., Strehler, B.L., 1983. Role of T cells and adherent cells in age-related decline in murine interleukin 2 production. J. Immunol. 129, 2426–2430.

Chaudhri, G., Hunt, N.H., Clark, J.A., Ceredig, R., 1988. Antioxidants inhibit proliferation and cell surface expression of receptors for interleukin-2 and transferrin in T lymphocytes stimulated with phorbol myristate acetate and ionomycin. Cell. Immunol 115, 204–215.

Chow, L.M.L., Fournel, M., Davidson, D., Veillette, A., 1993. Negative regulation of T-cell receptor signalling by tyrosine protein. Nature 365, 156–159.

Cochran, J.R., Aivazian, D., Cameron, T.O., Stern, L.J., 2001. Receptor clustering and transmembrane signaling in T cells. Trends Biochem. Sci. 26, 304–310.

Crabtree, G.R., 2001. Calcium, calcineurin, and the control of transcription. J. Biol. Chem. 276, 2313–2316.

Dasilva, D.R., Jones, E.A., Favata, M.F., Jaffee, B.D., Magolda, R.L., Trzaskos, J.M., Scherle, P.A., 1998. Inhibition of mitogen-activated protein kinase kinase blocks T cell proliferation but does not induce or prevent anergy. J. Immunol. 160, 4175–4181.

De Waart, F.G., Portengen, L., Doekes, G., Verwaal, C.J., Kok, F.J., 1997. Effect of 3 months vitamin E supplementation on indices of the cellular and humoral immune response in elderly subjects. Br. J. Nutr. 78, 761–774.

DeMartinis, M., Modesti, M., Loreto, M.F., Quaglino, D., Ginaldi, L., 2000a. Adhesion molecules on peripheral blood lymphocyte subpopulations in the elderly. Life Sci. 68, 139–151.

DeMartinis, M., Modesti, M., Profeta, V.F., Tullio, M., Loreto, M.F., Ginaldi, L., Quaglino, D., 2000b. CD50 and CD62L adhesion receptor expression on naive (CD45RA+) and memory (CD45RO+) T lymphocytes in the elderly. Pathobiology 68, 245–250.

Denisova, N.A., Erat, S.A., Kelly, J.F., Roth, G.S., 1998. Differential effect of aging on cholesterol modulation of carbachol-stimulated low-Km GTPase in striatal synaptosomes. Exp. Gerontol. 33, 249–265.

Dong, C., Juedes, A.E., Temann, U.A., Shresta, S., Allison, J.P., Ruddle, N.H., Flavell, R.A., 2001. ICOS co-stimulatory receptor is essential for T-cell activation and function. Nature 409, 97–101.

Douziech, N., Seres, I., Larbi, I., Szikszay, E., Roy, P.M., Arcand, M., Dupuis, G., Fulop, T., 2002. Modulation of human lymphocyte proliferative response with aging. Exp. Gerontol. 37, 369–387.

Dumont, F.J., Staruch, M.J., Fischer, P., Deasilva, C., Camacho, R., 1998. Inhibition of T cell activation by pharmacologic disruption of the MEK1/ERK MAP kinase or calcineurin signaling pathways results in differential modulation of cytokine production. J. Immunol. 160, 2579–2589.

Dustin, M.L., 2002. Membrane domains and the immunological synapse: keeping T cells resting and ready. J. Clin. Invest. 109, 155–160.

Effros, R.B., 2000. Costimulatory mechanisms in the elderly. Vaccine 18, 1661–1665.

Effros, R.B., Walford, R.L., 1983. The immune response of aged mice to influenza: diminished T-cell proliferation, interleukin 2 production and cytotoxicity. Cell. Immunol. 81, 298–305.

Eisenbraun, M.D., Tamir, A., Miller, R.A., 2000. Altered composition of the immunological synapse in anergic, age-dependentmemory T cell subset. J. Immunol. 164, 6105–6112.

Engwerda, C.R., Handwerger, B.S., Fox, B.S., 1996. An age-related decrease in rescue from T cell death following costimulation mediated by CD28. Cell. Immunol. 170, 141–148.

Erschler, W.B., Moore, A.L., Roessner, K., Ranges, G.E., 1985. Interleukin-2 and aging: decreased interleukin-2 production in healthy older people does not correlate with reduced helper cell numbers or antibody response to influenza vaccine and is not corrected in vitro by thymosin alpha 1. Immunopharmacology 10, 11–17.

Ershler, W.B., 1993. The influence of an aging immune system on cancer incidence and progression. J. Gerontol. 48, B3–B7.

Fagnoni, F.F., Vescovini, R., Passeri, G., Bologna, G., Pedrazzoni, M., Lavagetto, G., Casti, A., Franceschi, C., Passeri, M., Sansoni, P., 2000. Shortage of circulating naive CD8(+) T cells provides new insights on immunodeficiency in aging. Blood 95, 2860–2868.

Fagnoni, F.F., Vescovini, R., Mazzola, M., Bologna, G., Nigro, E., Lavagetto, G., Franceschi, C., Passeri, M., Sansoni, P., 1996. Expansion of cytotoxic CD8(+) CD28(−) T cells in healthy ageing people, including centenarians. Immunology 88, 501–507.

Fanger, G.R., 1999. Regulation of the MAPK family members: role of subcellular localization and architectural organization. Histol. Histopathol. 14, 887–894.

Fernandez Gutierrez, B., Jover, J.A., DeMiguel, S., Hernandez Garcia, C., Vidan, T., Ribera, J.M., Banares, A., Serra, J.A., 1999. Early lymphocyte activation in elderly humans: impaired T and T-dependent B cell responses. Exp. Gerontol. 34, 217–229.

Fraeyman, N., Vanscheeuwijck, P., De Wolf, M., Quatacker, J., 1993. Influence of aging on fluidity and coupling between receptors and g-proteins in rat lung membranes. Life Sci. 53, 153–160.

Franceschi, C., Bonafe, M., Valensin, S., Olivieri, F., De Luca, M., Ottaviani, E., De Benedictis, G., 2000. Inflamm-aging. An evolutionary perspective on immunosenescence. Ann. N.Y. Acad. Sci. 908, 244–254.

Frank, D.A., Robertson, M.J., Bonni, A., Ritz, J., Greenberg, M.E., 1995. Interleukin 2 signaling involves the phosphorylation of Stat proteins. Proc. Natl. Acad. Sci. USA 92, 7779–7783.

Franklin, R.A., Tordai, A., Patel, H., Gardner, A.M., Johnson, G.L., Gelfand, E.W., 1994. Ligation of the T cell receptor complex results in activation of the Ras/Raf-1/MEK/MAPK cascade in human T lymphocytes. J. Clin. Invest. 93, 2134–2140.

Fraser, J.D., Strauss, D.B., Weiss, A., 1993. Signal transduction events leading to T-cell lymphokine gene expression. Immunol. Today 14, 357–362.

Fülöp, T., Jacob, M.P., Robert, L., 1989. Biological effects of elastin peptides. In: L. Robert and W. Hornebeck, (Eds.), Elastin and Elastases 1 pp. 201–210. C.R.C. Press, Boca Raton.

Fülöp, T. Jr., 1994. Signal transduction changes in granulocytes and lymphocytes with aging. Immunol. Lett. 40, 259–268.

Fülöp, T. Jr., Douziech, N., Goulet, A.C., Desgeorges, S., Linteau, A., Lacombe, G., Dupuis, G., 2001. Cyclodextrin modulation of T lymphocyte signal transduction with aging. Mech. Ageing Dev. 152, 1413–1430.

Fülöp, T. Jr., Douziech, N., Jacob, M.P., Hauck, M., Wallach, J., Robert, L., 2001. Age-related alterations in the signal transduction pathways of the elastin-laminin receptor. Pathol. Biol. 49, 339–348.

Fülöp, T., Jr., Douziech, N., Larbi, A., Dupuis, G., 2002. Role of lipid rafts in T lymphocyte signal transduction with aging. Ann. N.Y. Acad. Sci. 973, 302–304.

Fülöp, T. Jr., Jacob, M.P., Varga, Z., Foris, G., Leovey, A., Robert, L., 1986. Effects of elastin peptides on human monocytes: Ca mobilization, stimulation of respiratory burst and enzyme secretion. Biochem. Biophys. Res. Commun. 141, 92–98.

Fülöp, T. Jr., Barabas, G., Varga, Zs., Csongor, J., Seres, I., Szucs, S., Mohacsi, A., Szikszay, E., Kékessy, D., Despont, J.P., Robert, L., Penyige, A., 1992. Transmembrane signalling changes with aging. Ann. N.Y. Acad. Sci. 673, 165–171.

Fülöp, T. Jr., Barabas, G., Varga, Zs., Csongor, J., Seres, I., Szucs, S., Mohacsi, A., Szikszay, E., Kékessy, D., Despont, J.P., Robert, L., Penyige, A., 1993. Alteration of transmembrane signalling with aging. Identification of G proteins in lymphocytes and granulocytes. Cell. Signal. 5, 593–603.

Fülöp, T. Jr., Foris, G., Worum, I., Paragh, G., Leovey, A., 1985a. Age-related variations of some PMNL functions. Mech. Ageing Dev. 29, 1–8.

Fülöp, T. Jr., Foris, G., Worum, I., Leovey, A., 1985b. Age-dependent alterations of Fc receptor mediated effector functions of human polymorphonuclear leukocytes. Clin. Exp. Immunol. 61, 425–432.

Fülöp, T. Jr., Foris, G., Worum, I., Leovey, A., 1986. Age-dependent variations of intralysosomal enzyme release from human PMN leukocytes under various stimuli. Immunobiology 171, 302–310.

Fülöp, T. Jr., Fouquet, C., Allaire, P., Perrin, N., Lacombe, G., Stankova, J., Rola-Pleszczinsky, M., Wagner, J.R., Khalil, A., Dupuis, G., 1997. Changes in apoptosis of human polumorphonuclear granulocytes with aging. Mech. Ageing Dev. 96, 15–31.

Fülöp, T. Jr., Gagné, D., Goulet, A.C., Desgeroges, S., Lacombe, G., Arcand, M., Dupuis, G., 1999. Age-related impairment of p56lck and ZAP70 activities in human T lymphocytes activated through the TCR/CD3 complex. Exp. Gerontol. 34, 197–216.

Fülöp, T. Jr., Kekessy, D., Foris, G., 1987. Impaired coupling of naloxone sensitive opiate receptors to adenylate cyclase in PMNLs of aged male subjects. Int. J. Immunopharmacol. 6, 651–659.

Fülöp, T. Jr., Leblanc, C., Lacombe, G., Dupuis, G., 1995. Determination of PKC isozymes in human T lymphocytes after CD3 stimulation with aging. FEBS Lett. 375, 69–74.

Fülöp, T. Jr., Utsuyama, M., Hirokawa, K., 1991. Determination of IL-2 receptor number of ConA stimulated human lymphocytes with aging. J. Clin. Lab. Immunol. 34, 31–36.

Fülöp, T. Jr., Varga, Z., Jacob, M.P., Robert, L., 1997. Effects of Lithium on superoxyde production and intracellular free calcium mobilization in elastin peptide (kappa-elastin) and FMLP stimulated human polymorphonuclear leukocytes. Effect of aging. Life Sci. 60, PL325–PL332.

Fülöp, T. Jr., Varga, Z., Csongor, J., Foris, G., Leovey, A., 1989. Age-related impairment of phosphatidylinositol breakdown of polymorphonuclear granulocytes. FEBS Lett. 245, 249–252.

Garcia, G.G., Miller, R.A., 1998. Increased Zap-70 association with CD3zeta in CD4 T cells from old mice. Cell. Immunol. 190, 91–100.

Garcia, G.G., Miller, R.A., 2001. Single-cell analyses reveal two defects in peptide-specific activation of naive T cells from aged mice. J. Immunol. 166, 3151–3157.

Geginat, J., Clissi, B., Moro, M., Dellabona, P., Bender, J.R., Pardi, R., 2000. CD28 and LFA-1 contribute to cyclosporin A-resistant T cell growth by stabilizing the IL-2 mRNA through distinct signaling pathways. Eur. J. Immunol. 30, 1136–1144.

Genot, E., Cantrell, D.A., 2000. Ras regulation and function in lymphocytes. Curr. Opin. Immunol. 12, 289–294.

Genot, E., Cleverley, S., Henning, S., Cantrell, D., 1996. Multiple p21ras effector pathways regulate nuclear factor of activated T cells. EMBO J. 15, 3923–3933.

Gesbert, F., Delespine-Carmagnat, M., Bertoglio, J., 1998. Recent advances in the understanding of Interleukin-2 signal transduction. J. Clin. Immunol. 18, 307–320.

Ghiotto-Ragueneau, M., Battifora, M., Truneh, A., Waterfield, M.D., Olive, D., 1996. Comparison of CD28-B7.1 and B7.2 functional interaction in resting human T cells: phosphatidylinositol 3-kinase association to CD28 and cytokine production. Eur. J. Immunol. 26, 34–41.

Ghosh, J., Miller, R.A., 1995. Rapid tyrosine phosphorylation of Grb2 and Shc in T cells exposed to anti-CD3, anti-CD4, and anti-CD45 stimuli: differential effects of aging. Mech. Ageing Dev. 80, 171–187.

Gillis, S., Kozak, R., Durante, M., Weksler, M.E., 1981. Immunological studies of aging. Decreased production of and response to T cell growth factor by lymphocytes from aged humans. J. Clin. Invest. 67, 931–942.

Gilman, S.C., Rosenberg, J.S., Feldman, J.D., 1982. T lymphocytes of young and aged rats. II. Functional defects and the role of interleukin-2. J. Immunol. 128, 644–650.

Gilmour, K.C., Pine, R., Reich, N.C., 1995. Interleukin 2 activates STAT5 transcription factor (mammary gland factor) and specific gene expression in T lymphocytes. Proc. Natl. Acad. Sci. USA 92, 10,772–10,776.

Ginaldi, L., DeMartinis, M., Dostilio, A., Marini, L., Loreto, F., Modesti, M., Quaglino, D., 2001. Changes in the expression of surface receptors on lymphocyte subsets in the elderly: quantitative flow cytometric analysis. Am. J. Hematol. 67, 63–72.

Gorgas, G., Butch, E.R., Guan, K.L., Miller, R.A., 1997. Diminished activation of the MAP kinase pathway in CD3-stimulated T lymphocytes from old mice. Mech. Ageing Dev. 94, 71–83.

Granja, C., . Lin, L.L., . Yunis, E.J., Relias, V., Dasgupta, J.D., 1991. PLC gamma 1, a possible mediator of T cell receptor function. J. Biol. Chem. 266, 16,277–16,280.

Grossmann, A., Ledbetter, J.A., Rabinovitch, P.R., 1989. Reduced proliferation in T lymphocytes in aged humans is predominantly in the CD8+ subset, and is unrelated to defects in transmembrane signaling which are predominantly in the CD4+ subset. Exp. Cell. Res. 180, 367–382.

Grossmann, A., Ledbetter, J.A., Rabinovitch, P.S., 1990. Aging-related deficiency in intracellular calcium response to anti-CD3 or concanavalin A in murine T-cell subsets. J. Gerontol. 45, B81–B86.

Grossmann, A., Rabinovitch, P.S., Kavanagh, T.J., Jinneman, J.C., Gilliland, L.K., Ledbetter, J.A., Kanner, S.B., 1995. Activation of murine T-cells via phospholipase-C gamma 1-associated protein tyrosine phosphorylation is reduced with aging. J. Gerontol. 50A, B205–B212.

Gupta, S., 1989. Membrane signal transduction in T cells in aging humans. Ann. N.Y. Acad. Sci. 568, 277–282.

Halliwell, B., Gutteridge, J.M.C. (Eds.), 1989. Free Radicals in Biology and Medicine. Clarendon Press, Oxford.

Harada, Y., Tanabe, E., Watanabe, R., Weiss, B.D., Matsumoto, A., Ariga, H., Koiwai, O., Fukui, Y., Kubo, M., June, C.H., Abe, R., 2001. Novel role of phosphatidylinositol 3-kinase in CD28-mediated costimulation. J. Biol. Chem. 276, 9003–9008.

Hardy, K., Chaudhri, G., 1997. Activation and signal transduction via mitogen-activated protein (MAP) kinases in T lymphocytes. Immunol. Cell Biol. 75, 528–545.

Harman, D., 1956. Aging: a theory based on free radical and radiation chemistry. J. Gerontol. 11, 298–300.

Hermiston, M.L., Xu, Z., Majeti, R., Weiss, A., 2002. Reciprocal regulation of lymphocyte activation by tyrosine kinases and phosphatases. J. Clin. Invest. 109, 9–14.

Heuertz, R.M., Tricomi, S.M., Ezekiel, U.R., Webster, R.O., 1999. C-reactive protein inhibits chemotactic peptide-induced p38 mitogen-activated protein kinase activity and human neutrophil movement. J. Biol. Chem. 274, 17,968–17,974.

Hirokawa, K., 1998. Immunity and aging. In: M.S.J. Pathy, (Ed.), Principles and Practice of Geriatric Medicine 1 pp. 35–48. John Wiley and Sons Chichester, England.

Hirokawa, K., Utsuyama, M., Katura, Y., Sado, T., 1988. Influence of age on the proliferation and peripheralization of thymic T cells. Arch. Pathol. Lab. Med. 112, 13–21.

Holdorf, A.D., Kanagawa, O., Shaw, A.S., 2000. CD28 and T cell co-stimulation. Rev. Immunogenet. 2, 175–184.

Hombach, A., Sent, D., Schneider, C., Heuser, C., Koch, D., Pohl, C., Seliger, B., Abken, H., 2001. T-cell activation by recombinant receptors: CD28 costimulation is required for interleukin 2 secretion and receptor-mediated T-cell proliferation but does not affect receptor-mediated target cell lysis. Cancer Res. 61, 1976–1982.

Hu, H., Rudd, C.E., Schneider, H., 2001. Src kinases Fyn and Lck facilitate the accumulation of phosphorylated CTLA-4 and its association with PI-3 kinase in intracellular compartments of T-cells. [Journal Article] Biochem. Biophys. Res. Commun. 288, 573–578.

Hubbard, S.R., Till, J.H., 2000. Protein tyrosine kinase structure and function. Ann. Rev. Biochem. 69, 373–398.

Hunt, A.E., Lali, F.V., Lord, J.D., Nelson, B.H., Miyazaki, T., Tracey, K.J., Foxwell, B.M.J., 1999. Role of interleukin (IL)-2 receptor beta-chain subdomains and Shc in p38 mitogen-activated protein (MAP) kinase and p54 MAP kinase (stress-activated protein kinase c-Jun N-terminal kinase) activation-IL-2-driven proliferation is independent of p38 and p54 MAP kinase activation. J. Biol. Chem. 274, 7591–7597.

Hutter, D., Yo, Y., Chen, W., Liu, P., Holbrook, N.J., Roth, G.S., Liu, Y., 2000. Age-related decline in Ras/ERK mitogen-activated protein kinase cascade is linked to a reduced association between Shc and EGF receptor. J. Gerontol. 55A, B125–B134.

Ilangumaran, S., Hoessli, D.C., 1998. Effects of cholesterol depletion by cyclodextrin on the sphingolipid microdomains of the plasma membrane. Biochem. J. 335, 433–440.

Isakov, N., 1997. Immunoreceptor tyrosine-based activation motif (ITAM), a unique module linking antigen and Fc receptors to their signaling cascades. J. Leuk. Biol. 61, 6–16.

Iwai, H., Fernandes, G., 1989. Immunological functions in food-restricted rats: enhanced expression of high-affinity interleukin-2 receptors on splenic T cells. Immunol. Lett. 23, 125–132.

Izquierdo, D., Pastor, M., Reif, K., Cantrell, D., 1995. The regulation and function of p21ras during T-cell activation and growth. Immunol. Today 16, 159–164.

Jabado, N., Pallier, A., Jauliac, S., Fischer, A., Hivroz, C., 1997. gp160 of HIV or anti-CD4 monoclonal antibody ligation of CD4 induces inhibition of JNK and ERK-2 activities in human peripheral CD4+ T lymphocytes. Eur. J. Immunol. 27, 397–404.

Janes, P.W., Ley, S.C., Magee, A.I., Kabouridis, P.S., 2000. The role of lipid rafts in T cell antigen receptor (TCR) signalling. Semin. Immunol. 12, 23–34.

Janes, P.W., Ley, S.C., Magee, A.I., 1999. Aggregation of lipid rafts accompagnies signaling via the T cell antigen receptor. J. Cell Biol. 147, 447–461.

June, C.H., Fletcher, M.C., Ledbetter, J.A., Schieven, G.L., Siegel, J.N., Philips, A.N., Samelson, L.E., 1990. Inhibition of tyrosine phosphorylation prevents T-cell receptor-mediated signal transduction. Proc. Natl. Acad. Sci. USA. 87, 7722–7726.

Kabouridis, P.S., Janzen, J., Magee, A.L., Ley, S.G., 2000. Cholesterol depletion disrupts lipid rafts and modulates the activity of multiple signaling pathways in T lymphocytes. Eur. J. Immunol. 30, 954–963.

Kane, L.P., Lin, J., Weiss, A., 2000. Signal transduction by the TCR for antigen. Curr. Opin. Immunol. 12, 242–249.

Karanfilov, C.I., Liu, B.Q., Fox, C.C., Lakshmanan, R.R., Whisler, R.L., 1999. Age-related defects in Th1 and Th2 cytokine production by human T cells can be dissociated from altered frequencies of CD45RA+ and CD45RO+ T cell subsets. Mech. Ageing Dev. 109, 97–112.

Kawanishi, H., 1993. Activation of calcium (Ca)-dependent protein kinase C in aged mesenteric lymph node T and B cells. Immunol. Lett. 35, 25–32.

Kawanishi, H., Joseph, K., 1992. Effects of phorbol myristate and ionomycin on in vitro growth of aged Peyer's patch T and B cells. Mech. Ageing Dev. 65, 289–300.

Kirchhoff, S., Muller, W.W., Krueger, A., Schmitz, I., Krammer, P.H., 2000a. TCR-mediated up-regulation of c-FLIPshort correlates with resistance toward CD95-mediated apoptosis by blocking death-inducing signaling complex activity. J. Immunol. 165, 6293–6300.

Kirchhoff, S., Muller, W.W., Li-Weber, M., Krammer, P.H., 2000b. Up-regulation of c-FLIPshort and reduction of activation-induced cell death in CD28-co-stimulated human T cells. Eur. J. Immunol. 30, 2765–2774.

Kirk, C.J., Freilich, A.M., Miller, R.A., 1999. Age-sensitive and -insensitive pathways leading to JNK activation in mouse CD4(+) T-cells. Cell. Immunol. 197, 75–82.

Kroemer, G., Dallaporta, B., Resche-Rigon, M., 1998. The mitochondrial death/life regulator in apoptosis and necrosis. Annu. Rev. Physiol. 60, 619–642.

Kukreja, A., Cost, G., Marker, J., Zhang, C., Sun, Z., Lin-Su, K., Ten, S., Sanz, M., Exley, M., Wilson, B., Porcelli, S., Maclaren, N., 2002. Multiple immuno-regulatory defects in type-1 diabetes. J. Clin. Invest. 109, 131–140.

Kyriakis, J.M., Avrach, J., 1996. Sounding the alarm: protein kinase cascades activated by stress and inflammation. J. Biol. Chem. 271, 24,313–24,316.

Lanzavecchia, A., Sallusto, F., 2000. Dynamics of T lymphocyte responses intermediates, effectors and memeory cells. Science 290, 92–97.

Leo, A., Wienands, J., Baier, G., Horejsi, V., Schraven, B., 2002. Adapters in lymphocyte signaling. J. Clin. Invest. 109, 301–309.

Leupin, O., Zaru, R., Laroche, T., Muller, S., Valitutti, S., 2000. Exclusion of CD45 from the T-cell receptor signaling area in antigen-stimulated T lymphocytes. Curr. Biol. 10, 277–280.

Li, M., Walter, R., Torres, C., Sierra, F., 2000a. Impaired signal transduction in mitogen activated rat splenic lymphocytes during aging. Mech. Ageing Dev. 113, 85–99.

Li, W.Q., Dehnad, F., Zafarullah, M., 2000b. Thiol antioxidant, N-acetylcysteine, activates extracellular signal-regulated kinase signaling pathway in articular chondrocytes. Biochem. Biophys. Res. Commun. 275, 789–794.

Linton, P., Thoman, M.L., 2001. T cell senescence. Front. Biosci. 6, D248–D261.

Lio, D., Candore, G., Cigna, D., Danna, C., Dilorenzo, G., Giordano, C., Lucania, G., Mansueto, P., Melluso, M., Modica, M.A., Caruso, C., 1996. In vitro T cell activation in elderly individuals: failure in CD69 and CD71 expression. Mech. Ageing Dev. 89, 51–58.

Lio, D., D'Anna, C., Gervasi, F., Scola, L., Potestio, M., Di Lorenzo, G., Listi, F., Colombo, A., Candore, G., Caruso, C., 1998. Interleukin-12 release by mitogen-stimulated mononuclear cells in the elderly. Mech. Ageing Dev. 102, 211–219.

Liparoto, S.F., Myszka, D.G., Wu, Z., Goldstein, B., Laue, T.M., Ciardelli, T.L., 2002. Analysis of the role of the interleukin-2 receptor gamma chain in ligand binding. Biochemistry 41, 2543–2551.

Liu, J., Farmer, J.J., Lane, W.S., Friedman, J., Weissman, I., Schreiber, S.L., 1991. Calcineurin is a common target of cyclophilin-cyclosporin A and FKBP-FK506 complexes. Cell 66, 807–815.

Lord, J.M., Butcher, S., Killampali, V., Lascelles, D., Salmon, M., 2001. Neutrophil ageing and immunesenescence. Mech. Ageing Dev. 122, 1521–1535.

Macian, F., Lopez-Rodriguez, C., Rao, A., 2001. Partners in transcription: NFAT and AP-1. Oncogene 20, 2476–2489.

Makinodan, T., Kay, M.M.B., 1980. Age influence on the immune system. Adv. Immunol. 29, 287–300.

Mamor, M.D., Julius, M., 2001. Role of lipid rafts in regulating interleukin 2 receptor signaling. Blood 98, 1489–1497.

Masuda, E.S., Imamura, R., Amasaki, Y., Arai, K., Arai, N., 1998. Signalling into the T-cell nucleus: NFAT regulation. Cell. Signal. 10, 599–611.

Mcleish, K.R., Gierschik, P., Jacobs, K.H., 1989a. Desensitization uncouples the formyl peptide receptor-guanine nucleotide-binding protein interaction in HL60 cells. Mol. Pharmacol. 36, 384–390.

Mcleish, K.R., Gierschik, P., Schepers, T., Sidiropoulos, D., Jakobs, K.H., 1989b. Evidence that activation of a common G-protein by receptors for leukotriene B4 and N-formylmethionyl-leucyl-phenylalanine in HL-60 cells occurs by different mechanisms. Biochem. J. 260, 427–434.

Miller, R.A., 2000. Effect of aging on T lymphocyte activation. Vaccine 18, 1654–1660.

Miller, R.A., 1986. Immunodeficiency of aging: restoration effects of phorbol ester combined with calcium ionophore. J. Immunol. 137, 805–808.

Miller, R.A., 1989. Defective calcium signal generation in a T cell subset that accumulates in old mice. Ann. N.Y. Acad. Sci. 568, 271–276.

Miller, R.A., 1991a. Accumulation of hyporesponsive, calcium extruding memory T cells as a key feature of age-dependent immune dysfunction. Clin. Immunol. Immunopathol. 58, 305–317.

Miller, R.A., 1991b. Aging and immune function. Int. Rev. Cytol. 124, 187–215.

Miller, R.A., Jacobson, B., Weil, G., Simons, E.R.J., 1987. Diminished calcium influx in lectin-stimulated T cells from old mice. J. Cell. Physiol. 132, 337–342.

Miller, R.A., Stutman, O., 1981. Decline in aging mice of the anti-2,4,6 trinitrophenyl (TNP) cytotoxic T cell response attributable to loss of Lyt-2-, interleukin-2 producing helper cell function. Eur. J. Immunol. 11, 751–756.

Minami, Y., Kono, R., Yamada, K., Kobayashi, N., Kawahara, A., Perlmutter, R.M., Taniguchi, T., 1993. Association of p56lck with IL-2 receptor chain is critical for the IL-2-induced activation of p56lck. EMBO J. 12, 759–768.

Monks, C.R., Freiberg, B.A., Kupfer, H., Sciaky, N., Kupfer, A., 1998. Three-dimensional segregation of supramolecular activation clusters in T cells. Nature 395, 82–86.

Montixi, C., Langlet, C., Bernard, A.M., Thimonier, J., Dubois, C., Wurbel, M.A., Chauvin, J.P., Pierres, M., He, H.T., 1998. Engagement of T cell receptor triggers its recruitment to low-density detergent-insoluble membrane domains. EMBO J. 17, 5334–5348.

Moran, M., Miceli, M.C., 1998. Engagement of GPI-linked CD48 contributes to TCR signals and cytoskeletal reorganization: a role for lipid rafts in T cell activation. Immunity 9, 787–796.

Moulias, R., Meaume, S., Raynaud-Simon, A., 1999. Sarcopenia, hypermetabolism and aging. Z. Gerontol. Geriat. 32, 425–432.

Nagel, J.E., Chopra, R.K., Chrest, F.J., McCoy, M.T., Schneider, E.L., Holbrook, N.J., Adler, W.H., 1988. Decreased proliferation, interleukin 2 synthesis, and interleukin 2 receptor expression are accompanied by decreased mRNA expression in phytohemagglutinin-stimulated cells from elderly donors. J. Clin. Invest. 81, 1096–1102.

Nagel, J.E., Chopra, R.K., Dowers, D.C., Adler, W.H., 1989. Effect of age on the human high affinity interleukin 2 receptor of phytohaemagglutinin stimulated peripheral blood lymphocytes. Clin. Exp. Immunol. 75, 286–291.

Nishizuka, Y., 1988. The molecular heterogeneity of protein kinase C and its implications for cellular regulation. Nature 334, 661–665.

Norian, L.A., Koretzky, G.A., 2000. Intracellular adapter molecules. Sem. Immunol. 12, 43–54.

Ohkusu, K., Du, J., Isobe, K.I., Akhand, A.A., Koto, A., Suzuki, H., Hidoka, H., Nakashima, I., 1997. Protein kinase C alpha-mediated chronic signal transduction for immunosenescence. J. Immunol. 159, 2082–2084.

Pahl, H.L., Bauerle, P.A., 1994. Oxygen and the control of gene expression. Bioessays 16, 497–502.

Pahlavani, M.A., Harris, M.D., Richardson, A., 1995. Activation of p21(ras)/MAPK signal transduction molecules decreases with age in mitogen-stimulated T cells from rats. Cell. Immunol. 165, 84–91.

Pahlavani, M.A., Harris, M.D., 1997. Effect of in vitro generation of oxygen free radicals on T cell function in young and old rats. Free Rad. Biol. Med. 25, 903–913.

Palmer, H.J., Paulson, K.E., 2000. Reactive oxygen species and anti-oxidants in signal transduction and gene expression. Nutr. Rev. 55, 353–361.

Park, C.S., Park, W.R., Sugimoto, N., Nakahira, M., Ahn, H.J., Hamaoka, T., Ohta, T., Kurimoto, M., Fujiwara, H., 2000. Differential effects of N-acetyl-l-cysteine on IL-2- vs IL-12-driven proliferation of a T cell clone: implications for distinct signalling pathways. Cytokine 12, 1419–1422.

Pascale, A., Govoni, S., Battaini, F., 1998. Age-related alteration of PKC, a key enzyme in memory processes physiological and pathological examples. Mol. Neurobiol. 16, 49–62.

Pastor, M.I., Woodrow, M., Cantrell, D., 1995. Regulation and function of p21ras in T lymphocytes. Cancer Surv. 22, 75–83.

Pawelec, G., Adibzadeh, M., Solana, R., Beckman, I., 1997. The T cell in the ageing individual. Mech. Ageing Dev. 93, 35–45.

Pawelec, G., Hirokawa, K., Fülöp, T., 2001. Altered T cell signalling in ageing. Mech. Ageing Dev. 122, 1613–1657.

Pawelec, G., Mariani, E., Bradley, B., Solana, R., 2000. Longevity in vitro of human CD4+ T helper cell clones derived from young donors and elderly donors, or from progenitor cells: age-associated differences in cell surface molecule expression and cytokine secretion. Biogerontology 1, 247–254.

Peterson, E.J., Koretzky, G.A., 1999. Signal transduction in T lymphocytes. Clin. Exp. Rheumatol. 17, 107–114.

Pimentel-Muinos, F.X., Mazana, J., Fresno, M., 1995. Biphasic control of nuclear factor-kappa B activation by the T cell receptor complex: role of tumor necrosis factor alpha. Eur. J. Immunol. 25, 179–186.

Pitha, J., Irie, T., Sklar, P.B., Nye, J.S., 1988. Drug solubilizers to aid pharmacologists: amorphous cyclodextrin derivatives. Life Sci. 43, 493–502.

Pivniouk, V.I., Geha, R.S., 2000. The role of SLP-76 and LAT in lymphocyte development. Curr. Opin. Immunol. 12, 173–178.

Ponnappan, U., Trebilcock, G.U., Zheng, M.Z., 1999. Studies into the effect of tyrosine phosphatase inhibitor phenylarsine oxide on NFkappaB activation in T lymphocytes during aging: evidence for altered IkappaB-alpha phosphorylation and degradation. Exp. Gerontol. 34, 95–107.

Posnett, D.N., Sinha, R., Kabak, S., Russo, C., 1994. The T cell equivalent to benign monoclonal gammapathy. J. Exp. Med. 179, 609–618.

Proust, J.J., Fiburn, C.R., Harrison, S.A., Buchholz, M.A., Nordin, A.A., 1987. Age-related defect in signal transduction during lectin activation of murine T lymphocytes. J. Immunol. 139, 1472–1478.

Quadri, R.A., Plastre, O., Phelouzat, M.A., Arbogast, A., Proust, J.J., 1996. Age-related tyrosine-specific protein phosphorylation defect in human T lymphocytes activated through CD3, CD4, CD8 or the IL-2 receptor. Mech. Ageing Dev. 88, 125–138.

Rao, A., Luo, C., Hogan, P.G., 1997. Transcription factors of the NFAT family: regulation and function. Ann. Rev. Immunol. 15, 707–747.

Rink, L., Cakman, I., Kirchner, H., 1998. Altered cytokine production in the elderly. Mech. Ageing Dev. 102, 199–209.

Ron, D., Kazanietz, M.G., 1999. New insights into the regulation of protein kinase C and novel phorbol ester receptors. FASEB J. 13, 1658–1676.

Roth, G.S., 1989. Calcium homeostasis and aging: role in altered signal transduction. Ann. N.Y. Acad. Sci. 568, 68–72.

Sansom, D.M., 2000. CD28, CTLA-4 and their ligands: who does what and to whom? Immunology 101, 169–177.

Sato, N., Sakamaki, K., Terada, N., Arai, K., Miyajima, A., 1993. Signal transduction by the high-affinity GM-CSF receptor: two distinct cytoplasmic regions of the common beta subunit responsible for different signaling. EMBO J. 12, 4181–4189.

Schaeffer, E.M., Schwartzberg, P.I., 2000. Tec family kinases in lymphocyte signaling and function. Curr. Opin. Immunol. 12, 282–288.

Schrum, A.G., Wells, A.D., Turka, L.A., 2000. Enhanced surface TCR replenishment mediated by CD28 leads to greater TCR engagement during primary stimulation. Int. Immunol. 12, 833–842.

Sen, C.K., Packer, L., 1996. Anti-oxidant and redox regulation of gene transcription. FASEB J. 10, 709–720.

Serafini, M., 2000. Dietary vitamin E and T cell-mediated function in the elderly: effectiveness and mechanism of action. Int. J. Dev. Neurosci. 18, 401–410.

Seres, I., Csongor, J., Mohacsi, A., Leovey, A., Fulop, T., 1993. Age-dependent alterations of human recombinant GM-CSF effects on human granulocytes. Mech. Ageing Dev. 71, 143–154.

Serfling, E., Avots, A., Neumann, M., 1995. The architecture of the interleukin-2 promoter: a reflection of T lymphocyte activation. Biochim. Biophys. Acta 1263, 181–200.

Shi, J., Miller, R.A., 1992. Tyrosine specific protein phosphorylation in response to anti-CD3 antibody is diminished in old mice. J. Gerontol. 47, B147–B153.

Shreck, R., Albermann, K., Bauerle, P.A., 1992. Nuclear factor kappa B: an oxidative stress-responsive transcription factor of eukaryotic cells (a review). Free Rad. Res. Commun. 17, 221–237.

Sikora, E., Kaminska, B., Radziszewska, E., Kaczmarek, L., 1992. Loss of transcription factor AP-1 DNA binding activity during lymphocyte aging in vivo. FEBS Lett. 312, 179–182.

Simons, K., . Ikonen, E., 1997. Functional rafts in cell membranes. Nature 387, 569–572.

Song, L., Stephens, J.M., Kittur, S., Collins, G.D., Nagel, J.E., Pekala, P.H., Adler, W.H., 1992. Expression of c-fos, c-jun and jun B in peripheral blood lymphocytes from young and elderly adults. Mech. Ageing Dev. 65, 149–156.

Spencer, N.F., Poynter, M.E., Im, S.Y., Daynes, R.A., 1997. Constitutive activation of NF-kappa B in an animal model of aging. Int. Immunol. 9, 1581–1588.

Stahl, W., Sies, H., 1997. Antioxidant defense: vitamins E and C and carotenoids. Diabetes 46(Suppl. 2), S14–S18.

Steimann, G.G., 1986. Changes in the human thymus during ageing. In: Muller-Hermelink, H.K. (Ed.), The Human Thymus. Heinemann, London, pp. 43–88.

Suresh, M., Whitmire, J.K., Harrington, L.E., Larsen, C.P., Pearson, T.C., Altman, J.D., Ahmed, R., 2001. Role of CD28-B7 interactions in generation and maintenance of CD8 T cell memory. J. Immunol. 167, 5565–5573.

Tamir, A., Eisenbraun, M.D., Garcia, G.G., Miller, R.A., 2000. Age-dependent alterations in tha assembly of signal transduction complexes at the site of T cell/APC interaction. J. Immunol. 165, 1243–1251.

Taniguchi, T., Minami, Y., 1993. The IL-2/IL-2 receptor system: a current overview. Cell 73, 5–8.

Thoman, M.L., Weigle, W.O., 1989. The cellular and subcellular bases of immunosenescence. Adv. Immunol. 46, 221–261.

Thoman, M.L., Weigle, W.O., 1988. Partial restoration of Con A-induced proliferation, IL-2 receptor expression, and IL-2 synthesis in aged murine lymphocytes by phorbol myristate acetate and ionomycin. Cell. Immunol. 114, 1–11.

Tomlinson, M.G., Lin, J., Weiss, A., 2000. Lymphocyte with a complex: adapter proteins in in antigem receptor signaling. Immunol. Today 21, 584–591.

Tortorella, C., Piazzolla, G., Spaccavento, F., Jirillo, E., Antonaci, S., 1999. Age-related effects of oxidative metabolism and cyclicAMP signaling on neutrophil apoptosis. Mech. Ageing Dev. 110, 195–205.

Utsuyama, M., Varga, Z., Fukami, K., Homma, Y., Takenawa, T., Hirokawa, K., 1993. Influence of age on the signal transduction of T-Cells in mice. Int. Immunol. 5, 1177–1182.

Utsuyama, M., Wakikawa, A., Tamura, T., Nariuchi, H., Kirokawa, K., 1997. Impairment of signal transduction in T cells from old mice. Mech. Ageing Dev. 93, 131–144.

Vallejo, A.N., Nestel, A.R., Schirmer, M., Weyand, C.M., Goronzy, J.J., 1998. Aging-related deficiency of CD28 expression in CD4+ T cells is associated with the loss of gene-specific nuclear factor binding activity. J. Biol. Chem. 273, 8119–8129.

Varga, Z., Jacob, M.P., Csongor, J., Robert, L., Leovey, A., Fulop, T. Jr., 1990a. Altered phosphatidylinositol breakdown after K-elastin stimulation in PMNLs of elderly. Mech. Ageing Dev. 52, 61–70.

Varga, Z., Jacob, M.P., Robert, L., Fülöp, T. Jr., 1989. Identification and signal transduction mechanism of elastin peptide receptor in human leukocytes. FEBS Lett. 258, 5–8.

Varga, Z., Kovacs, E.M., Paragh, G., Jacob, M.P., Robert, L., Fulop, T. Jr., 1988. Effects of Kappa elastin and FMLP on polymorphonuclear leukocytes of healthy middle-aged and elderly. K-elastin induced changes in intracellular free calcium. Clin. Biochem. 21, 127–130.

Varga, Zs., Bressani, N., Zaid, A.M., Bene, L., Fulop, T. Jr., Leovey, A., Fabris, N., Damjanovich, S., 1990b. Cell surface markers, Inositol phosphate levels and membrane potential of lymphocytes from young and old human patients. Immunol. Lett. 23, 275–280.

Varga, Zs., Jacob, M.P., Robert, L., Fulop, T. Jr., 1997. Studies on effector functions of soluble elastin receptors of phagocytic cells at various ages. Exp. Gerontol. 32, 653–662.

Viani, P., Cervato, G., Fiorelli, A., Cestaro, B., 1991. Age-related differences in synaptosomal peroxidative damage and membrane properties. J. Neurochem. 56, 253–258.

Viola, A., Schroeder, S., Sakakibara, Y., Lanzavecchia, A., 1999. T lymphocyte costimulation mediated by reorganization of membrane microdomains. Science 283, 680–682.

Vlahos, C.J., Matter, W.F., 1992. Signal transduction in neutrophil activation. Phosphatidylinositol 3-kinase is stimulated without tyrosine phosphorylation. FEBS Lett. 309, 242–248.

Wada, T., Seki, H., Konno, A., Ohta, K., Nunogami, K., Kaneda, H., Kasahara, Y., Yachie, A., Koizumi, S., Taniguchi, N., Miyawaki, T., 1998. Developmental changes and functional properties of human memory T cell subpopulations defined by CD60 expression. Cell. Immunol. 187, 117–123.

Wang, C.R., Hashimoto, K., Kubo, S., Yokochi, T., Kubo, M., Suzuki, M., Suzuki, K., Tada, T., Nakayama, T., 1995. T cell receptor-mediated signaling events in CD4+CD8+ thymocytes undergoing thymic selection: requirement of calcineurin activation for thymic positive selection but not negative selection. J. Exp. Med. 181, 927–941.

Wang, E., 1997. Regulation of apoptosis resistance and ontogeny of age-dependent diseases. Exp. Gerontol. 52, 471–484.

Watanabe, S., Itoh, T., Arai, K., 1997. Roles of JAK kinase in human GM-CSF receptor signals. Leukemia 11(Suppl. 3), 76–78.

Weiss, A., Littman, D.R., 1994. Signal transduction by lymphocyte antigen receptors. Cell 76, 263–274.

Weissman, A.M., 1994. The T cell antigen-receptor a multisubunit signalling complex. In: Samelson (Ed.). Chem. Immunol., Vol. 59. Basel, Karger, pp. 1–18.

Wenisch, C., Patruta, S., Daxbock, F., Krause, R., Horl, W., 2000. Effect of age on human neutrophil function. J. Leuk. Biol. 67, 40–45.

Weyand, C.M., Brandes, J.C., Schmidt, D., Fulbright, J.W., Goronzy, J.J., 1998. Functional properties of CD4+CD28− T cells in the aging immune system. Mech. Ageing Dev. 102, 131–147.

Whisler, R.L., Newhouse, Y.G., Donnerberg, R.L., Tobin, C.M., 1991a. Aging: Immunol. Infect. Dis. 3, 27–36.

Whisler, R.L., Bagenstose, S.E., Newhouse, Y.G., Carle, K.W., 1997. Expression and catalytic activites of protein tyrosine kinases (PTKs) Fyn and Lck in peripheral blood T cells from elderly humans stimulated through the T cell receptor (TCR)/CD3 complex. Mech. Ageing Dev. 98, 57–73.

Whisler, R.L., Karanfilov, C.I., Newhouse, Y.G., Fox, C.C., Lakshmanan, R.R., Liu, B.Q., 1998. Phosphorylation and coupling of zeta-chains to activated T-cell receptor (TCR)/CD3 complexes from peripheral blood T-cells of elderly humans. Mech. Ageing Dev. 105, 115–135.

Whisler, R.L., Liu, B.Q., Newhouse, Y.G., Walters, J.D., Breckenridge, M.B., Grant, I.S., 1991b. Signal transduction in human B cells during aging: alterations in stimulus-induced phosphorylations of tyrosine and serine/threonine substrates and in cytosolic calcium responsiveness. Lymph. Cytokine Res. 10, 463–473.

Whisler, R.L., Newhouse, Y.G., Bagenstose, S.E., 1996. Age-related reductions in the activation of mitogen-activated protein kinases p44mapk/ERK1 and p42mapk/ERK2 in human T cells stimulated via ligation of the T cell receptor complex. Cell. Immunol. 168, 201–210.

Whyte, M., Renshaw, S., Lawson, R., Bingle, C., 1999. Apoptosis and the regulation of neutrophil lifespan. Biochem. Soc. Trans. 27, 802–807.

Wick, G., Grubeck-Loebenstein, B., 1997. The aging immune system: primary and secondary alterations of immune reactivity in the elderly. Exp. Gerontol. 32, 401–413.

Wilkinson, S.E., Nixon, J.S., 1998. T-cell signal transduction and the role of protein kinase C. Cell. Mol. Life Sci. 54, 1122–1144.

Wu, W.T., Pahlavani, M.A., Cheung, H.T., Richardson, A., 1986. The effect of aging on the expression of interleukin 2 messenger ribonucleic acid. Cell. Immunol. 100, 224–231.

Wu, D., Meydani, M., Beharka, A.A., Serafini, M., Martin, K.R., Meydani, S.N., 2000. In vitro supplementation with different tocopherol homologues can affect the function of immune cells in old mice. Free Rad. Biol. Med. 28, 643–651.

Xavier, R., Brennan, T., Li, Q., McCormack, C., Seed, B., 1998. Membrane compartmentation is required for efficient T cell activation. Immunity 8, 723–732.

Yagisawa, M., Saeki, K., Okuma, E., Kitamura, T., Kitagawa, S., Hirai, H., Yazaki, Y., Takaku, F., Yuo, A., 1999. Signal transduction pathways in normal human monocytes stimulated by cytokines and mediators: comparative study with normal human neutrophils or transformed cells and the putative roles in functionality and cell biology. Exp. Hematol. 27, 1063–1076.

Yamamoto, K., Sasada, M., 1999. Neutrophil function. Japanese J. Clin. Med. 57 (Suppl.), 524–526.

Zatz, M.M., Goldstein, A.L., 1985. Thymosins, lymphokines, and the immunology of aging. Gerontology. 31, 263–277.

Zeng, Y.-X., Hirokawa, K., 1994. Age change in signal transduction of T cells. Aging: Immunol. Infect. Dis. 5, 147–158.

Zhang, W., Samelson, L.E., 2000. The role of membrane-associated adaptors in T cell receptor signalling. Semin. Immunol. 12, 35–41.

Zhang, W., Sloan-Lancester, J., Kitchen, J., Trible, R.P., Samelson, L.E., 1989. LAT: the ZAP-70 tyrosine kinase substrate that links T cell receptor to cellular activation. Cell 92, 83–92.

Zhang, W., Trible, R.P., Samelson, L.E., 1998. LAT palmitoylation: its essential role in membrane microdomain targeting and tyrosine phosphorylation during T cell activation. Immunity 9, 239–246.

Zs-Nagy, I., Kitani, K., Ohta, M., Zs-Nagy, V., Imahori, K., 1986. Age-dependent decrease of the lateral diffusion constant of proteins in the plasma membrane of hepatocytes as revealed by fluorescent recovery after photobleaching in tissue smears. Arch. Gerontol. Geriatr. 5, 131–146.

Zu, Y.L., Qi, J., Gilchrist, A., Fernandez, G.A., Vazquez-Abad, D., Kreutze, D.L., Huang, C.K., Sha'afi, R.I., 1998. p38 mitogen-activated protein kinase activation is required for human neutrophil function triggered by TNF-alpha or FMLP stimulation. J. Immunol. 160, 1982–1989.

Advances in
Cell Aging and
Gerontology

CD28 downregulation and expression of NK-associated receptors on T cells in aging and situations of chronic activation of the immune system

Javier G. Casado[1], Olga DelaRosa[1], Esther Peralbo[1],
Raquel Tarazona and Rafael Solana

Department of Immunology, Faculty of Medicine, "Reina Sofía" University Hospital,
Av. Menendez Pidal s/n, 14004, Córdoba, Spain
Correspondence address: E-mail: rsolana@uco.es (R.S.)

Contents

1. Introduction
2. CD28 expression in aging and situations of chronic activation of the immune system
3. NK-R expression in healthy elderly donors
4. Expression of NK-Rs on T lymphocytes in HIV infection
5. NK-Rs on cytotoxic T lymphocytes from melanoma patients
6. Expression of CD28 and NK-Rs on T lymphocytes aged in long-term cultures

Abbreviations

NK-R: NK associated receptors; CTLs: cytotoxic T lymphocytes; HIV: human immunodeficiency virus; IL-2: interleukin-2.

1. Introduction

Decreased expression of the co-stimulatory molecule CD28 may be considered as a biomarker of aging in human T lymphocytes. Several studies show that the proportion and density of CD28+ T cells decreases in elderly individuals, as well as in certain conditions associated with chronic activation of the immune system. It has

---

[1]J.G.C., O.D.R., and E.P. contributed equally to this work.

Advances in Cell Aging and Gerontology, vol. 13, 123–132

been demonstrated that at birth almost all human peripheral blood CD8+ T cells are CD28+, and that the proportion of CD8+CD28− T cells increase with age (Dennett et al., 2002; Solana and Mariani, 2000). We have previously proposed that the expression of NK-associated receptors (NK-Rs) on T cells is mainly restricted to the CD8+CD28− subset and, hence, may be considered a marker of cytotoxic effector T cells that are expanded in vivo after antigenic activation leading to extensive proliferation (Tarazona et al., 2000).

Some of the age-associated phenotypic and functional alterations observed in T cells isolated from aged donors have also been observed in vitro in T cells maintained in long-term cultures. These alterations include, though not exclusively, loss of proliferative capacity and decreased IL-2 production in response to mitogens (Adibzadeh et al., 1996; Grubeck-Loebenstein et al., 1994). Also described is a decrease in CD28 co-stimulatory molecule expression (Effros et al., 1994a), loss of telomerase activity (Pawelec, 2000a), telomere shortening (Monteiro et al., 1996), reduced response to stress (Effros et al., 1994b) or resistance to apoptosis (Spaulding et al., 1999). Therefore, it has been suggested that the maintenance of T lymphocytes in long-term cultures, in which these cells are stimulated repeatedly to divide, can offer an in vitro model for studying processes associated with in vivo aging in T lymphocytes (Pawelec et al., 2000b, 1999; Effros and Pawelec, 1997).

In this review, we analyse the expression of CD28 co-stimulatory molecules and NK-Rs in several situations that involve chronic activation of the immune system, including in vivo and in vitro studies.

2. CD28 expression in aging and situations of chronic activation of the immune system

CD28+ T cells are known to have longer telomeres than their CD28− counterparts (Speiser et al., 2001). It has also been observed that T cells maintained in long-term culture reach a state of replicative senescence after multiple rounds of cell division (Pawelec et al., 2000b). These cells have shortened telomeres and undetectable levels of telomerase, respond poorly to stimuli and lack CD28 expression; suggesting that the accumulation of CD28− T cells observed in vivo may reflect those cells that have undergone the process of immunosenescence after chronic stimulation (Effros, 1997; Spaulding et al., 1999; Yi-qun et al., 1997; Pitcher et al., 1999). Moreover, T cells lacking CD28 gene expression cannot undergo clonal expansion since CD28/B7 interactions are required for T cell proliferation (Effros and Pawelec, 1997).

Senescent T cells also show lower expression of CD95 (Fas), and are less prone to spontaneous apoptosis ex vivo, and more resistant to in vitro activation-induced cell death. Thus, the persistence of expanded clones in vivo in the CD8+CD28− T cell subset may be the consequence of antigen-driven differentiation from their CD8+CD28+ precursors and their relative resistance to apoptosis (Laux et al., 1998; Spaulding et al., 1999).

These results further support the previous data confirming that as seen in aging, in some clinical conditions of chronic activation of the immune system (e.g. Human

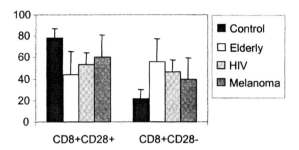

Fig. 1. *Comparison of CD28 expression on CD8+ cells in different clinical situations.* Peripheral blood lymphocytes were analysed in controls (*n* = 7), healthy elderly donors (*n* = 10), HIV infected (*n* = 5) and melanoma patients (*n* = 12) by immunofluorescence using CD8-PerCP in combination with CD28-PE or FITC. Viable cells were gated on forward scatter/side scatter in a contour plot.

Immunodeficiency Virus (HIV) infection or tumour patients) we observe downregulation of CD28 expression on T lymphocytes (Fig. 1). In addition, we and others have shown that in diseases that involve chronic antigenic stimulation, the expression of NK-Rs is mainly restricted to the CD8+CD28− T cell subpopulation (Scognamiglio et al., 1999; Mittrucker et al., 2001; Hislop et al., 2001; Crucian et al., 1995; Azuma et al., 1993; Borthwick et al., 1994; Effros et al., 1996; Fagnoni et al., 1996; Posnett et al., 1999; Speiser et al., 1999).

3. NK-R expression in healthy elderly donors

In humans, several markers associated with NK cells, termed NK-Rs, have also been found on a subset of CD8+ T cells. It has been shown that the expression of certain NK-Rs on CD3+ cells is mainly restricted to the CD8+CD28− T cell subset both in young and elderly donors. We have analysed the expression of CD56, CD16, CD161, NKB1, CD94, NKG2A, KIR2DL1, KIR2DL2, and ILT2 on T cells in healthy elderly patients and in young controls. Our results showed a significant increase in the expression of CD56, CD16, CD94 and ILT2 on CD3+ cells from elderly donors compared with young, but no significant differences were found in the expression of the other markers analysed. Moreover, the analysis of NK-R expression in relation to the CD28 phenotype shows a significant increase of CD56, CD16, CD161, ILT2, and KIR2DL1 in the CD28− T cell subset in elderly donors compared with young individuals (Fig. 2). In contrast, the expression of NK-Rs on the CD28+ T cell subset is less than 10%, in both young and elderly donors.

These results confirm and support our previous findings showing that the expression of several NK-Rs is increased in elderly individuals (Borrego et al., 1999). Our results also indicate the preferential expression of NK-Rs in the CD28− T cell subpopulation; a subset that is expanded in elderly individuals, and consequently, the increased expression of NK-Rs observed in the elderly may reflect the accumulation of CD3+CD28− cells (Fig. 1). CD3+CD28− cells expressing NK-Rs may represent cytotoxic effectors or effector/memory cells that have undergone

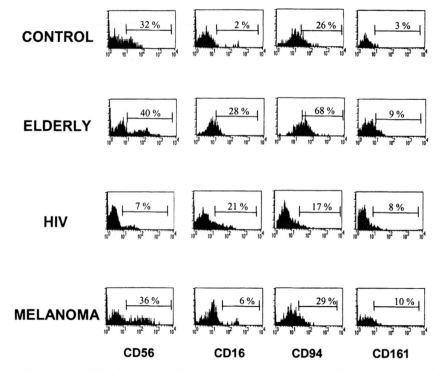

Fig. 2. *Expression of NK cell markers on T cells in a representative control, elderly donor, HIV infected individual and melanoma patient.* Peripheral blood lymphocytes were analysed by multiparametric immunofluorescence using CD8-PerCP in combination with CD28-PE or FITC and the following NK cell receptors: CD56, CD16, CD94 and CD161 conjugated with PE or FITC. Viable cells were gated on forward scatter/side scatter in a contour plot. The histograms show the staining for the NK receptors gated in the CD8 + CD28− subpopulation. The percentages refer to the total number of CD8 + CD28− T cells.

extensive proliferation and replicative senescence after in vivo chronic activation and that accumulate in vivo, probably as a consequence of being resistant to apoptosis.

The increased percentage of these cells with characteristics of replicative senescence in the elderly may be responsible, at least in part, for the decreased lymphoproliferative response and altered cytotoxicity observed in the elderly. We propose here that downregulation of CD28 and the increased expression of NK-Rs on T cells in the elderly may be considered as markers for senescent T cells that have been expanded in vivo either by chronic antigenic activation or by cytokine dependent proliferation.

4. Expression of NK-Rs on T lymphocytes in HIV infection

Cytotoxic T lymphocytes (CTLs) play a central role in the control of persistent HIV infection in humans. Thus, it has been shown that HIV-infected individuals have a high frequency of HIV-specific CD8 T lymphocytes in peripheral blood including both CTL precursors and effectors, and that the HIV-specific cytotoxic response has a protective role in the course of HIV infection (Pantaleo and Fauci, 1996; Bariou et al., 1997; Ogg et al., 1998; Wilson et al., 2000).

However, a defect in HIV-specific cytotoxicity has frequently been found in HIV-infected individuals. Several possibilities that include clonal exhaustion of responder cells (Pantaleo et al., 1997), decreased levels of expression of CD3 ζ chain in these cells (Trimble et al., 2000), increased expression of cytotoxicity inhibitory receptors (De Maria and Moretta, 2000) or blockade of differentiation to cytotoxic effectors (Appay et al., 2000), have been shown to be involved in the decreased cytotoxicity observed in HIV-infected individuals. Furthermore, a progressive increase in the expression of activation markers on CD8+ cells has also been extensively documented in HIV infection (Peritt et al., 1999; Sherman et al., 2002). A significant increase in the expression of CD94, CD244 and intracellular perforin has been found on CD8+ T cells in HIV infected individuals (Tarazona et al., 2002).

It has been shown that the expression of CD28 on T cells is significantly decreased in HIV-infected individuals. In particular, we found an increase in the percentage of CD8+CD28− T cells (Fig. 1). When the expression of NK-Rs on T cells from healthy controls were analysed in relation to their CD28 phenotype, our results showed a preferential expression of NK-Rs on CD8+CD28− T cells, rather than on CD8+CD28+ cells, with the only exception being CD161, which was expressed at similar levels in both subsets (Tarazona et al., 2002). The CD8+CD28− T cell subset includes cytotoxic effector and memory T cells.

The analysis of the percentage of CD8+CD28− T cells expressing NK-Rs showed that only the expression levels of CD56 were significantly decreased in HIV-infected individuals (Fig. 2), whereas the expression of other markers studied was comparable between HIV-infected individuals and healthy control values (Tarazona et al., 2002).

In conclusion, the phenotypic analysis of the expression of NK-Rs on T cells from HIV-infected individuals reveals that the increased expression of CD94, CD244 or perforin on T cells from HIV-infected individuals can be explained by the redistribution of CD8 T cell subsets, namely the expansion of CD8+CD28− cells, observed in these patients. The defective expression of CD56 on CD8+ T cells in HIV infected individuals may be involved in the decreased peripheral blood T cell cytotoxicity observed in HIV infection.

5. NK-Rs on cytotoxic T lymphocytes from melanoma patients

T lymphocytes play an important role in protection against tumours. CD8+ T cells can kill tumour cells through the recognition of tumour cell-derived peptides presented by cell surface Major Histocompatibility Complex (MHC) class I molecules (Robbins and Kawakami, 1996; Boon et al., 1994). The achievement of an efficient immune protection requires an adequate activation of tumour-specific T cells (Speiser et al., 1997). In this sense, cytotoxicity inhibitory receptors that regulate effector function have been described on CD8+ T cells from melanoma patients (Mingari et al., 1998; Speiser et al., 1999) including a subset of melanoma-specific T lymphocytes (Huard and Karlsson, 2000). Furthermore, we have analysed the expression of several NK-Rs (KIR2DL1, KIR2DL2, CD94/NKG2A, ILT-2 and NKB1) on CD8+ T cells from melanoma patients and healthy controls. These analyses showed a high variability in the expression of these molecules between patients as previously described on CD3+ cells from melanoma patients (Speiser et al., 1999). We also showed that CD8+ T cells from melanoma patients had a significant decrease in CD28 expression (Fig. 1). Further analysis of NK-R expression on CD8+ T cells, together with the co-stimulatory molecule CD28, showed a significant increase in the presence of KIR2DL1/KIR2DS1 on CD8+CD28− T cells in melanoma patients. Although no significant differences were found in the expression of other NK-Rs, we observed that some patients had an increased expression of certain NK-Rs (Fig. 2). Together, these results support the hypothesis that T cells from melanoma patients have undergone a process of chronic activation leading to a downregulation of CD28 molecules and the expression of NK-Rs, similar to that observed in elderly and HIV-infected individuals, that could impair tumour rejection. Nevertheless, the role of NK-R expression in melanoma progression needs to be further analysed.

6. Expression of CD28 and NK-Rs on T lymphocytes aged in long-term cultures

It has been demonstrated that the expression of the CD28 molecule decreases in aged CD8+ T cells in long-term culture after repeated stimulation (Effros, 1997). It has also been reported that CD28 expression decreases on CD4+ T cells undergoing multiple rounds of cell division in long-term culture (Pawelec et al., 2000b).

Using long-term cultures we have studied the expression of several NK-Rs on CD8+ T cells in relation to the CD28 phenotype. T cell cultures were stimulated weekly with PHA in the presence of exogenous Interleukin 2 (IL-2). Our results indicated that the percentage of CD8+CD28− T cells decreased dramatically in the first week of culture, probably due to apoptosis, and then we observed a gradual decrease in the percentage of CD8+CD28+ and an increase of CD8+CD28− T cells with the duration of the culture. In addition, we analysed the expression of CD161, CD56 and CD94 NK cell markers and the CD244 activation marker on CD8+ T cells together with their CD28 phenotype. The expression of CD56, CD94 and CD244 was higher in the CD8+CD28− T cell subset than in the CD8+CD28+ equivalent, whereas CD161 expression was similar on both subsets (Fig. 3). These results are comparable with those found in vivo in the elderly, with HIV-infection and melanoma patients (Figs. 1 and 2) (Tarazona et al., 2000).

In conclusion, several in vivo and in vitro studies indicate that T cells expressing NK-Rs are expanded in elderly individuals and in certain conditions that involve chronic activation of the immune system. Both in vivo and in vitro studies show that NK-R+ T cells are preferentially found within the CD8+CD28− T cell subset. Changes in the expression of different NK-Rs have been defined in the elderly, with HIV-infection and in melanoma patients, suggesting that CD8+CD28− T cells expressing NK-Rs have undergone a process of senescence after chronic stimulation with the cognate antigen. Because it has been shown that CD8+CD28− T cells have a lower proliferative capacity and shorter telomeres than their CD8+CD28+

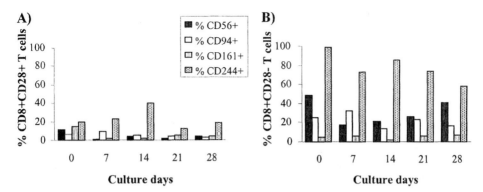

Fig. 3. *CD56, CD94 and CD244 are preferentially expressed in the CD8 + CD28 − T cell subset in long-term culture.* Peripheral blood T lymphocyte bulk cultures from healthy young donors were stimulated in vitro with PHA and exogenous IL-2. Expression of NK-Rs on CD8+ T cells in relation to their CD28 phenotype was analysed every week by multiparametric immunofluorescence using monoclonal antibodies against human surface molecules CD8, CD28 CD56, CD94, CD161, CD244 conjugated to an appropriate fluorochrome. Percentages refer to the total numbers of CD8 + CD28 + and CD8 + CD28 − T cells.

counterparts, our results demonstrating a higher expression of NK-Rs further support the hypothesis that the CD28− phenotype is associated with a state of replicative senescence.

Acknowledgements

This work was supported by grants FIS01/478 (to R.S.), and FIS00/0853 (to R.T.) from the Ministry of Health and PM98/0165 (to R.S.) from the Ministry of Education. R.T. is supported by a research contract from FIS (98/3082) until May 2002, and from May 2002 by a Ramon y Cajal Contract from MCYT. This work was carried out under the aegis of the European Union Thematic Network ImAginE (Immunology and Aging in Europe, QLK6-CT-1999-02031).

References

Adibzadeh, M., Mariani, E., Bartoloni, C., Beckman, I., Ligthart, G., Remarque, E., Shall, S., Solana, R., Taylor, G.M., Barnett, Y., Pawelec, G., 1996. Lifespans of T lymphocytes. Mech. Ageing Dev. 91, 145–154.

Appay, V., Nixon, D.F., Donahoe, S.M., Gillespie, G.M., Dong, T., King, A., Ogg, G.S., Spiegel, H.M., Conlon, C., Spina, C.A., Havlir, D.V., Richman, D.D., Waters, A., Easterbrook, P., McMichael, A.J., Rowland-Jones, S.L., 2000. HIV-specific CD8(+) T cells produce antiviral cytokines but are impaired in cytolytic function. J. Exp. Med. 192, 63–75.

Azuma, M., Phillips, J.H., Lanier, L.L., 1993. CD28− T lymphocytes. Antigenic and functional properties. J. Immunol. 150, 1147–1159.

Bariou, C., Genetet, N., Ruffault, A., Michelet, C., Cartier, F., Genetet, B., 1997. Longitudinal study of HIV-specific cytotoxic lymphocytes in HIV type 1-infected patients: relative balance between host immune response and the spread of HIV type 1 infection. AIDS Res. Hum. Retroviruses 13, 1301–1312.

Boon, T., Cerottini, J.C., Van den Eynde, B., van der Bruggen, P., Van Pel, A., 1994. Tumor antigens recognized by T lymphocytes. Annu. Rev. Immunol. 12, 337–365.

Borrego, F., Alonso, M.C., Galiani, M.D., Carracedo, J., Ramirez, R., Ostos, B., Pena, J., Solana, R., 1999. NK phenotypic markers and IL2 response in NK cells from elderly people. Exp. Gerontol. 34, 253–265.

Borthwick, N.J., Bofill, M., Gombert, W.M., Akbar, A.N., Medina, E., Sagawa, K., Lipman, M.C., Johnson, M.A., Janossy, G., 1994. Lymphocyte activation in HIV-1 infection. II. Functional defects of CD28− T cells. AIDS 8, 431–441.

Crucian, B., Dunne, P., Friedman, H., Ragsdale, R., Pross, S., Widen, R., 1995. Alterations in levels of CD28−/CD8+ suppressor cell precursor and CD45RO+/CD4+ memory T lymphocytes in the peripheral blood of multiple sclerosis patients. Clin. Diagn. Lab. Immunol. 2, 249–252.

De Maria, A., Moretta, L., 2000. HLA-class I-specific inhibitory receptors in HIV-1 infection. Hum. Immunol. 61, 74–81.

Dennett, N.S., Barcia, R.N., McLeod, J.D., 2002. Age associated decline in CD25 and CD28 expression correlate with an increased susceptibility to CD95 mediated apoptosis in T cells. Exp. Gerontol. 37, 271–283.

Effros, R.B., 1997. Loss of CD28 expression on T lymphocytes: a marker of replicative senescence. Dev. Comp. Immunol. 21, 471–478.

Effros, R.B., Allsopp, R., Chiu, C.P., Hausner, M.A., Hirji, K., Wang, L., Harley, C.B., Villeponteau, B., West, M.D., Giorgi, J.V., 1996. Shortened telomeres in the expanded CD28−CD8+ cell subset in HIV disease implicate replicative senescence in HIV pathogenesis. AIDS 10, F17–F22.

Effros, R.B., Boucher, N., Porter, V., Zhu, X., Spaulding, C., Walford, R.L., Kronenberg, M., Cohen, D., Schachter, F., 1994a. Decline in CD28+ T cells in centenarians and in long-term T cell cultures: a possible cause for both in vivo and in vitro immunosenescence. Exp. Gerontol. 29, 601–609.

Effros, R.B., Pawelec, G., 1997. Replicative senescence of T cells: does the Hayflick Limit lead to immune exhaustion? Immunol. Today 18, 450–454.

Effros, R.B., Zhu, X., Walford, R.L., 1994b. Stress response of senescent T lymphocytes: reduced hsp70 is independent of the proliferative block. J. Gerontol. 49, B65–B70.

Fagnoni, F.F., Vescovini, R., Mazzola, M., Bologna, G., Nigro, E., Lavagetto, G., Franceschi, C., Passeri, M., Sansoni, P., 1996. Expansion of cytotoxic CD8+ CD28− T cells in healthy ageing people, including centenarians. Immunology 88, 501–507.

Grubeck-Loebenstein, B., Lechner, H., Trieb, K., 1994. Long-term in vitro growth of human T cell clones: can postmitotic 'senescent' cell populations be defined? Int. Arch. Allergy Immunol. 104, 232–239.

Hislop, A.D., Gudgeon, N.H., Callan, M.F., Fazou, C., Hasegawa, H., Salmon, M., Rickinson, A.B., 2001. EBV-specific CD8+ T cell memory: relationships between epitope specificity, cell phenotype, and immediate effector function. J. Immunol. 167, 2019–2029.

Huard, B., Karlsson, L., 2000. A subpopulation of CD8+ T cells specific for melanocyte differentiation antigens expresses killer inhibitory receptors (KIR) in healthy donors: evidence for a role of KIR in the control of peripheral tolerance. Eur. J. Immunol. 30, 1665–1675.

Laux, I., Khoshnan, A., Tindell, C., Bae, D., Zhu, X., June, C.H., Effros, R.B., Nel, A., 1998. Response differences between human CD4(+) and CD8(+) T-cells during CD28 costimulation: implications for immune cell-based therapies and studies related to the expansion of double-positive T-cells during aging. Clin. Immunol. 96, 187–197.

Mingari, M.C., Moretta, A., Moretta, L., 1998. Regulation of KIR expression in human T cells: a safety mechanism that may impair protective T-cell responses. Immunol. Today 19, 153–157.

Mittrucker, H.W., Kursar, M., Kohler, A., Hurwitz, R., Kaufmann, S.H., 2001. Role of CD28 for the generation and expansion of antigen-specific CD8(+) T lymphocytes during infection with Listeria monocytogenes. J. Immunol. 167, 5620–5627.

Monteiro, J., Batliwalla, F., Ostrer, H., Gregersen, P.K., 1996. Shortened telomeres in clonally expanded CD28−CD8+ T cells imply a replicative history that is distinct from their CD28+CD8+ counterparts. J. Immunol. 156, 3587–3590.

Ogg, G.S., Jin, X., Bonhoeffer, S., Dunbar, P.R., Nowak, M.A., Monard, S., Segal, J.P., Cao, Y., Rowland-Jones, S.L., Cerundolo, V., Hurley, A., Markowitz, M., Ho, D.D., Nixon, D.F., McMichael, A.J., 1998. Quantitation of HIV-1-specific cytotoxic T lymphocytes and plasma load of viral RNA. Science 279, 2103–2106.

Pantaleo, G., Fauci, A.S., 1996. Immunopathogenesis of HIV infection. Annu. Rev. Microbiol. 50, 825–854.

Pantaleo, G., Soudeyns, H., Demarest, J.F., Vaccarezza, M., Graziosi, C., Paolucci, S., Daucher, M., Cohen, O.J., Denis, F., Biddison, W.E., Sekaly, R.P., Fauci, A.S., 1997. Evidence for rapid disappearance of initially expanded HIV-specific CD8+ T cell clones during primary HIV infection. Proc. Natl. Acad. Sci. USA 94, 9848–9853.

Pawelec, G., 2000a. Hypothesis: loss of telomerase inducibility and subsequent replicative senescence in cultured human T cells is a result of altered costimulation. Mech. Ageing Dev. 121, 181–185.

Pawelec, G., Adibzadeh, M., Rehbein, A., Hahnel, K., Wagner, W., Engel, A., 2000b. In vitro senescence models for human T lymphocytes. Vaccine 18, 1666–1674.

Pawelec, G., Wagner, W., Adibzadeh, M., Engel, A., 1999. T cell immunosenescence in vitro and in vivo. Exp. Gerontol. 34, 419–429.

Peritt, D., Sesok-Pizzini, D.A., Schretzenmair, R., Macgregor, R.R., Valiante, N.M., Tu, X., Trinchieri, G., Kamoun, M., 1999. C1.7 antigen expression on CD8+ T cells is activation dependent: increased proportion of C1.7+CD8+ T cells in HIV-1-infected patients with progressing disease. J. Immunol. 162, 7563–7568.

Pitcher, C.J., Quittner, C., Peterson, D.M., Connors, M., Koup, R.A., Maino, V.C., Picker, L.J., 1999. HIV-1-specific CD4+ T cells are detectable in most individuals with active HIV-1 infection, but decline with prolonged viral suppression. Nat. Med. 5, 518–525.

Posnett, D.N., Edinger, J.W., Manavalan, J.S., Irwin, C., Marodon, G., 1999. Differentiation of human CD8 T cells: implications for in vivo persistence of CD8+ CD28− cytotoxic effector clones. Int. Immunol. 11, 229–241.

Robbins, P.F., Kawakami, Y., 1996. Human tumor antigens recognized by T cells. Curr. Opin. Immunol. 8, 628–636.

Scognamiglio, P., Accapezzato, D., Casciaro, M.A., Cacciani, A., Artini, M., Bruno, G., Chircu, M.L., Sidney, J., Southwood, S., Abrignani, S., Sette, A., Barnaba, V., 1999. Presence of effector CD8+ T cells in hepatitis C virus-exposed healthy seronegative donors. J. Immunol. 162, 6681–6689.

Sherman, G.G., Scott, L.E., Galpin, J.S., Kuhn, L., Tiemessen, C.T., Simmank, K., Meddows-Taylor, S., Meyers, T.M., 2002. CD38 expression on CD8(+) T cells as a prognostic marker in vertically HIV-infected pediatric patients. Pediatr. Res. 51, 740–745.

Solana, R., Mariani, E., 2000. NK and NK/T cells in human senescence. Vaccine 18, 1613–1620.

Spaulding, C., Guo, W., Effros, R.B., 1999. Resistance to apoptosis in human CD8+ T cells that reach replicative senescence after multiple rounds of antigen-specific proliferation. Exp. Gerontol. 34, 633–644.

Speiser, D.E., Migliaccio, M., Pittet, M.J., Valmori, D., Lienard, D., Lejeune, F., Reichenbach, P., Guillaume, P., Luscher, I., Cerottini, J.C., Romero, P., 2001. Human CD8(+) T cells expressing HLA-DR and CD28 show telomerase activity and are distinct from cytolytic effector T cells. Eur. J. Immunol. 31, 459–466.

Speiser, D.E., Miranda, R., Zakarian, A., Bachmann, M.F., McKall-Faienza, K., Odermatt, B., Hanahan, D., Zinkernagel, R.M., Ohashi, P.S., 1997. Self antigens expressed by solid tumors Do not efficiently stimulate naive or activated T cells: implications for immunotherapy. J. Exp. Med. 186, 645–653.

Speiser, D.E., Valmori, D., Rimoldi, D., Pittet, M.J., Lienard, D., Cerundolo, V., MacDonald, H.R., Cerottini, J.C., Romero, P., 1999. CD28-negative cytolytic effector T cells frequently express NK receptors and are present at variable proportions in circulating lymphocytes from healthy donors and melanoma patients. Eur. J. Immunol. 29, 1990–1999.

Tarazona, R., DelaRosa, O., Alonso, C., Ostos, B., Espejo, J., Pena, J., Solana, R., 2000. Increased expression of NK cell markers on T lymphocytes in aging and chronic activation of the immune system reflects the accumulation of effector/senescent T cells. Mech. Ageing Dev. 121, 77–88.

Tarazona, R., DelaRosa, O., Casado, J.G., Torre-Cisneros, J., Villanueva, J.L., Galiani, M.D., Pena, J., Solana, R., 2002. NK-associated receptors on CD8 T cells from treatment-naive HIV-infected individuals: defective expression of CD56. AIDS 16, 197–200.

Trimble, L.A., Shankar, P., Patterson, M., Daily, J.P., Lieberman, J., 2000. Human immunodeficiency virus-specific circulating CD8 T lymphocytes have down-modulated CD3zeta and CD28, key signaling molecules for T-cell activation. J. Virol. 74, 7320–7330.

Wilson, J.D., Ogg, G.S., Allen, R.L., Davis, C., Shaunak, S., Downie, J., Dyer, W., Workman, C., Sullivan, S., McMichael, A.J., Rowland-Jones, S.L., 2000. Direct visualization of HIV-1-specific cytotoxic T lymphocytes during primary infection. AIDS 14, 225–233.

Yi-qun, Z., Lorre, K., de Boer, M., Ceuppens, J.L., 1997. B7-blocking agents, alone or in combination with cyclosporin A, induce antigen-specific anergy of human memory T cells. J. Immunol. 158, 4734–4740.

Advances in
Cell Aging and
Gerontology

Characterisation of NK cells in the elderly

Erminia Mariani and Andrea Facchini

*Dipartimento di Medicina Interna e Gastroenterologia, Laboratorio di Immunologia e Genetica, Istituto di
Ricerca Codivilla-Putti, IOR, Università di Bologna, Bologna, Italy
Correspondence address: Tel.: +39-051-6366803; fax: +39-051-6366807.
E-mail: marianie@alma.unibo.it (E.M.)*

Contents

1. Definition and identification of human NK cells

Human Natural Killer (NK) cells together with phagocytes, such as neutrophils and macrophages are central components of the innate immune system. Although some NK functions overlap with those delivered by classical T lymphocytes, NK cells represent a primary host immuno-surveillance system and have the ability to both spontaneously lyse tumour cells and virus-infected cells and provide an early source of immunoregulatory cytokines (Robertson and Ritz, 1990; Seaman, 2000; Cooper et al., 2001b).

NK cells had been originally described as "large granular lymphocytes" because their extensive cytoplasm contains granules that are used in cytolytic response. Almost all these cells express, in humans, CD16 (Fcγ RIIIA), a low affinity receptor for IgG Fc fragment (Perussia et al., 1984) and also CD56 (N-CAM), an isoform of the human neural-cell adhesion molecule, with unknown function on human NK cells (Lanier et al., 1989). Although early studies suggested that this molecule might mediate homotypic adhesion between NK and target cells, CD56 does not appear to play an important role in cytolytic activity, except for target cells that express CD56 themselves (Lanier et al., 1991; Takasaki et al., 2000). Together CD16 and CD56 are commonly used to identify human NK cells.

The relative expression of CD16 and CD56 is variable and about 90% of human NK cells have low levels (dim) of CD56 and high levels (bright) of CD16, whereas remaining NK cells are CD56 bright and CD16 dim or negative (Cooper et al., 2001b). This small subset also differs from the bulk of NK cells by expressing higher levels of CD94 (another NK associated molecule) (Campbell et al., 2001). In addition, NK cells are characterised as non-T cells, lacking the expression of T cell antigen receptor and CD3 complex at the cell surface, except for the zeta chain that is coupled to the CD16 receptor.

NK cells display two alternative mechanisms of cell killing: a direct spontaneous cytotoxicity (MHC-unrestricted) against a variety of susceptible neoplastic and virus infected target cells, and an indirect cytotoxicity which requires the activation and engagement of the CD16 Fcγ RIIIA by antibody-coated target cells (ADCC; Antibody-Dependent Cell-mediated Cytotoxicity) (Lanier et al., 1986; Cooper et al., 2001b).

The mechanisms by which NK cells recognize target cells and discriminate between normal and transformed and/or foreign cells are complex and not entirely understood. Recently three molecules (NKp46, NKp30 and NKp44) of a new group of receptors, different from the adhesion molecules previously established to play a role in the recognition process such as $\beta2$ integrins and CD2 (Melero et al., 1993; Tangye et al., 2000), have been identified on NK cells. These receptors, termed NCR (Natural Cytotoxicity Receptors), belong to the Ig-superfamily and are responsible for NK-cell triggering in the process of non-HLA-restricted natural cytotoxicity and a direct correlation has been shown between the surface density of NCR and the ability of NK cells to kill various tumours (Moretta et al., 2000). NKp46 and NKp30 are expressed exclusively by resting and activated NK cells, while NKp44 is progressively induced in vitro on IL-2 activated NK cells (Cantoni et al., 1999; Pende et al., 1999; Sivori et al., 1999; Biassoni et al., 2001). NCR are coupled to different transmembrane signalling molecules, including CD3zeta, FcεRIγ and KARAP/DAP-12 (Biassoni et al., 2001). Another NK surface receptor, termed 2B4 and structurally related to the CD2 subfamily, appears to induce natural cytotoxicity only when co-engaged with a triggering receptor, similar to NKp80 which preferentially activates NCR bright cells (Nakajima and Colonna, 2000).

Evidence obtained in the last 10 years supports the idea that the cytolytic activity of NK cells is the result of a balance between opposite signals delivered by the recognition of class I MHC molecules, the engagement of which can activate or inhibit killing, depending on the length of the cytoplasmic tails (Biassoni et al., 2001). Two major inhibitory/activatory receptor superfamilies have been described in humans: the killer immunoglobulin-like receptor (KIR), which primarily discriminates among the HLA-A, B and C allelic determinant of the classic class I molecule; and the C-type lectin which includes a receptor recognizing HLA-E.

KIR are characterised by two to four extra-cellular Ig-like domains. The anti-p58 (CD158a and 158b) monoclonal antibodies interact with HLA-Cw4 and Cw3, and related alleles, whereas p70 (CD159) appears to be the NK receptor for the HLA-Bw4 determinant (Biassoni et al., 1996, 2001). Close to KIR is another related family: the ILT (Ig-like transcripts) or LIR (leukocyte Ig-like receptors) family, whose products are found also on T and B cells, macrophages and dendritic cells (Colonna et al., 1997; Cosman et al., 1997).

Only a limited number of lectin-like receptors have been identified. One of them is formed by the association of CD94 with NKG2A (a member of the NKG2 family), which delivers an inhibitory signal to NK cells when CD94 recognizes HLA-E (Lopez-Botet et al., 1998). CD94/NKG2A inhibitory receptors are expressed on KIR-negative NK cells. The other members of this family appear to be activatory receptors. Moreover, the activating counterpart of the HLA-specific receptor p50 (CD158c), and CD94/NKG2C, may play a role in triggering the NK-mediated cytotoxicity against HLA-class I-positive target cells (Braud et al., 1998). Another NK triggering receptor, NKG2D, appears to play either a complementary or a synergistic role with NCR (Bauer et al., 1999; Biassoni et al., 2001).

Other surface molecules also expressed on human NK cells, or NK subpopulations, and useful for their definition in health and disease are: CD7, CD11a, CD11b, CD18, CD27, CD29, CD31, CD44, CD45RA, CD49d, CD49e, CD50, CD54, CD57, CD58, CD62L, CD69, CD81, p38 (C1.7.1), 2B4, PEN5 and p75/AIRM-1. (Robertson, 1997; Frey et al., 1998; André et al., 2000; Miller, 2001; Sedlmayr et al., 2002).

NK cells express receptors for a number of cytokines and chemokines, and therefore, they respond to a variety of stimuli. All NK cells express a functional heterodimeric interleukin-2 receptor (IL-2R $\beta\gamma$) with intermediate affinity, while CD56 bright NK cells appear to be the only lymphocyte with constitutive expression of the high-affinity heterotrimeric IL-2R (IL-2R $\alpha\beta\gamma$) (Caligiuri et al., 1990; Nagler et al., 1990; Voss et al., 1992). In addition, NK cells constitutively express several receptors for the monocyte-derived cytokines IL-1, IL-10, IL-12, IL-15 and IL-18, and probably receive some of their earliest activation signals from monocytes during the innate immune response (Carson et al., 1994, 1995a,b; Kunikata et al., 1998; Fehniger et al., 1999; Fitzgerald and O'Neill, 2000; Wang et al., 2000; Cooper et al., 2001a). The β and γ subunits of the IL-2R are shared by the receptor for IL-15, while the receptor for IL-18, a member of the IL-1 cytokine family, does not share components with IL-2 or IL-15 (Sims, 2002). Furthermore, NK cells express a number of chemokine receptors that not only favour NK cell migration, but also NK-target cell binding and NK recruitment. CD56-dim CD16-positive NK cells uniformly express CXCR1, CXCR2, CXCR3, CXCR4 and CX3CR1, but negligible levels of CC chemokine receptors (CCR2–6) on the cell surface (Robertson et al., 2000; Campbell et al., 2001; Inngjerdingen et al., 2001; Robertson, 2002; Mariani et al., 2002c). In contrast, CD56-bright, CD16-negative NK cells have been reported to express various CCR2, CCR5, CCR7, CXCR3 and CXCR4 (Campbell et al., 2001; Robertson, 2002). In addition, IL-2-activated NK cells express CCR4 and CCR8 (Inngjerdingen et al., 2000), while IL-2 or IL-12 stimulation was ineffective on CCR3, CCR5 and CXCR1 expression (Mariani et al., 2002b).

NK cells can also produce cytokines with antiviral functions and through the release of cytokines, NK cells have the capacity to regulate the activity of other cells, particularly cells of the immune system. The pattern of cytokines released by NK cells varies, in part, with the monokine induced during the early pro-inflammatory response to infection, and by the subset of NK cells present at the site of inflammation. Freshly isolated CD56-bright human NK cells are the primary source of NK-cell-derived immunoregulatory cytokines; including IFN-γ, TNF-α, TGF-β, IL-3, IL-5, IL-10, IL-13 and GM-CSF. These cells appear to have an intrinsic capacity for high-level production of cytokine compared with CD56-dim NK cells, both spontaneously and after activation in culture (Cooper et al., 2001b). These data suggest that the major function of CD56-bright NK cells during the innate immune response in vivo might be to provide macrophages and other antigen-presenting cells with early IFN-γ and other cytokines, promoting a positive cytokine feed back loop and efficient control of infection. NK cells can also produce chemokines, such as MIP-1α, Mip-1β, RANTES and IL-8 (Fehniger et al., 1999; Mariani et al., 2001, 2002b,c). These factors have important chemo-attractant and pro-inflammatory functions.

A variety of cytokines (Table 1) and chemokines (Table 2) have been demonstrated to activate particular NK cell responses and/or to regulate NK cell functions. Accumulating evidence indicates that the host, following identification of particular pathogen characteristics, responds with a different innate cytokine profile to elicit defence mechanisms most effective against the infective agent. Moreover, the chemokines are also likely to be important factors. In fact, these soluble factors promoting recruitment and accumulation of NK cells into critical sites, and then converting the innate NK-cell response into a T cell response through the activity of IFN-γ inducible chemokines, constitute an important link between the innate and adaptive immune response (Cooper et al., 2001a,b; Luster, 2002).

2. Phenotypic properties of human NK cells in the elderly

The study of the human immune response in healthy elderly donors has shown that immunosenescence not only affects the T cell response, but also different aspects of innate immunity, and probably the most extensive studies on age-associated alterations in the innate immune system have been directed toward NK cells (Pawelec et al., 1998; Solana and Mariani, 2000).

In 1987, a quantitative analysis of cells expressing the NK cell phenotype demonstrated for the first time that circulating NK cells increase in the peripheral blood of healthy individuals over 70 years old (Facchini et al., 1987), compared to young or middle-aged people. This observation was subsequently confirmed by other groups of investigators (Vitale et al., 1992; Sansoni et al., 1993; Borrego et al., 1999; Potestio et al., 1999; Fahey et al., 2000; Ginaldi et al., 2000; Miyaji et al., 2000; Ginaldi et al., 2001). The increase in NK cell numbers in the peripheral blood from elderly people is directly correlated with age (Ogata et al., 1997), and is associated with an increased percentage of CD3 cells co-expressing mainly CD57 and CD94 (Borrego et al., 1999), and also with a decreased number of CD3-positive T cells

Table 1
Cytokine regulation of Natural Killer cell functions

Cytokine	Effects on NK cells	References
IFN-α/-β	stimulates proliferation; increases cytolytic activity; induces cytolysis of NK resistant targets; induces cell trafficking; inhibits IL-12 production; elicits expression of IL-15 mRNA	Trinchieri et al., 1978; Biron et al., 1984; Kasaian and Biron, 1990; Salazar-Mather et al., 1996; Biron, 1997; Cousens et al., 1997; Zhang et al., 1998; Biron et al., 1999
IFN-γ	increases granule and conjugate formation; increases degranulation and cytolytic activity; increases chemotaxis and chemokinesis; increases interactions with endothelial cells	Taub, 1999
IL-1α/-1β	synergises with IL-12 for IFN-γ production	Hunter et al., 1997; Biron et al., 1999; Cooper et al., 2001a
IL-2	stimulates proliferation; increases granule and conjugate formation; increases degranulation and cytolytic activity; induces cytolysis of NK resistant targets; increases chemotaxis and chemokinesis; induces cytokine secretion; increases MIP-1α, RANTES, IL-8 secretion; increases interactions with endothelial cells	Baume et al., 1992; Rabinovich et al., 1993; Taub, 1999; Mariani et al., 2000b, 2001, 2002b,c
IL-4	down regulates LAK activity induced by IL-15; modulate cell adhesion and chemotaxis; stimulates production of IL-5 and IL-13	Salvucci et al., 1996; Taub 1999
IL-10	induces cytolysis of NK resistant targets; inhibits IL-12 production; synergises with IL-2 for proliferation and cytolytic activity; synergises with IL-2 for IFN-γ, TNF-α, GM-CSF production	D'Andrea et al., 1993; Hunter et al., 1994; Carson et al., 1995a; Biron et al., 1999
IL-12	increases granule and conjugate formation; increases degranulation and cytolytic activity; critical for the response during certain infections; increases interactions with endothelial cells; stimulates IFN-γ production; increases MIP-1α, RANTES, IL-8 secretion; synergises with IL-15 for IL-10 and MIP-1α production; synergises with IL-18 for IFN-γ production	Chan et al., 1991; Robertson et al., 1992; Rabinovich et al., 1993; Tripp et al., 1993; Allavena et al., 1994; Hunter et al., 1994; Carson et al., 1995; Bluman et al., 1996; Orange and Biron 1996a,b; Biron, 1997; Monteiro et al., 1998; Biron et al., 1999; Taub, 1999; Aste-Amezaga et al., 2002; Mariani et al., 2001, 2002b,c
IL-15	promotes cell growth and maturations; synergises with IL-12 for IFN-γ, TNF-α, GM-CSF, IL-10, MIP-1α production; synergises with IL-18 for GM-CSF production	Carson et al., 1994; Puzanov et al., 1996; Warren et al., 1996; Biron et al., 1999; Fehniger et al., 1999; Taub, 1999
IL-18	induces cytolytic activity in culture and following in vivo administration; stimulates IFN-γ production; induces IL-13 and/or IL-4 production; synergises with IL-12 for IFN-γ production and cytolytic activity; synergises with IL-15 for GM-CSF production	Nakamura et al., 1993; Biron et al., 1999; Fehninger, 1999; Akira, 2000; Nakanishi et al., 2001
IL-21	induces proliferation and maturation of NK from bone marrow	Parrish-Novak et al., 2000
TGF-β	inhibits proliferation and cytolytic activity; increases chemotaxis and chemokinesis; inhibits FN-γ and IL-12 production	Rook et al., 1986; Maghazachi and Al-Aoukaty, 1993; Bellone et al., 1995; Biron et al., 1999
TNF-α	increases granule and conjugate formation; increases degranulation and cytolytic activity; increases chemotaxis and increases chemokinesis; increases interactions with endothelial cells; synergises with IL-12 for IFN-γ production	Tripp et al., 1993; Hunter et al., 1994; Pilaro et al., 1994; Biron et al., 1999; Taub, 1999

Table 2
Chemokine regulation of Natural Killer cell functions

Chemokine	Effects on NK cells	References
MIP-1α	increases proliferation; increases adhesion, degranulation and cytolytic activity; increases chemotaxis	Maghazachi et al., 1994; Taub et al., 1995a,b; Loetscher et al., 1996; Maghazachi et al., 1996; Taub et al., 1996; Biron et al., 1999; Taub 1999; Robertson, 2002
MIP-1β	increases proliferation; increases adhesion, degranulation, and cytolytic activity; increases chemotaxis	Taub et al., 1995a,b; Loetscher et al., 1996; Maghazachi et al., 1996; Taub et al., 1996; Biron et al., 1999; Taub, 1999; Campbell et al., 2001; Robertson 2002
RANTES	increases proliferation; increases adhesion, degranulation and cytolytic activity; increases chemotaxis	Maghazachi et al., 1994; Taub et al., 1995a,b; Loetscher et al., 1996; Maghazachi et al., 1996; Taub et al., 1996; Biron et al., 1999; Taub, 1999; Campbell et al., 2001; Robertson, 2002
MCP-1	increases proliferation; increases adhesion, degranulation and cytolytic activity; increases chemotaxis	Maghazachi et al., 1994; Taub et al., 1995a,b; Loetscher et al., 1996; Taub et al., 1996; Biron et al., 1999; Taub, 1999; Inngjerdingen et al., 2001; Robertson, 2002
MCP-2/-3	increases degranulation and cytolytic activity; increases chemotaxis	Maghazachi et al., 1994; Taub et al., 1995a,b; Loetscher et al., 1996; Taub et al., 1996; Taub, 1999; Biron et al., 1999
Ltn	increases degranulation and cytolytic activity; decreases adhesion; increases chemotaxis	Hedrik et al., 1997; Maghazachi et al., 1997; Taub, 1999
Exodus-1	increases chemotaxis	Kim et al., 1999
Exodus-2	stimulates migration of policlonally activated NK; co-stimulates proliferation induced by IL-2	Robertson et al., 2000; Robertson, 2002
Exodus-3	increases chemotaxis; stimulates migration of policlonally activated NK; co-stimulates proliferation induced by IL-1	Kim et al., 1999; Robertson et al., 2000; Robertson, 2002
IL-8	decreases adhesion, degranulation, and cytolytic activity; increases chemotaxis	Taub, 1999; Campbell et al., 2001
GRO-α/-β	decreases degranulation and cytolytic activity; increases chemotaxis	Taub, 1999; Inngjerdingen et al., 2001
IP-10	increases proliferation; increases adhesion, degranulation and cytolytic activity; increases chemotaxis	Taub et al., 1995a,b; Maghazachi et al., 1997; Taub, 1999; Campbell et al., 2001; Robertson, 2002
PF4	decreases proliferation; decrease adhesion, degranulation and cytolytic activity; increases chemotaxis	Taub, 1999
SDF-1α	increases cytolytic activity; increases chemotaxis	Taub, 1999; Inngjerdingen et al., 2001; Robertson, 2002
SDF-1β	increases chemotaxis	Taub, 1999
Fractalkine	increases chemotaxis and cytolytic activity	Yoneda et al., 2000; Campbell et al., 2001; Robertson, 2002

(Mariani et al., 1999; Miyaji et al., 2000). The analysis of the absolute number of CD56-positive cells confirms that also this subset can be increased in healthy elderly donors (Vitale et al., 1992; Borrego et al., 1999). Furthermore, the increased percentage of NK cells observed in the elderly, more attributable to the expansion of the CD56-dim subpopulation (the mature NK subset) (Baume et al., 1992) than to the decrease of a CD56-bright one, suggests a phenotypic and functional shift in the maturity status of NK cells with ageing (Krishnaraj and Svanborg, 1992; Krishnaraj, 1997).

The analysis of healthy elderly NK subpopulations defined by their relative expression of CD56, does not reveal phenotypic changes in the CD56-bright subset, whereas CD56-dim NK cells expressed higher levels of HLA-DR and CD95 (Apo1/Fas), and lower levels of CD69, than CD56-dim from young donors (Borrego et al., 1999).

2.1. NK receptors

Although very little is known about the effect of senescence on the expression of NK receptors other than CD56 and CD16, some data demonstrated that NK cells from the elderly present a pattern of expression and density of some KIR similar to the young population, suggesting a preserved NK repertoire for the recognition of allogeneic cells (Mariani et al., 1994). In particular, p58 molecules as defined by CD158 a and b monoclonal antibodies are widely distributed on NK cells of both young and old subjects, and their membrane antigenic density is also preserved (Figure 1) (Mariani et al., 1997). Furthermore, preliminary data seem to indicate a similarly preserved distribution and density on NK cells from old subjects for CD158c, p70 (CD159), CD160, CD94 and NKG2A and LIR1 receptors (Figure 1) (Mariani et al., 2000a and personal unpublished observations).

Neoplastic and virally-infected cells may change in their expression of class I molecules, modifying the susceptibility of these cells to appropriate recognition.

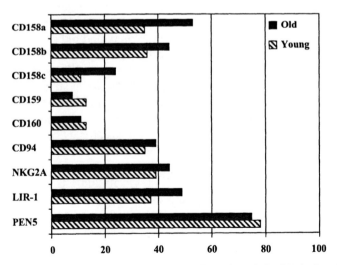

Fig. 1 Distribution of NK receptors on CD16 positive cells in the elderly (Mariani et al., 1994, 1997, 2000a,c).

Therefore, preserved distribution of receptors on NK cells able to recognise hetero-geneous alleles is particularly important in the elderly and relevant for the develop-ment of an optimal signal after the interaction with specific HLA alleles. Although the best characterized activating receptor is CD16, other molecules (such as NK p46, NKRP1, NKp44 and CD94/NKG2C) can also be involved in the activation of NK cytolysis, but the expression of these receptors has not been examined thus far, in the elderly. PEN5 is an NK cell-restricted glycoprotein expressed on the functionally differentiated CD56-dim NK cells (Vivier et al., 1993). Its expression on the surface of NK cells is coordinated with the disappearance of L-selectin and the up-regulation of KIR (André et al., 2000). PEN5 is widely and similarly expressed on NK cells from young and old subjects (Figure 1) with no modifications of antigenic density and is not detectable on T cells from either young or old donors (Mariani et al., 2000c). The selectivity of PEN 5 expression on NK cells suggests that specific NK cell functions depend on the acquisition of this developmentally regulated L-selectin ligand. Therefore, the control of NK cell interactions via PEN5 may have important implications for the induction and maintenance of immunity to tumours and infections (André et al., 2000). Thus, the maintenance of this epitope on NK cells of elderly people suggests a preserved ability to interact with L-selectin-positive cells, of innate and adoptive immunity, in amplifying the immune response to inflammation.

2.2. Adhesion molecules

A major role in the binding between effector cells and their targets is played by several families of adhesion receptors (Bakker et al., 2000). Among these, 2B4 is a cell surface glycoprotein involved in the regulation of NK and T lymphocyte func-tion, through the modulation of other receptor–counter receptor interaction. (Bakker et al., 2000; Nakajima and Colonna, 2000). The 2B4 molecule is expressed on the majority of NK cells, similarly in both young and old subjects (Figure 2), but the low density of this molecule on the membrane of NK cells obtained from the old subjects suggests a smaller number of specific receptors (Mariani et al., 2002a). The low antigen density of 2B4 observed on NK cells from old subjects may be involved in the inefficient transduction of the activating signal and in the impaired recognition of virus-infected cells and contribute to the increased incidence of viral infections observed during ageing. In fact, 2B4 dysfunction has already been associated with an inability to control EBV infections in a severe form of immunodeficiency (Moretta et al., 2001).

The expression of CD11a, CD11b, CD18, CD50 and CD58 adhesion molecules (Mariani et al., 1992, 1998b; De Martinis et al., 2000) as well as CD49d (participat-ing in endothelium rolling) (De Martinis et al., 2000) and CD29 expression (Mariani et al., 1992) is unchanged on the majority of NK cells in the elderly, while CD2 and CD54 molecules, usually expressed on a low percentage of NK cells, are represented more in the old group than in the young (Figure 2), but with a similar surface density (Mariani et al., 1992, 1998b), consistent with the maintained capability of single NK cells to bind tumour target cells (Facchini et al., 1987; Ligthart et al., 1989).

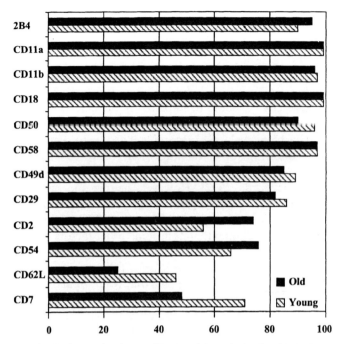

Fig. 2 Distribution of adhesion molecules on CD16 positive cells in the elderly (Mariani et al., 1992, 1998b; De Martinis et al., 2000; Ginaldi et al., 2000; Mariani et al., 2002a).

In contrast, a decreased proportion of NK cells bearing CD62L (Figure 2) (preferentially expressed on the less mature CD56-bright subset) was found, that could result in impaired NK cell homing (De Martinis et al., 2000). Similarly, the CD7 homing receptor was down-regulated (Figure 2) (Ginaldi et al., 2000).

3. Functional responses of NK cells in the elderly

3.1. Cytolytic activity

The "spontaneous" lytic activity of NK cells has been fully analysed in elderly people and original reports were contradictory, indicating that the overall spontaneous NK cytolytic activity developed by the bulk of peripheral blood lymphocytes could be either unaffected (Facchini et al., 1987), decreased (Levy et al., 1991; Sansoni et al., 1993), or increased in the elderly (Krishnaraj and Blandford, 1987). Discrepancies in the data from different groups may be partially explained by the use of peripheral blood samples in which the relative distribution of the effector lymphocytes, and of the other mononuclear cells, can differently modulate NK activity via cytokines and chemokines. In addition, the use of poorly selected, apparently healthy, individuals or of blood donors, without gathering sufficient information on disease history or drug use, may have influenced the results.

The introduction of strict criteria for the selection of only very healthy elderly subjects, for evaluating the effects of senescence on spontaneous cytolytic activity, demonstrated that this activity is not significantly affected by ageing in the peripheral blood lymphocytes (Sansoni et al., 1993; Kutza and Murasako, 1994, 1996). However, the results indicating a lack of correlation between the increase in number of peripheral NK cells and the preserved lytic activity of these cells, suggested that NK-cell activity should be impaired in the elderly when considered on a per-cell basis, as was demonstrated using NK cells purified by sorting, or cloned by limiting dilution (Facchini et al., 1987; Mariani et al., 1990). This was consistent with the finding that, following target binding, each CD16-positive cell from elderly donors exerts only about half the cytolytic activity displayed by NK cells from young donors (Vitale et al., 1992). However, neither the binding of effector to target cells (Facchini et al., 1987), nor the effector cells' perforin content, perforin distribution or perforin utilization was significantly different between young and old groups (Mariani et al., 1996), and therefore some other factors must account for the decreased lytic activity found in NK cells from aged people.

The triggering of some receptors on NK cells, following recognition of a susceptible target cell, activates intracellular signals that result in many different responses (Trinchieri, 1989). However, the ability of NK cells to transform a receptor-mediated signal into an effector response, which is linked to the ability to generate second messengers after K562 stimulation, does show a pronounced decrease with ageing. The major biochemical defect appears to be an age-related delay in PIP_2 hydrolysis and decrease in the levels of IP_3 formation in old NK cells after K562 stimulation (spontaneous cytotoxicity) (Mariani et al., 1998a).

Because the density of membrane receptors involved in adhesion among cells was essentially preserved on NK cells with increasing donor age (Mariani et al., 1998b), intracellular signal transduction may be impaired at a step distal to the binding of the receptor. Ageing seems to affect phospholipase C (PLC) activity, which is up-regulated together with adenylate cyclase, by signals transduced from surface NK receptors through G-protein. Modified levels of G-protein expression are probably involved in the intracellular cAMP increases that can functionally inhibit spontaneous cytolytic activity of NK cells, through a selective activation of protein kinase (Borzì et al., 1992; Fülöp et al., 1993). Since another activity of cAMP also appears to be the regulation of spreading of cells and cytoskeletal rearrangement, intracellular increases in cAMP found in aged subjects may also differently regulate the adhesion to endothelial and tissue localization of NK cells (Poggi et al., 1996).

Antibody-dependent cellular cytotoxicity (ADCC) is mediated by NK cells when triggered via the FcγRIII (CD16). ADCC has also been studied in the peripheral blood lymphocytes of aged subjects. The results from different studies have shown that ADCC was comparable between young and ageing subjects (Edwards and Avis, 1979; Fernandes and Gupta, 1981). Furthermore, the PI-3-kinase pathway triggered by CD16 cross-linking is not significantly affected in NK cells from elderly people, indicating that the transduction pathways involved in spontaneous, or CD16-dependent NK cytotoxicity, are differentially affected by ageing (Mariani et al., 1998a).

3.2. Response to cytokines

Cytokine activation of NK cells results in enhanced cytolytic activity against both NK-susceptible and NK-resistant target lines, cell proliferation, synthesis and release of cytokines and chemokines. The data accumulated so far indicate that anti-tumour MHC-unrestricted cytolytic activity of NK cells, induced by cytokines, is preserved in the healthy elderly; in fact, incubation with IFNα, β, γ, IL2 or IL12 (over a wide range of doses and incubation times) (Kutza and Murasako, 1996) up-regulated lytic activity of peripheral blood lymphocytes, or isolated NK subsets, from the elderly to the corresponding levels of the stimulated cells derived from young controls (Krishnaraj and Svanborg, 1992; Kutza and Murasako, 1994, 1996; Krishnaraj, 1997). Thus, NK treatment with these cytokines enhances cytolytic activity against the NK-sensitive target K562 in both young and elderly samples. However, the activity of lymphokine-activated killer (LAK) cells in the elderly, as defined by the ability of peripheral blood mononuclear cell samples to lyse the normally NK-resistant tumour cell lines, is significantly decreased against Daudi targets, in the elderly, when compared to young people (Kutza and Murasako, 1994, 1996). Comparable age-associated differences are observed in LAK activity after induction with IL-2, IL-12 and IFN-α at varying doses and incubation times (Kutza and Murasako, 1996). Thus, in contrast to NK activity, LAK activity induced by IL-2, IL-12 and IFN-α appears to be impaired in the elderly. Therefore, these results suggest that NK cells have an age-associated defect in their response to cytokines with a subsequent deficit in their capacity to develop LAK activity against NK-resistant cell lines. In this respect, the type of cells used as targets may become particularly important because NK cells can use different killing mechanisms against different targets. The differential susceptibility of target cells to perforin, TNF, or Fas-mediated cytolysis may result in differences in NK cell killing after cytokine stimulation. The effects of other cytokines that have been shown to be very important in NK cell differentiation and function, (i.e. IL-15, IL-18) or chemokines has not been studied in the elderly.

Analysis of proliferative response of purified NK cells from healthy elderly donors to IL-2 shows a decrease when compared to proliferation of NK cells from young individuals. However, a broad variability in the proliferation, from nearly normal to negligible, was observed (Borrego et al., 1999). Proliferative response paralleled the modulation of the CD69 antigen following IL-2 activation, indicating that the decreased proliferation is due to a general defect of the NK cells to IL-2 activation.

Immunological changes associated to senescence are particularly dramatic when the profile of cytokine production by different cell populations is considered. Thus whereas IFN-α and IFN-γ production is decreased in the elderly compared to the young, the release of IL-4 and IL-10 is increased. Furthermore, production of pro-inflammatory cytokines such as IL-1, IL-6, IL-8 and TNF-α is increased in the elderly. A major function of NK cells is their capacity to synthesize and release a broad range of cytokines such as IFN-γ or TNF-α that will participate in the initiation of the Th1-dependent adaptive immune response. The IL-2 inducible secretion of IFN-γ by NK cells shows an early significant decrease in the elderly when compared to the production in young individuals, which can be overcome by prolonging the incubation time of NK cells with IL-2 (Krishnaraj and Bhooma, 1996; Rink et al., 1998). A partial loss of high affinity IL-2 α,β and γ receptors on NK cells may be involved in this decreased response, probably as a consequence of the decline in the level of IL-2 during ageing (Krishnaraj, 1997). IL-2 treatment also induces the release of TNF-α to a similar extent to that observed in the young (Borrego et al., 1999) and there remains an adequate synthesis of perforin by NK cells in culture following PHA stimulation (Mariani et al., 1996) even in those individuals with a decreased spontaneous perforin expression (Rukavina et al., 1998). In addition, IL-2 and IL-12 stimulation significantly up-regulated the production of MIP1-α, RANTES and IL-8 from NK cells of old subjects. However, the total amount is lower than that obtained under the same conditions from young NK cells (Mariani et al., 2001, 2002b,c) and is not or is only marginally modified by extension of the incubation time. The ability of IL-12 and particularly IL-2 to induce high levels of chemokines from NK cells is of interest in view of the potential for NK-derived chemokines to interfere with virus replication before the onset of antigen-specific immune responses and the decreased production of MIP1-α, RANTES and IL-8 chemokines during ageing may be involved in the lower response of old subjects to infections.

4. Role of NK cells against infections in the elderly

The role of NK cell responses and functions against viral infections, originally evaluated in murine models, has also been extensively studied in humans. In general, innate NK responses to the primary viral infection peak during the first three days of infection (Welsh, 1981; Welsh and Vargas-Cortes, 1992) and are gradually replaced by adaptive specific responses that B and T cells developed in more than a week. When T cell responses do not occur, such as in SCID and athymic nude mice, the increased NK-cell response is maintained for a long time to defend the host (Welsh et al., 1991; Welsh and Vargas-Cortes, 1992; Su et al., 1993). Increased NK cytolytic activity and IFN-γ production are demonstrated during different viral infections and the role played by NK cells is due to both these functions. The induction of NK-cell cytotoxicity is common during many viral infections; however the relative contribution of cytotoxicity to NK mediated defence is not yet clear and NK cell cytolytic function might be uniquely required in particular compartments. In contrast, the importance of IFN-γ produced by NK cells in antiviral defence has been definitively established. These infections, eliciting the production of IFN-γ by NK cells, also induce endogenous synthesis of biologically active IL-12. Thus, the early production of IFN-γ is a consequence of virus-induced IL-12 and contributes to inhibition of virus replication (Biron et al., 1999; Biron and Brossay, 2001). In addition, the production of IFN-γ (which is probably the most important cytokine with anti-infective activity produced by NK cells) has a central role by promoting downstream adaptive response against infections.

Besides the production of these cytokines, NK cells can make certain chemokines. These factors have important chemoattractant and pro-inflammatory functions; moreover, under certain circumstances (such as HIV infection) chemokines can also mediate direct antiviral effects probably through MIP-1α, contributing to control of the infection. More definitive proof that MIP-1α may be a direct chemo-attractant for NK cells is derived from CMV infection of MIP-1α KO mice that do not present an NK cell inflammatory infiltrate in the liver (Salazar-Mather et al., 1998; Biron et al., 1999). In humans, the contribution of NK cells to early antiviral immune defence mechanisms is supported by clinical observations that rare patients with low or absent NK cell cytolytic activity, but without derangements of other immunological functions, suffer from particular susceptibility to infective diseases, primarily herpes viruses (Fleisher et al., 1982; Biron et al., 1989). In addition to their antiviral functions, NK cells may play a role in eliminating various other pathogens such as bacteria and fungi (Garcia et al., 1989; Hidore and Murphy, 1989).

It has been reported that the number and the function of T cells are reduced in healthy elderly subjects compared to young people (Sansoni et al., 1993; Miller, 1996 Ogata et al., 1997). Considering the inverse correlation between T and NK cell number (Mariani et al., 1999) and the functional link between these two cell populations, the role of NK cells against infections may be more important in the elderly than in young people. Previous studies (Levy et al., 1991; Ogata et al., 1997) suggested that low NK cell activity might be related to increased susceptibility to common infectious diseases such as respiratory infections, in immunologically normal elderly people. In addition, a low NK activity was associated with the development of severe infections, correlated significantly with a history of recurrent infections and virus-associated diseases (common cold, pneumonia and cystitis, being the most prevalent) and with death due to pneumonia in the follow-up period (Ogata et al., 2001).

Furthermore, a follow-up study (Remarque and Pawelec, 1998) demonstrated that elderly people (over 85-year-old) with low numbers of NK cells have a threefold greater mortality risk in the first 2 years following laboratory determination than those with high NK cell numbers. Further evidence supporting the importance of NK cells in healthy ageing comes from studies in centenarians, that can be considered the best example of successful ageing, as they have escaped the major age-associated diseases: these subjects, indeed, show a very well-preserved NK cell cytolytic activity (Sansoni et al., 1993; Franceschi et al., 1995).

5. Effect of hormones and micronutrients on NK cells in the elderly

Different hormones affect the homeostasis of the immune system, either directly by interacting with specific receptors on immunocompetent cells and the corresponding signalling pathways, or indirectly by regulating cellular respiration and many aspects of carbohydrate, fat and amino acid metabolism (Manolagas et al., 1985; Kiess and Belohradsky, 1986; Mariotti and Pinchera, 1990; Angeli et al., 1994; Paavonen, 1994). The decrease of both intake and synthesis of vitamin D, the decline of growth hormone secretion and insulin-like growth factor levels and the impairment of thyroid, gonadal and adrenal function are well established with advancing age (Morley, 1995). These hormonal systems are important in maintaining the integrity of muscle and bone mass, and together with malnutrition and deficit of vitamins and micronutrients, are a critical determinant of immunocompetence in the elderly (Lesourd and Meaume, 1994; Chandra, 1995; Straub et al., 2001). A positive association among NK cell number and function, and vitamin D metabolism, was established in a group of people over 90 years old: in fact those individuals with the highest NK cell numbers and the best preserved cytolytic activity, also showed a high serum vitamin D level, were well nourished, had a balanced basal metabolism and had better ability to develop an autonomous life style (Mariani et al., 1999). A poor vitamin D status is frequently found in elderly people (Ravaglia et al., 1994). The significant association between NK cell number and activity, and vitamin D levels is consistent with the observations that vitamin D deficiency is associated with a reduction of non-specific immune responses in vitro, and that supplementation of $1,25(OH_2)D3$ to elderly subjects in vivo significantly increases the circulating levels of IFN-α, one of the cytokines that modulates NK activity (Manolagas et al., 1989). Furthermore, since vitamin D acts as regulator of cell growth and differentiation (Hansen et al., 2001) and, together with retinoids and IL-12, is a potent inhibitor of angiogenesis induced by tumour cells (Majewski et al., 1996), it is possible that preserved vitamin D levels and NK number and activity, protects these subjects from the development of tumours until extreme old age.

Many data have been accumulated on the frequent immune alterations in physiological and pathological fluctuations in thyroid hormones or the existence of receptors for thyroid hormones on lymphocytes (Pekonen and Weintraub, 1978). Thyrotropin-stimulating hormone restores the reduced IL-2-activated NK activity of lymphocytes from old subjects. Low triiodothyronine serum levels were correlated with reduced NK cell number (Mariani et al., 1999; Kmiec et al., 2001) and were associated with a reduced mitogen response and altered T cell subset distribution (frequently observed in old subjects). In addition, thyroxin decreased NK cell activity (Fabris et al., 1995) and serum levels of free-thyroxin correlated with NK cell number (Mariani et al., 1999).

Glucocorticoids are the most important inhibitors of immunological response (Holbrook et al., 1983). The effect of cortisol (the most representative endogenous glucocorticoid in humans) (Angeli et al., 1994) is apparent at physiological hormone concentrations and downregulates NK activity through the control of granzyme A synthesis. However, it was recently suggested that the glucocorticoid inhibitory effect on immune responses is reduced in the elderly, consistent with the development of peripheral glucocorticoid resistance with advancing age, with the lack of effect of plasma cortisol on NK cell activity in vivo (Bodner et al., 1998) and with the lack of correlation between serum level and NK cell number and activity in vitro, in old subjects (Mariani et al., 1999).

Both protein and specific micronutrient deficits are very common in the elderly (Lesourd, 1995), leading to or paralleling a metabolic insufficiency that involves lymphocytes and mononuclear cells and impairs specific host defences, consistent with data showing that healthy subjects over 65 years old improved NK cell number and activity after one year supplementation with physiological amounts of micro-nutrients (Chandra, 1992). A significant positive correlation of NK cell cytolytic activity with vitamin E and ubiquinone-10 concentration were found (Ravaglia et al., 2000). Plasma concentration of vitamin E is affected by ageing (Hallfrisch et al., 1994; Campbell et al., 1989). Several animal and human studies showed that adequate vitamin E intake is essential for immune function and that vitamin E supplementation can successfully improve some aspects of the age related decline in immunity (Meydani, 1995; Meydani and Beharka, 1996), probably regulating prostaglandin synthesis and/or decreasing free radical formation (Folkers et al., 1993). By contrast, little is known about the interaction between ubiquinone-10 and the immune response. Ubiquinone-10 is a redox component in the mitochondrial respiratory chain and is a membrane antioxidant. It is a vitamin-like substance, found in small amounts in a wide variety of foods but also synthesized in human tissues (Littarru et al., 1991). Chronic malnutrition is presumed to affect ubiquinone-10 status either directly (by reducing its dietary intake) or indirectly (by reducing the intake of other vitamins involved in ubiquinone-10 biosynthesis) (Littarru et al., 1991). The basal plasma concentration of ubiquinone-10 has been described as either being increased (Di Palma et al., 1986) or decreased (Ravaglia et al., 2000) with ageing.

In contrast with some studies suggesting that vitamin A (Semba, 1998), vitamin B-6 (Baum et al., 1991), folate (Rosenthal et al., 1987) and vitamin B-12 (Kubota et al., 1992) may affect NK cell activity in deficient subjects with various diseases, Ravaglia et al. (2000) did not find any association between NK cell number, cytolytic activity and blood concentrations of A, B-6, B-12 vitamins and folate in a population of healthy free-living nonagenarian subjects.

A further remarkable finding is the strong correlation between the number of NK cells and serum concentrations of zinc (Ravaglia et al., 2000), deficiency of which is known to play a central role in multiple aspects of the immune system. In fact, several studies in animals and humans describe decreased NK cell activity and increased susceptibility to a variety of pathogens, in zinc-deficient states. Furthermore, the production or the biological activity of multiple cytokines influencing the development and function of T and B-lymphocytes, monocytes and NK cells is affected by zinc deficiency (see Ibs et al., this volume). Zinc influences the activity of multiple enzymes involved in activation, replication and pro-grammed cell death of lymphocytes and also acts as an antioxidant, protecting cells from the effects of oxygen radicals generated during immune activation (Shankar and Prasad, 1998). The preferential binding of zinc to metallothioneins in ageing can account for the continuous zinc sequestration inside lymphocytes responsible for the low zinc ion bioavailability and the resulting impaired immune response (Mocchegiani et al., 2002). Selenium is another trace element crucial for both specific and non-specific immune responses and, similar to zinc, its serum concentration correlated with NK cell number. The up-regulation of interleukin-2 receptors on the surface of activated lymphocytes, resulting in enhanced prolif-eration and clonal expansion of cytotoxic precursor cells (Kiremidijan-Schumacher and Roy, 1998) and, perhaps, the antioxidant role of selenoproteins appear to be the main mechanisms by which selenium affects cell-mediated immunity (Ravaglia et al., 2000).

6. Conclusions

In conclusion, the ageing process is a multifactorial process where not only the immune system, but also numerous other physiological systems decline, resulting in the loss or decrease of adaptation capacity. In fact, metabolic decompensation that would be easily tolerated in a young subject, can lead to problems for older people, although long-lived individuals are likely to maintain their residual homeostatic reserve capacity better. An improved understanding of the causes of immunosenescence may offer the possibility of therapeutic intervention and may result in a significant reduction in the cost of medical care in old age.

Acknowledgements

This work was partially supported by grants from MURST (60% fund), Ricerca Corrente IOR, FP Health Ministry from Italy and the EU ImAginE framework (QLK6-CT-1999-02031). The authors would like to thank Patrizia Rappini and Graziella Salmi for their typing assistance.

References

Akira, S., 2000. The role of IL-8 in innate immunity. Curr. Opin. Immunol. 12, 59–63.
Allavena, P., Paganin, C., Zhou, D., Bianchi, G., Sozzani, S., Mantovani, A., 1994. Interleukin-12 is chemotactic for natural killer cells and stimulates their interaction with vascular endothelium. Blood 84, 2261–2268.
André, P., Spertini, O., Guia, S., Rihet, P., Dignat-George, F., Brailly, H., Sampol, J., Anderson, P.J., Vivier, E., 2000. Modification of P-selectin glycoprotein ligand-1 with a natural killer cell-restricted sulphated lactosamine creates and alternate ligand for L-selectin. Proc. Natl. Acad. Sci. USA 97, 3400–3405.
Angeli, A., Masera, R.G., Staurenghi, A.H., Solomon, S., Bateman, A., Sartori, M.L., Lazzero, A., Griot, G., 1994. The expanding field of hypothalamic–pituitary–adrenal modulation of human natural killer cell activity. Ann. N.Y. Acad. Sci. 719, 328–342.
Aste-Amezaga, M., D'Andrea, A., Kubin, M., Trinchieri, G., 2002. Cooperation of natural killer cell stimulatory factor/interleukin-12 with other stimuli in the induction of cytokines and cytotoxic cell-associated molecules in human T and NK cells. Cell. Immunol. 156, 480–492.
Bakker, A.B., Wu, J., Phillips, J.H., Lanier, L.L., 2000. NK cell activation: distinct stimulatory pathways counterbalancing inhibitory signals. Hum. Immunol. 61, 18–27.
Bauer, S., Groh, V., Wu, J., Steinle, A., Phillips, J.H., Lanier, L.L., Spies, T., 1999. Activation of NK cells and T cells by NKG2D, a receptor for stress-inducible MICA. Science 285, 727–729.
Baum, M.K., Mantero-Atienza, E., Shor-Posner, G., Fletcher, M.A., Morgan, R., Eisdorfer, C., Sauberlich, H.E., Cornwell, P.E., Beach, R.S., 1991. Association of vitamin B6 status with parameters of immune function in early HIV-1 infection. J. Acquir. Immune Defic. Syndr. 4, 1122–1132.
Baume, D.M., Robertson, M.J., Levine, H., Manley, T.J., Schow, P.W., Ritz, D., 1992. Differential responses to interleukin 2 define functionally distinct subsets of human natural killer cells. Eur. J. Immunol. 22, 1–6.

Bellone, G., Aste-Amezaga, M., Trinchieri, G., Rodeck, U., 1995. Regulation of NK cell functions by TGF-β1. J. Immunol. 155, 1066–1073.

Biassoni, R., Cantoni, C., Falco, M., Verdiani, S., Bottino, C., Vitale, M., Conte, R., Poggi, A., Moretta, A., Moretta, L., 1996. The human leukocyte antigen (HLA)-C-specific "activatory" or "inhibitory" natural killer cell receptors display highly homologous extra cellular domains but differ in their transmembrane and intracytoplasmic portions. J. Exp. Med. 183, 645–650.

Biassoni, R., Cantoni, C., Pende, D., Sivori, S., Parolini, S., Vitale, M., Bottino, C., Moretta, A., 2001. Human natural killer cell receptors and co-receptors. Immunol. Rev. 181, 203–214.

Biron, C.A., 1997. Activation and function of natural killer cell responses during viral infections. Curr. Opin. Immunol. 9, 24–34.

Biron, C.A., Brossay, L., 2001. NK cells and NKT cells in innate defence against viral infections. Curr. Opin. Immunol. 13, 458–464.

Biron, C.A., Byron, K.S., Sullivan, J.L., 1989. Severe herpes virus infections in an adolescent without natural killer cells. N. Engl. J. Med. 320, 1731–1735.

Biron, C.A., Nguyen, K.B., Pien, G.C., Cousens, L.P., Salazar-Mather, T.P., 1999. Natural killer cells in antiviral defence: function and regulation by innate cytokines. Annu. Rev. Immunol. 17, 189–220.

Biron, C.A., Sonnenfeld, G., Welsh, R.M., 1984. Interferon induces natural killer cell blastogenesis in vivo. J. Leukocyte Biol. 35, 31–37.

Bluman, E.M., Bartynski, K.J., Avalos, B.R., Caligiuri, M.A., 1996. Human natural killer cells produce abundant macrophage inflammatory protein-1α in response to monocyte-derived cytokines. J. Clin. Invest. 97, 2722–2727.

Bodner, G., Ho, A., Kreek, M.J., 1998. Effect of endogenous cortisol levels on natural killer cell activity in healthy humans. Brain Behav. Immun. 12, 285–296.

Borrego, F., Alonso, C., Galiani, M., Carracedo, J., Ramirez, R., Ostos, B., Pena, J., Solana, R., 1999. NK phenotypic markers and IL2 response in NK cells from elderly people. Exp. Gerontol. 34, 253–265.

Borzi, R.M., Dal Monte, P., Uguccioni, M., Meliconi, R., Facchini, A., 1992. Intracellular nucleotides of lymphocytes and granulocytes from normal ageing subjects. Mech. Ageing Dev. 64, 1–11.

Braud, V.M., Allan, D.S.J., O'Callaghan, C.A., Soderstrom, K., D'Andrea, A., Ogg, G.S., Lazetic, S., Young, N.T., Bell, J.I., Phillips, J.H., Lanier, L.L., McMichael, A.J., 1998. HLA-E binds to natural killer cell receptors CD94/NKG2A, B and C. Nature 391, 795–799.

Caligiuri, M.A., Zmuidzinas, A., Manley, T.J., Levine, H., Smith, K.A., Ritz, J., 1990. Functional consequences of interleukin 2 receptor expression on resting human lymphocytes. Identification of a novel natural killer cell subset with high affinity receptors. J. Exp. Med. 171, 1509–1526.

Campbell, D., Bunker, W., Thomas, A.J., Clayton, B.E., 1989. Selenium and vitamin E status of healthy and institutionalised elderly subjects: analysis of plasma, erythrocytes and platelets. Br. J. Nutr. 62, 221–227.

Campbell, J.J., Qin, S., Unutmaz, D., Soler, D., Murphy, K.E., Hodge, M.R., Wu, L., Butcher, E.C., 2001. Unique subpopulations of CD56 + NK and NK-T peripheral blood lymphocytes identified by chemokine receptor expression repertoire. J. Immunol. 166, 6477–6482.

Cantoni, C., Bottino, C., Vitale, M., Pessino, A., Augugliaro, R., Malaspina, A., Parolini, S., Moretta, L., Moretta, A., Biassoni, R., 1999. NKp44, a triggering receptor involved in tumour cell lysis by activated human natural killer cells, is a novel member of the immunoglobulin superfamily. J. Exp. Med. 189, 787–796.

Carson, W.E., Giri, J.G., Lindemann, M.J., Linett, M.L., Ahdieh, M., Paxton, R., Anderson, D., Eisenmann, J., Grabstein, K., Caligiuri, M.A., 1994. Interleukin (IL) 15 is a novel cytokine that activates human natural killer cells via components of the IL-2 receptor. J. Exp. Med. 180, 1395–1403.

Carson, W.E., Lindemann, M.J., Baiocchi, R.A., Linett, M., Tan, J.C., Chou, C.C., Narula, S., Caligiuri, M.A., 1995a. The functional characterization of interleukin-10 receptor expression on human natural killer cells. Blood 85, 3577–3585.

Carson, W.E., Ross, M.E., Baiocchi, R.A., Marien, M.J., Boiani, N., Grabstein, K., Caligiuri, M.A., 1995b. Endogenous production of interleukin 15 by activated human monocytes is critical for optimal production of interferon-γ by natural killer cells in vitro. J. Clin. Invest. 96, 2578–2582.

Chan, S.H., Perussia, B., Gupta, J.W., Kobayashi, M., Pospisil, M., Young, H.A., Wolf, S.F., Young, D., Clark, S.C., Trinchieri, G., 1991. Induction of interferon γ production by natural killer cell stimulatory factor: characterization of the responder cells and synergy with other inducers. J. Exp. Med. 173, 869–879.

Chandra, R.K., 1992. Effect of vitamin and trace-element supplementation on immune responses and infection in elderly subjects. Lancet 52, 927–932.

Chandra, R.K., 1995. Nutrition and immunity in the elderly: clinical significance. Nutr. Rev 53, S80–S85.

Colonna, M., Navarro, F., Bellon, T., Llano, M., Garcia, P., Samaridis, J., Angman, L., Cella, M., Lopez-Botet, M., 1997. A common inhibitory receptor for major histocompatibility complex class I molecules on human lymphoid and myelomonocytic cells. J. Exp. Med. 186, 1809–1818.

Cooper, M.A., Fehniger, T.A., Ponnappan, A., Mehta, V., Wewers, M.D., Caligiuri, M.A., 2001a. Interleukin-1β co stimulates interferon-γ production by human natural killer cells. Eur. J. Immunol. 31, 792–801.

Cooper, M.A., Fehniger, T.A., Turner, S.C., Chen, K.S., Ghaheri, B.A., Ghayur, T., Carson, W.E., Caligiuri, M.A., 2001b. Human natural killer cells: a unique innate immunoregulatory role for the CD56^{bright} subset. Blood 97, 3146–3151.

Cosman, D., Fanger, N., Borges, L., Kubin, M., Chin, W., Peterson, L., Hsu, M.L., 1997. A novel immunoglobulin superfamily receptor for cellular and viral MHC class I molecules. Immunity 7, 273–282.

Cousens, L.P., Orange, J.S., Su, H.C., Biron, C.A., 1997. Interferon-α/β inhibition of interleukin 12 and interferon-γ production in vitro and endogenous during viral infection. Proc. Natl. Acad. Sci. USA 94, 634–639.

D'Andrea, A., Aste-Amezaga, M., Valiante, N.M., Ma, X., Kubin, M., Trinchieri, G., 1993. Interleukin 10 (IL-10) inhibits human lymphocyte interferon γ-production by suppressing natural killer cell stimulatory factor/IL-12 synthesis in accessory cells. J. Exp. Med. 178, 1041–1048.

De Martinis, M., Modesti, M., Loreto, M.F., Quaglino, D., Ginaldi, L., 2000. Adhesion molecules on peripheral blood lymphocyte subpopulations in the elderly. Life Sci. 68, 139–151.

Di Palma, A., Masini, M., Cimatti, G., Di Palma, P., Adamanti, S., 1986. Ubiquinone-10: a probable longevity marker. In: Folkers, K., Yamamura, Y. (Eds.), Biomedical and Clinical Aspects of Coenzyme Q. Vol. 5. Elsevier, Amsterdam, pp. 489–492.

Edwards, D.L., Avis, F.P., 1979. Antibody-dependent cellular cytotoxicity effector cell capability among normal individuals. J. Immunol. 123, 1887–1893.

Fabris, N., Mocchegiani, E., Provinciali, M., 1995. Pituitary–thyroid axis and immune system: a reciprocal neuroendocrine–immune interaction. Horm. Res. 43, 29–38.

Facchini, A., Mariani, E., Mariani, A.R., Papa, S., Vitale, M., Manzoli, F.A., 1987. Increased number of circulating Leu11+ (CD16) large granular lymphocytes and decreased NK activity during human ageing. Clin. Exp. Immunol. 68, 340–347.

Fahey, J.L., Schnelle, J.F., Boscardin, J., Thomas, J.K., Gorre, M.E., Aziz, N., Sadeghi, H., Nishanian, P., 2000. Distinct categories of immunologic changes in frail elderly. Mech. Ageing Dev. 115, 1–20.

Fehniger, T.A., Shah, M.H., Turner, M.J., VanDeusen, J.B., Whitman, S.P., Cooper, M.A., Suzuki, K., Wechser, M., Goodsaid, F., Caligiuri, M.A., 1999. Differential cytokine and chemokine gene expression by human NK cells following activation with IL-18 or IL-15 in combination with IL-12: implications for the innate immune response. J. Immunol. 162, 4511–4520.

Fernandes, G., Gupta, S., 1981. Natural killing and antibody-dependent cytotoxicity by lymphocyte subpopulations in young and ageing humans. J. Clin. Immunol. 1, 141–148.

Fitzgerald, K.A., O'Neill, L.A., 2000. The role of the interleukin-1/Toll-like receptor superfamily in inflammation and host defence. Microbes Infect. 2, 933–943.

Fleisher, G., Starr, S., Koven, N., Kamiya, H., Douglas, S.D., Henle, W., 1982. A non x-linked syndrome with susceptibility to severe Epstein–Barr virus infections. J. Pediatr. 100, 727–730.

Folkers, K., Morita, M., McRee, J. Jr., 1993. The activities of ubiquinone-10 and vitamin B6 for immune responses. Biochem. Biophys. Res. Commun. 193, 88–92.

Franceschi, C., Monti, D., Sansoni, P., Cossarizza, A., 1995. The immunology of exceptional individuals: the lesson of centenarians. Immunol. Today 16, 12–16.

Frey, M., Packianathan, N.B., Fehniger, T.A., Ross, M.E., Wang, W.C., Stewart, C.C., Caligiuri, M.A., Evan, S.S., 1998. Differential expression and function of L-selectin on CD56^{bright} and CD56^{dim} natural killer cell subsets. J. Immunol. 161, 400–408.

Fülöp, T., Barabas, G., Varga, Z., Josef, C., Csabina, S., Szucs, S., Seres, I., Szikszay, E., Jeney, Z., Penyige, A., 1993. Age-dependent changes in transmembrane signalling-identification of G-proteins in human lymphocytes and polymorphonuclear leukocytes. Cell. Signal. 5, 593–603.

Garcia, P.P., Koster, F.T., Kelley, R.O., McDowell, T.D., Bankhurst, A.D., 1989. Antibacterial activity of human natural killer cells. J. Exp. Med. 169, 99–113.

Ginaldi, L., De Martinis, M., D'Ostilio, A., Marini, L., Loreto, F., Modesti, M., Quaglino, D., 2001. Changes in the expression of surface receptors on lymphocyte subsets in the elderly: quantitative flow cytometric analysis. Am. J. Haematol. 67, 63–72.

Ginaldi, L., De Martinis, M., Modesti, M., Loreto, F., Corsi, M.P., Quaglino, D., 2000. Immunophenotypical changes of T lymphocytes in the elderly. Gerontology 46, 242–248.

Hallfrisch, J., Muller, D.C., Singh, V.N., 1994. Vitamin A and E intakes and plasma concentrations of retinol, beta-carotene and alpha-tocopherol in men and women of the Baltimore Longitudinal Study of Ageing. Am. J. Clin. Nutr. 60, 176–182.

Hansen, C.M., Binderup, L., Hamberg, K.J., Carlberg, C., 2001. Vitamin D and cancer: effect of 1,25(OH) 2D3 and its analogs on growth control and tumorigenesis. Front. Biosci. 6, D820–D848.

Hedrick, J.A., Saylor, V., Figueroa, D., Mizoue, L., Xu, Y., Menon, S., Abrams, J., Handel, T., Zlotnick, A., 1997. Lymphotactin is produced by NK cells and attracts both NK cells and T cells in vivo. J. Immunol. 158, 1533–1540.

Hidore, M.R., Murphy, J.W., 1989. Murine natural killer cell interactions with a fungal target, *Cryptococcus neoformans*. Infect. Immun. 57, 1990–1997.

Holbrook, N.J., Cox, W.I., Horner, H.C., 1983. Direct suppression of natural killer activity in human peripheral blood leukocyte culture by glucocorticoids and its modulation by interferon. Cancer Res. 43, 4019–4025.

Hunter, C.A., Subauste, C.S., Van Cleave, V.H., Remington, J.S., 1994. Production of γ interferon by natural killer cells from *Toxoplasma gondii*-infected SCID mice: regulation by interleukin-10, interleukin-12, and tumour necrosis factor α. Infect. Immun. 62, 2818–2824.

Hunter, C.A., Timans, J., Pisacane, P., Menon, S., Cai, G., Walker, W., Aste-Amezaga, M., Chizzonite, R., Bazan, J.F., Kastelein, R.A., 1997. Comparison of the effects of interleukin-1α, interleukin-1β, and interferon-γ-inducing factor on the production of interferon-γ by natural killer. Eur. J. Immunol. 27, 2787–2792.

Inngjerdingen, M., Damaj, B., Maghazachi, A.A., 2000. Human NK cells express CC chemokine receptors 4 and 8 and respond to thymus and activation-regulated chemokine, macrophage-derived chemokine, and I-309. J. Immunol. 164, 4048–4054.

Inngjerdingen, M., Damaj, B., Maghazachi, A.A., 2001. Expression and regulation of chemokine receptors in human natural killer cells. Blood 97, 367–375.

Kasaian, M.T., Biron, C.A., 1990. Cyclosporin A inhibition of interleukin 2 gene expression, but not natural killer cell proliferation, after interferon induction in vivo. J. Exp. Med. 171, 745–762.

Kiess, W., Belohradsky, B.H., 1986. Endocrine regulation of the immune system. Klin. Wochenschr. 64, 1–7.

Kim, C.H., Pelus, L.M., Appelbaum, E., Johanson, K., Anzai, N., Broxmeyer, H.E., 1999. CCR7 ligands, SLC/6Ckine/Exodus2/TCA4 and CKbeta-11/MIP-3beta/ELC, are chemoattractants for CD56(+) CD16(−) NK cells and late stage lymphoid progenitors. Cell. Immunol. 193, 226–235.

Kiremidjian-Schumacher, I., Roy, M., 1998. Selenium and immune function. Z. Ernahrungswiss 37, 50–56.

Kmiec, Z., Mysliwska, J., Rachon, D., Kotlaz, G., Sworczak, K., Mysliwski, A., 2001. Natural killer activity and thyroid hormone levels in young and elderly persons. Gerontology 47, 282–288.

Krishnaraj, R., 1997. Senescence and cytokines modulate the NK cell expression. Mech. Ageing Dev. 96, 89–101.

Krishnaraj, R., Bhooma, T., 1996. Cytokine sensitivity of human NK cells during immunosenescence. 2. IL2-induced interferon gamma secretion. Immunol. Lett. 50, 59–63.

Krishnaraj, R., Blandford, G., 1987. Age-associated alterations in human natural killer cells. I. Increased activity as per conventional and kinetic analysis. Clin. Immunol. Immunopathol. 45, 268–274.

Krishnaraj, R., Svanborg, A., 1992. Preferential accumulation of mature NK cells during human immunosenescence. J. Cell. Biochem. 50, 386–391.

Kubota, K., Kurabayashi, H., Kawada, E., Okamoto, K., Shirakura, T., 1992. Restoration of abnormally high Cd4/CD8 ratio and low natural killer cell activity by vitamin B12 therapy in patients with post-gastrectomy megaloblastic anaemia. Intern. Med. 31, 125–126.

Kunikata, T., Torigoe, K., Ushio, S., Okura, T., Ushio, C., Yamauchi, H., Ikeda, M., Ikegami, H., Kurimoto, M., 1998. Constitutive and induced IL-18 receptor expression by various peripheral blood cell subsets as determined by anti-hIL-18R monoclonal antibody. Cell. Immunol. 189, 135–143.

Kutza, J., Murasako, D.M., 1994. Effects of ageing on natural killer cell activity and activation by interleukin-2 and IFN-alpha. Cell. Immunol. 155, 195–204.

Kutza, J., Murasako, D.M., 1996. Age-associated decline in IL-2 and IL-12 induction of LAK cell activity of human PBMC samples. Mech. Ageing Dev. 90, 209–222.

Lanier, L.L., Chang, C., Azuma, M., Ruitenberg, J.J., Hemperly, J.J., Phillips, J.H., 1991. Molecular and functional analysis of human natural killer cell-associated neural cell adhesion molecule (N-CAM/CD56). J. Immunol. 146, 4421–4426.

Lanier, L.L., Le, A.M., Civin, C.I., Loken, M.R., Phillips, J.H., 1986. The relationship of CD16 (Leu-11) and Leu-19 (NKH-1) antigen expression on human peripheral blood NK cells and cytotoxic T lymphocytes. J. Immunol. 136, 4480–4486.

Lanier, L.L., Testi, R., Bindl, J., Phillips, J.H., 1989. Identity of Leu-19 (CD56) leukocyte differentiation antigen and neural cell adhesion molecule. J. Exp. Med. 169, 2233–2238.

Lesourd, B.M., Meaume, S., 1994. Cell mediated immunity changes in ageing, relative importance of cell subpopulation switches and of nutritional factors. Immunol. Lett. 40, 235–242.

Lesourd, B.M., 1995. Protein under nutrition as the major cause of decreased immune function in the elderly: clinical and functional implications. Nutr. Rev. 53, S86–S91.

Levy, S.M., Herberman, R.B., Lee, J., Whiteside, T., Beadle, M., Heiden, L., Simons, A., 1991. Persistently low natural killer cell activity, age, and environmental stress as predictors of infectious morbidity. Nat. Immun. Cell. Growth Reg. 10, 289–307.

Ligthart, G.J., Schuit, H.R., Hijmans, W., 1989. Natural killer cell function is not diminished in the healthy aged and is proportional to the number of NK cells in the peripheral blood. Immunology 68, 396–402.

Littarru, G.P., Lippa, S., Oradei, A., Fiorini, R.M., Mazzanti, L., 1991. Metabolic and diagnostic implications of blood CoQ10 levels. In: Folkers, K., Yamagami, T., Littarru, G.P. (Eds.), Biomedical and Clinical Aspects of Coenzyme Q. Vol. 6. Elsevier, Amsterdam, pp. 167–178.

Loetscher, P., Seitz, M., Clark-Lewis, I., Baggiolini, M., Moser, B., 1996. Activation of NK cells by CC chemokines: chemotaxis, Ca+ mobilization, and enzyme release. J. Immunol. 156, 322–327.

Lopez-Botet, M., Carretero, M., Bellon, T., Perez-Villar, J.J., Llano, M., Navarro, F., 1998. The CD94/NKG2 C-type lectin receptor complex. Curr. Top. Microbiol. Immunol. 230, 41–52.

Luster, A.D., 2002. The role of chemokines in linking innate and adaptive immunity. Curr. Opin. Immunol. 14, 129–135.

Maghazachi, A.A., Al-Aoukaty, A., 1993. Transforming growth factor-1 is chemotactic for interleukin-2-activated natural killer cells. Nat. Immunity 12, 57–61.

Maghazachi, A.A., Al-Aoukaty, A., Schall, T.J., 1994. C-C chemokines induce the chemotaxis of NK and IL-2 activated NK cells. J. Immunol. 153, 4969–4977.

Maghazachi, A.A., Al-Aoukaty, A., Schall, T.J., 1996. CC chemokines induce the generation of killer cells from CD56+ cells. Eur. J. Immunol. 26, 315–319.

Maghazachi, A.A., Skalhegg, B.S., Rolstad, B., Al-Aoukaty, A., 1997. Interferon-inducible protein-10 and lymphotactin induce the chemotaxis and mobilization of intracellular calcium in natural killer cells through pertussis toxin-sensitive and -insensitive heterotrimeric G-proteins. FASEB J. 11, 765–774.

Majewski, S., Skopinska, M., Marczak, M., Szmurlo, A., Bollag, W., Jablonska, S., 1996. Vitamin D3 is a potent inhibitor of tumour cell-induced angiogenesis. J. Investig. Dermatol. Symp. Proc. 1, 97–101.

Manolagas, S.C., Hustmyer, F.G., Yu, X.P., 1989. 1,25-Dihydroxyvitamin D3 and the immune system. Proc. Soc. Exp. Biol. Med. 191, 238–245.

Manolagas, S.C., Provvedini, D.M., Tsoukas, C.D., 1985. Interactions of 1,25-dihydroxivitamin D3 and the immune system. Mol. Cell. Endocrinol. 43, 113–122.

Mariani, E., Mariani, A.R., Meneghetti, A., Tarozzi, A., Cocco, L., Facchini, A., 1998a. Age-dependent decreases of NK cell phosphoinositide turnover during spontaneous but not Fc-mediated cytolytic activity. Int. Immunol. 7, 981–989.

Mariani, E., Meneghetti, A., Neri, S., Cattini, L., Facchini, A., 2000a. Do NK cells modify the pattern of specific receptors for HLA Class I with aging? Tissue Antigens 55, 79.

Mariani, E., Meneghetti, A., Neri, S., Cattini, L., Facchini, A., 2002a. Distribution of 2B4 surface glycoprotein on NK and T lymphocytes of elderly people. In: Leukocyte typing VII. Oxford University Press, in press.

Mariani, E., Meneghetti, A., Neri, S., Ravaglia, G., Forti, P., Cattini, L., Facchini, A., 2002b. Chemokine production by Natural Killer cells from nonagenarians. Eur. J. Immunol. 32, 1524–1529.

Mariani, E., Meneghetti, A., Tarozzi, A., Cattini, L., Facchini, A., 2000b. Interleukin-12 induces efficient lysis of natural killer-sensitive and natural killer-resistant human osteosarcoma cells: the synergistic effect of interleukin-2. Scand. J. Immunol. 51, 618–625.

Mariani, E., Monaco, M.C.G., Cattini, L., Sinoppi, M., Facchini, A. 1992. Distribution of adhesion molecules on CD4, CD8 and CD16 lymphocytes during aging. In: F. Licastro C.M. Caldarera (Eds.), Biomarkers of aging: expression and regulation, Editrice CLUEB, Bologna, pp. 341–348.

Mariani, E., Monaco, M.C.G., Cattini, L., Sinoppi, M., Facchini, A., 1994. Distribution and lytic activity of NK cell subsets in the elderly. Mech. Ageing Dev. 76, 177–187.

Mariani, E., Neri, S., Meneghetti, A., Cattini, L., Facchini, A., 2000c. PEN5 expression is preserved on NK cells from subjects over seventy years old. Tissue Antigens 55, 79–80.

Mariani, E., Pulsatelli, L., Meneghetti, A., Dolzani, P., Mazzetti, I., Neri, S., Ravaglia, G., Forti, P., Facchini, A., 2001. Different IL-8 production by T and NK lymphocytes in elderly subjects. Mech. Ageing Dev. 122, 1383–1395.

Mariani, E., Pulsatelli, L., Neri, S., Dolzani, P., Meneghetti, A., Silvestri, T., Ravaglia, G., Forti, P., Cattini, L., Facchini, A., 2002c. Rantes and MIP-1α production by T lymphocytes, monocytes and NK cells from nonagenarian subjects. Exp. Gerontol. 37, 219–226.

Mariani, E., Ravaglia, G., Forti, P., Meneghetti, A., Tarozzi, A., Maioli, F., Boschi, F., Pratelli, L., Pizzoferrato, A., Piras, F., Facchini, A., 1999. Vitamin D, thyroid hormones and muscle mass influence natural killer (NK) innate immunity in healthy nonagenarians and centenarians. Clin. Exp. Immunol. 116, 19–27.

Mariani, E., Ravaglia, G., Meneghetti, A., Tarozzi, A., Forti, P., Maioli, F., Boschi, F., Facchini, A., 1998b. Natural immunity and bone and muscle remodelling hormones in the elderly. Mech. Ageing Dev. 102, 279–292.

Mariani, E., Roda, P., Mariani, A.R., Vitale, M., Degrassi, A., Papa, S., Facchini, A., 1990. Age-associated changes in CD8+ and CD16+ cell reactivity: clonal analysis. Clin. Exp. Immunol. 81, 479–484.

Mariani, E., Sgobbi, S., Meneghetti, A., Tadolini, M., Tarozzi, A., Sinoppi, M., Cattini, L., Facchini, A., 1996. Perforins in human cytolytic cells: the effect of age. Mech. Ageing Dev. 92, 195–209.

Mariani, E., Sgobbi, S., Meneghetti, A., Tarozzi, A., Cattini, L., Facchini, A., 1997. CD158a and b Workshop: distribution of p58 molecules on peripheral blood lymphocytes during aging. In: Kishimoto, T., Kikutani, H., von dem Borne, A.E.G.K., Goyert, S.M., Mason, D.Y., Miyasaka, M., Moretta, L., Okumura, K., Shaw, S., Springer, T.A., Sugamura, K., Zola, H. (Eds.), Leukocyte Typing VI. White Cell Differentiation Antigens. Garland Publishing Inc., New York and London, pp. 304–306.

Mariotti, S., Pinchera, A., 1990. Role of the immune system in the control of thyroid function. In: Greer, M.A. (Ed.), Comprehensive Endocrinology: The Thymus Gland. Raven Press, New York, pp. 147–219.

Melero, I., Balboa, M.A., Alonso, J.L., Yague, E., Pivel, J.P., Sanchez-Madrid, F., Lopez-Botet, M., 1993. Signalling through the LFA-1 leukocyte integrin actively regulates intercellular adhesion and tumour necrosis factor-alpha production in natural killer cells. Eur. J. Immunol. 23, 1859–1865.

Meydani, M., 1995. Vitamin E. Lancet 345, 170–175.

Meydani, S.N., Beharka, A.A., 1996. Recent development in vitamin E and immune response. Nutr. Rev. 565, 49S–58S.

Miller, J.S., 2001. The biology of natural killer cells in cancer, infection, and pregnancy. Exp. Gerontol. 29, 1157–1168.

Miller, R.A., 1996. The aging immune system: primer and prospectus. Science 273, 70–74.

Miyaji, C., Watanabe, H., Toma, H., Akisaka, M., Tomiyama, K., Sato, Y., Abo, T., 2000. Functional alteration of granulocytes, NK cells, and natural killer T cells in centenarians. Hum. Immunol. 61, 908–916.

Mocchegiani, E., Giacconi, R., Cipriano, C., Muzzioli, M., Gasparini, N., Moresi, R., Stecconi, R., Suzuki, H., Cavalieri, E., Mariani, E., 2002. MtmRNA gene expression, via IL-6 and glucocorticoids, as potential genetic marker of immunosenescence: lessons from very old mice and humans. Exp. Gerontol. 37, 349–357.

Monteiro, J.M., Harvey, C., Trinchieri, G., 1998. Role of interleukin-12 in primary influenza virus infection. J. Virol. 72, 4825–4831.

Moretta, A., Biassoni, R., Bottino, C., Mingari, M.C., Moretta, L., 2000. Natural cytotoxicity receptors that trigger human NK-cell-mediated cytolysis. Immunol. Today 21, 228–234.

Moretta, A., Bottino, C., Vitale, M., Pende, D., Cantoni, C., Zingari, M.C., Biassoni, R., Moretta, L., 2001. Activating receptors and co receptors involved in human natural killer cell-mediated lysis. Annu. Rev. Immunol. 19, 197–223.

Morley, J.E., 1995. Hormones, aging, and endocrine disorders in the elderly. In: Felig, P., Baxter, J.D., Frhoman, L.A. (Eds.), Endocrinology and Metabolism. McGraw Hill, New York, pp. 1813–1836.

Nagler, A., Lanier, L.L., Phillips, J.H., 1990. Constitutive expression of high affinity interleukin 2 receptors on human CD16-natural killer cells in vivo. J. Exp. Med. 171, 1527–1533.

Nakajima, H., Colonna, M., 2000. 2B4: an NK cell activating receptor with unique specificity and signal transduction mechanism. Hum. Immunol. 61, 39–43.

Nakamura, K., Okamura, H., Nagata, K., Komatsu, T., Tamura, T., 1993. Purification of a factor which provides a co stimulatory signal for γ interferon production. Infect. Immun. 61, 64–70.

Nakanishi, K., Yoshimoto, T., Tsutsui, H., Okamura, H., 2001. Interleukin-18 is a unique cytokine that stimulates both Th1 and Th2 responses depending on its cytokine milieu. Cytokine Growth Factor Rev. 12, 53–72.

Ogata, K., An, E., Shioi, Y., Nakamura, K., Luo, S., Yokose, N., Minami, S., Dan, K., 2001. Association between natural killer cell activity and infection in immunologically normal elderly people. Clin. Exp. Immunol. 124, 392–397.

Ogata, K., Yokose, N., Tamura, H., An, E., Nakamura, K., Dan, K., Nomura, T., 1997. Natural killer cells in the late decades of human life. Clin. Immunol. Immunopathol. 84, 269–275.

Orange, J.S., Biron, C.A., 1996a. An absolute and restricted requirement for IL-12 in natural killer cell interferon-γ production and antiviral defence: studies of natural killer and T cell responses in contrasting viral infections. J. Immunol. 156, 1138–1142.

Orange, J.S., Biron, C.A., 1996b. Characterization of early IL-12, IFN-α/β, and TNF effects on antiviral state and NK cell responses during murine cytomegalovirus infection. J. Immunol. 156, 4746–4756.

Paavonen, T., 1994. Hormonal regulation of immune responses. Ann. Med. 26, 255–258.

Parrish-Novak, J., Dillon, S.R., Nelson, A., Hammond, A., Sprecher, C., Gross, J.A., Johnston, J., Madden, K., Xu, W., West, J., Schrader, S., Burkhead, S., Heipel, M., Brandt, C., Kuijper, J.L., Kramer, J., Conklin, D., Presnell, S.R., Berry, J., Shiota, F., Bort, S., Hambly, K., Mudri, S., Clegg, C., Moore, M., Grant, F.J., Lofton-Day, C., Gilbert, T., Rayond, F., Ching, A., Yao, L., Smith, D., Webster, P., Whitmore, T., Maurer, M., Kaushansky, K., Holly, R.D., Foster, D., 2000. Interleukin 21 and its receptor are involved in NK cell expansion and regulation of lymphocyte function. Nature 408, 57–63.

Pawelec, G., Solana, R., Remarque, E., Mariani, E., 1998. Impact of aging on innate immunity. J. Leukoc. Biol. 64, 703–712.

Pekonen, F., Weintraub, B.D., 1978. Thyrotropin binding to cultured lymphocytes and thyroid cells. Endocrinology 103, 1668–1677.

Pende, D., Parolini, S., Pessino, A., Sivori, S., Augugliaro, R., Morelli, L., Marcenaro, E., Accame, L., Malaspina, A., Biassoni, R., Bottino, C., Moretta, L., Moretta, A., 1999. Identification and molecular characterization of NKp30, a novel triggering receptor involved in natural cytotoxicity mediated by human natural killer cells. J. Exp. Med. 190, 1505–1516.

Perussia, B., Trinchieri, G., Jackson, A., Warner, N.L., Faust, J., Rumpold, H., Kraft, D., Lanier, L.L., 1984. The Fc receptor for IgG on human natural killer cells: phenotypic, functional, and comparative studies with monoclonal antibodies. J. Immunol. 133, 180–189.

Pilaro, A.M., Taub, D.D., McCormick, K.L., Williams, H.M., Sayers, T.J., Fogler, W.E., Wiltrout, R.H., 1994. TNF-α is a principal cytokine involved in the recruitment of NK cells to liver parenchyma. J. Immunol. 153, 333–342.

Poggi, A., Panzeri, M.C., Moretta, L., Zocchi, M.R., 1996. CD31-triggered rearrangement of the actin cytoskeleton in human natural killer cells. Eur. J. Immunol. 26, 817–824.

Potestio, M., Pawelec, G., Di Lorenzo, G., Candore, G., D'Anna, C., Gervasi, F., Lio, D., Tranchida, G., Caruso, C., Colonna Romano, G., 1999. Age-related changes in the expression of CD95 (AP01/FAS) on blood lymphocytes. Exp. Gerontol. 34, 659–673.

Puzanov, I.J., Bennett, M., Kumar, V., 1996. IL-15 can substitute for the marrow microenvironment in the differentiation of natural killer cells. J. Immunol. 157, 4282–4285.

Rabinowich, H., Herberman, R.B., Whiteside, T., 1993. Differential effects of IL-12 and IL-2 in expression and function of cellular adhesion molecules on purified NK cells. Cell. Immunol. 152, 481–498.

Ravaglia, G., Forti, P., Maioli, F., Bastagli, L., Facchini, A., Mariani, E., Savarino, L., Sassi, S., Cucinotta, D., Lenaz, G., 2000. Effect of micronutrient status on natural killer cell immune function in healthy free-living subjects aged $\geq$90 y. Am. J. Clin. Nutr. 71, 590–598.

Ravaglia, G., Forti, P., Pratelli, L., Maioli, F., Scali, C.R., Bovini, A.M., Rtediolo, S., Marasti, N., Pizzoferrato, A., Gasbarrini, G., 1994. The association of ageing with calcium active hormone status in men. Age Ageing 23, 127–131.

Remarque, E., Pawelec, G., 1998. T cell immunosenescence and its clinical relevance in man. Rev. Clin. Gerontol. 8, 5–14.

Rink, L., Cakman, I., Kirchner, H., 1998. Altered cytokine production in the elderly. Mech. Ageing Dev. 102, 199–210.

Robertson, M.J., 1997. Natural Killer Cell clinical studies: surface antigens of human natural killer cells in health and disease. In: Kishimoto, T., Kikutani, H., von dem Borne, A.E.G.K., Goyert, S.M., Mason, D.Y., Miyasaka, M., Moretta, L., Okumura, K., Shaw, S., Springer, T.A., Sugamura, K., Zola, H. (Eds.), Leukocyte Typing VI. White Cell Differentiation Antigens, pp. 327–329.

Robertson, M.J., 2002. Role of chemokines in the biology of natural killer cells. J. Leukoc. Biol. 71, 173–183.

Robertson, M.J., Ritz, J., 1990. Biology and clinical relevance of human natural killer cells. Blood 76, 2421–2438.

Robertson, M.J., Soiffer, R.J., Wolf, S.F., Manley, T.J., Donahue, C., Young, D., Herrmann, S.H., Ritz, J., 1992. Response of human natural killer (NK) cells to NK cell stimulatory factor (NKSF): cytolytic activity and proliferation of NK cells are differentially regulated by NKSF. J. Exp. Med. 175, 779–788.

Robertson, M.J., Williams, B.T., Christopherson, K.2., Brahmi, Z., Hromas, R., 2000. Regulation of human natural killer cell migration and proliferation by the exodus subfamily of CC chemokines. Cell. Immunol. 199, 8–14.

Rook, A.H., Kehrl, J.H., Wakefield, L.M., Roberts, A.B., Sporn, M.B., Burlington, D.B., Lane, H.C., Fauci, A.S., 1986. Effects of transforming growth factor β on the functions of natural killer cells: depressed cytolytic activity and blunting of interferon responsiveness. J. Immunol. 136, 3916–3920.

Rosenthal, G.J., Germolec, D.R., Lamma, K.R., Ackermann, M.F., Luster, M.I., 1987. Comparative effects on the immune system of methotrexate and trimetrexate. Int. J. Immunopharmacol. 9, 793–801.

Rukavina, D., Laskarin, G., Rubesa, G., Strbo, N., Bedenicki, I., Manestar, D., Glavas, M., Christmas, S.E., Podack, E.R., 1998. Age-related decline of perforin expression in human cytotoxic T lymphocytes and natural killer cells. Blood 92, 2410–2420.

Salazar-Mather, T.P., Ishikawa, R., Biron, C.A., 1996. Natural killer (NK) cell trafficking and cytokine expression in splenic compartments after interferon-induction and viral infection. J. Immunol. 157, 3054–3064.

Salazar-Mather, T.P., Orange, J.S., Biron, C.A., 1998. Early murine cytomegalovirus (MCMV) infection induces liver natural killer (NK) cell inflammation and protection through macrophage inflammatory protein 1α (MIP-1α)-dependent pathways. J. Exp. Med. 187, 1–14.

Salvucci, O., Mami-Chouaib, F., Moreau, J.L., Theze, J., Chehimi, J., Chouaib, S., 1996. Differential regulation of interleukin-12- and interleukin-15-induced natural killer cell activation by interleukin-4. Eur. J. Immunol. 26, 2736–2741.

Sansoni, P., Cossarizza, A., Brianti, V., Fagnoni, F., Snelli, G., Monti, D., Marcato, A., Passeri, G., Ortolani, C., Forti, E., Fagiolo, U., Passeri, M., Franceschi, C., 1993. Lymphocyte subsets and natural killer cell activity in healthy old people and centenarians. Blood 82, 2767–2773.

Seaman, W.E., 2000. Natural killer cells and natural killer T cells. Arthritis Rheum. 43, 1204–1217.

Sedlmayr, P., Schallhammer, L., Hammer, A., Wilders-Trusching, M., Wintersteiger, R., Dohr, G., 2002. Differential phenotypic properties of human peripheral blood CD56^{dim+} and CD56$^{bright+}$ natural killer cell subpopulations. Int. Arch. Allergy Immunol. 110, 308–313.

Semba, R.D., 1998. The role of vitamin A and related retinoids in immune function. Nutr. Rev. 56, 38S–48S.

Shankar, A.H., Prasad, A.S., 1998. Zinc and immune function: the biological basis of altered resistance to infection. Am. J. Clin. Nutr. 68, 447S–463S.

Sims, J.E., 2002. IL-1 and IL-18 receptors, and their extended family. Curr. Opin. Immunol. 14, 117–122.

Sivori, S., Pende, D., Bottino, C., Marcenaro, E., Pessino, A., Biassoni, R., Moretta, L., Moretta, A., 1999. NKp46 is the major triggering receptor involved in the natural cytotoxicity of fresh or cultured human NK cells. Correlation between surface density of NKp46 and natural cytotoxicity against autologous, allogeneic or xenogeneic target cells. Eur. J. Immunol. 29, 1656–1666.

Solana, R., Mariani, E., 2000. NK and NK/T cells in human senescence. Vaccine 18, 1613–1620.

Straub, R.H., Cutolo, M., Zietz, B., Scholmerich, J., 2001. The process of ageing changes the interplay of the immune, endocrine and nervous system. Mech. Ageing Dev. 122, 1591–1611.

Su, H.C., Ishikawa, R., Biron, C.A., 1993. Transforming growth factor-β expression and natural killer cell responses during virus infection of normal, nude, and SCID mice. J. Immunol. 151, 4874–4890.

Takasaki, S., Hayashida, K., Morita, C., Ishibashi, H., Niho, Y., 2000. CD56 directly interacts in the process of NCAM + target cell-killing by NK cells. Cell Biol. Int. 24, 101–108.

Tangye, S.G., Phillips, J.H., Lanier, L.L., 2000. The CD2-subset of the Ig superfamily of cell surface molecules: receptor-ligand pairs expressed by NK cells and other immune cells. Semin. Immunol. 12, 149–157.

Taub, D.D., 1999. Natural killer cell-chemokine interactions. Biologic effects on natural killer cell trafficking and cytolysis. In: Rollins, B.J. (Ed.), Chemokines and Cancer. Humana Press Inc, Totowa, NJ, pp. 73–93.

Taub, D.D., Ortaldo, J.R., Turcovski-Corrales, S.M., Key, M.L., Longo, D.L., Murphy, W.J., 1996. β Chemokines co-stimulate lymphocyte cytolysis, proliferation and lymphokine production. J. Leukoc. Biol. 59, 81–89.

Taub, D.D., Sayers, T.J., Carter, C.R.D., Ortaldo, J.R., 1995a. α and β chemokines induce NK cell migration and enhance NK-mediated cytolysis. J. Immunol. 155, 3877–3888.

Taub, D.D., Sayers, T.J., Carter, C.R., Ortaldo, J.R., 1995b. Chemokines induce NK cell migration and enhance NK cell cytolytic activity via cellular degranulation. J. Immunol. 155, 3877–3888.

Trinchieri, G., 1989. Biology of natural killer cells. Adv. Immunol. 47, 187–376.

Trinchieri, G., Santoli, D., Koprowski, H., 1978. Spontaneous cell-mediated cytotoxicity in humans: role of interferon and immunoglobulins. J. Immunol. 120, 1849–1855.

Tripp, C.S., Wolf, S.F., Unanue, E.R., 1993. Interleukin 12 and tumour necrosis factor α are co stimulators of interferon γ production by natural killer cells in severe combined immunodeficiency mice with listeriosis, and interleukin 10 is physiologic antagonist. Proc. Natl. Acad. Sci. USA 90, 3725–3729.

Vitale, M., Zamai, L., Neri, L.M., Galanzi, A., Facchini, A., Rana, R., Cataldi, A., Papa, S., 1992. The impairment of natural killer function in the healthy aged is due to a post binding deficient mechanism. Cell. Immunol. 145, 1–10.

Vivier, E., Sorrell, J.M., Ackerly, M., Robertson, M.J., Rasmussen, R.A., Levine, H., Anderson, P., 1993. Developmental regulation of a mucin like glycoprotein selectively expressed on natural killer cells, J. Exp. Med. 2033.

Voss, S.D., Sondel, P.M., Robb, R.J., 1992. Characterization of the interleukin 2 receptors (IL-2R) expressed on human natural killer cells activated in vivo by IL-2: association of the p64 IL-2R γ chain with the IL-2R β chain in functional intermediate-affinity IL-2R. J. Exp. Med. 176, 531–541.

Wang, K.S., Frank, D.A., Ritz, J., 2000. Interleukin-2 enhances the response of natural killer cells to interleukin-12 through up-regulation of the interleukin-12 receptor and STAT4. Blood 95, 3183–3190.

Warren, H.S., Kinnear, B.F., Kastelein, R.L., Lanier, L.L., 1996. Analysis of the co stimulatory role of IL-2 and IL-15 in initiating proliferation of resting (CD56^{dim}) human NK cells. J. Immunol. 156, 3254–3259.

Welsh, R.M., 1981. Natural cell-mediated immunity during viral infections. Curr. Top. Microbiol. Immunol. 92, 83–106.

Welsh, R.M., Brubaker, J.O., Vargas-Cortes, M., O'Donnell, C.L., 1991. Natural killer (NK) cell response to virus infections in mice with severe combined immunodeficiency. The stimulation of NK cells and the NK cell-dependent control of virus infections occur independently of T and B cell function. J. Exp. Med. 173, 1053–1063.

Welsh, R.M., Vargas-Cortes, M., 1992. Natural killer cells in viral. In: Lewis, C.E., McGee, J.O. (Eds.), The Natural Killer Cell. Oxford University Press, Oxford, pp. 107–150.

Yoneda, O., Imai, T., Goda, S., Inoue, H., Yamauchi, A., Okazaki, T., Imai, H., Yoshie, O., Bloom, E.T., Domae, N., Umehara, H., 2000. Fractalkine-mediated endothelial cell injury by NK cells. J. Immunol. 164, 4055–4062.

Zhang, X., Sun, S., Hwang, I., Tough, D.F., Sprent, J., 1998. Potent and selective stimulation of memory-phenotype CD8 + T cells in vivo by IL-15. Immunity 8, 591–599.

Advances in
Cell Aging and
Gerontology

T cell ageing and immune surveillance

Julie McLeod

Centre for Research in Biomedicine, University of the West of England, Bristol BS16 1QY, England, UK
Correspondence address: Tel.: +44-117-344-2531; fax: +44-117-344-2906.
E-mail: Julie.mcleod@uwe.ac.uk

Contents

Abbreviations

TNF: tumour necrosis factor; CD95-L: CD95 ligand; IL: interleukin; TGFβ: tumour grouth factor β; AICD: activation induced cell death; TAA's: tumour associated antigen's; RIPc: RIP cleaved; Pgp: P-glycoprotein; MDR-1: multi drug resistance-1; MM: multiple myeloma.

1. Introduction

As we age the cells of the immune system show clear signs of 'ageing' termed immunosenescence. This leads to increased rates of infection and mortality, predisposition to an autoimmune profile and increased risk of cancer. Although all the cells, which constitute the immune system, are affected by age, the T cells show specific dysfunction that has implications throughout the immune system.

Advances in Cell Aging and Gerontology, vol. 13, 159–172

In this chapter we will discuss the interaction of T cells with 'immune privileged' sites – in particular the immune surveillance which occurs between T cells and tumour cells.

2. What is immune privilege?

For over a century it has been recognised that there are sites in the body where a reduced or non-existent immune response occurs. These sites were specifically targeted for their good transplantation capability as no major histocompatability response occurs. Such sites include the brain, ovary, testis, pregnant uterus and the eye. Due to their apparent isolation from the immune system they were termed 'immune privileged'. In addition, certain tumours have been recognised as also utilising the concept of immune privilege to escape the surveillance of the immune system (Green and Ferguson, 2001).

It is known that T cells, among others, are able to recognise and respond to antigenic epitopes on immune privileged sites. However, this recognition does not lead to activation and inflammation but rather to the 'attacking' cells being immunosuppressed. The factors that are involved in this suppressive effect are still under debate, however the death receptor, CD95 is a strong candidate (Green and Ferguson, 2001).

CD95 is a cysteine-rich transmembrane protein, which is expressed on the cells of the lymphoid system, heart and in various places throughout embryonic and adult life (Miyawaki et al., 1992). It is a member of the tumour necrosis factor (TNF) receptor family and, through its intracellular death domain, initiates apoptosis. Apoptosis occurs through trimerisation of the CD95 molecule upon binding of its natural ligand, CD95 ligand (CD95-L). CD95-L is expressed on the cell surface and is ultimately cleaved to a soluble form by metalloproteinases, both products having differential apoptotic capability (Tanaka et al., 1998). The membrane form is the most effective; however, soluble CD95-L has been reported to both induce apoptosis (Tanaka et al., 1998) and to have a neutralising role particularly at the immune : tumour interface (Hohlbaum et al., 2000). Indeed, immune-privileged sites, e.g. corneal epithelial cells, have been found to have constitutive expression of CD95-L through a *cis*-acting element in the CD95-L promoter (Zhang et al., 1999). In addition, intriguingly, membrane CD95-L has also been found to induce a bipolar signal, which can costimulate T cell responses, particularly in CD8 cells (Suzuki and Fink, 2000). Therefore, expression of the different forms of CD95-L will play a central role in determining the balance of anti- or pro-inflammatory responses occurring during immune surveillance.

3. CD95-L and immune privilege

The demonstration supporting a role for CD95-L in immune privilege has been undertaken in a wide array of sites (O'Connell et al., 1999) where suppression of the inflammatory response to a range of antigens was reversed in CD95-L-deficient *gld* mice. This was due to apoptosis of the CD95$^+$ infiltrating cells (Griffith et al.,

1996). The destruction of attacking immune cells through apoptosis at sites of immune-privilege is certainly one key way to prevent inflammation, although this might be rather simplistic in some immune-privileged areas. One central element of an immune-privileged site is the ability to induce tolerance to antigens delivered to these sites. Interestingly, CD95-induced apoptosis has been implicated in the induction of immune tolerance (Green and Ferguson, 2001). Work suggests that the production of apoptotic cells plus antigen can promote tolerance possibly through redirecting antigen presentation towards regulatory cell generation (Hill et al., 1999; Green and Ferguson, 2001). In addition to involving CD95:CD95-L interactions, the immunosuppressive cytokines IL-10 (Green and Ferguson, 2001) and TGFβ (Chen et al., 2001) have been implicated in directing the immune tolerant state. Therefore, CD95:CD95-L interactions are key players in an immune suppressive state although this appears to rely heavily on the physiological relevance of the site and level of expression. In this regard, engineered expression of CD95-L can lead to a pro-inflammatory state primarily involving neutrophils (Ottonello et al., 1999). However, the expression and presence of CD95-L in a natural context has only been reported to be immuno-suppressive (Green and Ferguson, 2001). The ability of cells comprising specific sites to induce a suppressive or tolerant state in attacking immune cells has been utilised by tumour cells as a protection mechanism.

4. Involvement of CD95-L in tumour 'counter attack'

Certain tumours are capable of generating CD95-L, suggesting that this expression would provide protection from infiltrating attacking cells. A range of tumours have been reported to express CD95-L (Hahne et al., 1996; O'Connell et al., 2001) and can lead to destruction of infiltrating lymphocytes through a CD95-dependant pathway. The protective effect of CD95-L expression also encompasses destruction of cytotoxic T cells and reduction in tumour-specific antibody production (Arai et al., 1997). However, there still remains contention as to the role of CD95:CD95-L interactions in tumour protection (Restifo, 2000). It has been generally thought that tumour cells would utilise an up-regulated expression of CD95-L to ligate CD95 on tumour infiltrating T cells, leading to their death and thus tumour protection. However, in a recent commentary, Restifo (2000) suggested that the simplicity of the idea may not be physiologically relevant and that attacking T cells may undergo apoptosis through Activation Induced Cell Death (AICD). In addition, as mentioned above, experimental use of CD95-L over-expression to induce immune-privilege has been found to be pro-inflammatory. A recent review by O'Connell et al. (2001) robustly interrogates the role of CD95-L in tumour protection. AICD, where activation through the T cell receptor complex leads to CD95-L induction and apoptosis of CD95[+] cells (Alderson et al., 1995), has been found to occur in tumour-infiltrating T cells (Restifo, 2000). However, although T cells will recognise tumour associated antigens (TAA's) the cellular activation will be muted through down-regulation of Major Histocompatability Complex molecules and the TAA's themselves (Seliger et al., 1997). However, AICD does not comprise the whole of the tumour protection response since T cell apoptosis is still observed when the T cell-derived CD95-L generation is abolished, therefore supporting a role for tumour-generated CD95-L in T cell apoptosis (O'Connell et al., 1996). Support for a direct role of tumour-derived CD95-L is also gained from in vivo studies which correlate apoptotic levels of tumour-infiltrating lymphocytes with site-specific expression of CD95-L (Manukata et al., 2000). In addition, the ratio of CD95-L: CD95 expression on tumour cells derived from patients is a strong prognostic indicator of tumour survival (O'Connell et al., 2001). Therefore, tumour-derived CD95-L can be argued to play a key role in tumour protection from immune surveillance, although AICD and a balance between the anti- and pro-inflammatory effects will also be involved. In addition, in the context of the aged immune system, dysregulation of the attacking T cells will be pivotal in the overall response to tumours.

5. T cell ageing

Immunosenescent T cells show several characteristics which can be seen in both in vivo and in vitro models of ageing. There is a block in cell division (Abidzadeh et al., 1996) and shortening of telomere length (Vaziri et al., 1993). Also there is a specific reduction in expression of the costimulatory receptors; CD28, CD154 (CD40 ligand), CD134 (OX-40) (Pawelec et al., 1999a) and CD25 (Dennett et al., 2002). Of these, CD28 and CD25 are very significant due to their pivotal role in generating a full proliferative T cell response. Stimulation through CD28 is known (Murtaza et al., 1999) to be the key balance in determining the T cell proliferative outcome. Indeed in aged T cells it is the expression level of CD28 that correlates with response (Tortorella et al., 1997; Boucher et al., 1998). CD25 expression is also particularly key in determining the overall T cell response since aged human T cells have been reported (Pawelec et al., 2002) to become more dependent on exogenous IL-2 for survival. However, not all receptors are absent from aged T cells, with the apoptotic receptor CD95 being up-regulated (Pawelec et al., 1999b), although there is some debate whether CD95 expression is retained at a high level or is reduced in very elderly individuals (Pawelec et al., 2002). In addition, aged CD4 T cell clones and T cells (CD4 and CD8) derived from the elderly show heightened apoptotic susceptibility, although long-term cultures of CD8 cells are more resistant to apoptotic stimuli (Pawelec et al., 2002). Therefore, in the aged T cell there is a clear change in the balance between proliferation and death. These alterations in the apoptotic capability of aged T cells may well have an important role in immune surveillance and tumour-derived counter attack.

6. Apoptotic capability in aged T cells

Apoptosis can be initiated through myriad factors including, cytokine withdrawal and ligation of members of the TNF family of receptors (Cohen et al., 1992; Smith et al., 1994; Dhein et al., 1995). The latter death receptors comprise CD95 (for a detailed account of apoptosis in aged T cells see Chapter 10 in this edition). CD95 is expressed on a variety of cell types including T and B cells (Miyawaki et al., 1992) and is a potent mediator of apoptosis in susceptible cells. However, susceptibility in T cells may be dependent on the state of differentiation and/or activation (Wesselborg et al., 1993). CD95-L exhibits a more restricted distribution, being found mainly on activated T cells (Suda et al., 1994) and, with age, T cell-derived CD95-L is upregulated (Aggarwal and Gupta, 1998). Therefore, aged T cells may have an advantage in the balance of activation and death signals.

Extracellular and intracellular factors are involved in regulating apoptotic capability. Intracellular regulation of CD95 apoptotic signals occurs through the proteins of the Bcl-2 family. Pro-apoptotic signals are derived through Bax and Bad, whereas $Bclx_L$ and Bcl-2 provide anti-apoptotic signals (Chao and Korsmeyer, 1998). Work by Shimizu et al. (1999) has shown that these proteins exercise their functions at the mitochondrial surface, which is known to be the defining point of whether apoptosis proceeds or not (Kroemer et al., 1997; Shimizu et al., 1999). Therefore, the relative levels of these intracellular proteins may determine the susceptibility of cells for apoptosis. Differences within cell type, as to the impact of specific intracellular regulators, have been reported (Scaffidi et al., 1998) where type 1 and type 2 cells have been proposed (not to be confused with type 1 and type 2 T cell cytokine profiles). Type 1 cells utilise the caspase pathway for apoptotic signalling, whereas the mitochondria play a predominant role in Type 2 cells. Therefore, the regulatory effects of $Bclx_L$, Bcl-2, Bax and Bad have more impact in selected (Type 2) cells. This may explain some of the discrepancies between apoptotic capability in human peripheral T cells and T cell-lines (Walker et al., 1999). In elderly T cells, both CD4 and CD8 cells show a decrease in the anti-apoptotic protein BcL-2 and an increase in Bax, which correlated with increased apoptotic susceptibility (Aggarwal and Gupta, 1998). Therefore, alterations in the ratio of intracellular regulators directly impacts on apoptotic capability in aged T cells.

The caspase pathway itself is also affected in aged T cells. Increases in constitutive and CD95-induced activity of caspase 3 and 8 are seen in aged cultures with heightened apoptotic capability (Aggarwal and Gupta, 1999), whereas reduced caspase 3 activity occurs in resistant senescent CD8 lymphocytes (Spaulding et al., 1999). Recent studies by our group (unpublished observations) have also investigated constitutive caspase activity in aged T cell clones where SENIEUR-derived clones showed a predominant balance towards caspase activity, whereas younger clones involved both the caspase and the novel RIP pathway. The RIP pathway is classically one that is proliferative, working through Ikβ and NFKβ (Ting et al., 1996; Kelliher et al., 1998). However, recent (Lin et al., 1999; Martinon et al., 2000) reports suggest that RIP can induce apoptosis by CD95 through interaction with caspase 8 and subsequent generation of a cleaved RIP product (RIPc). Our group have found that T cell clones derived from young and CD34+ cells and aged in culture show heightened expression of RIPc, which correlates with apoptotic capability. Therefore, the balance between the caspase and RIP pathways are implicated as central regulators of the apoptotic capability of aged lymphocytes.

7. The aged immune : tumour interface

One major facet of an efficient immune response against tumours are effective and responsive CD4 and CD8 T cells. As described above, aged T cells show distinct characteristics which would not support the idea of an effective T cell attack against tumours. In particular their heightened apoptotic susceptibility would tip the balance in favour of tumour counter attack. There are a number of factors which influence the balance of aged T cells towards apoptosis and these will be discussed in the context of the immune : tumour interface.

7.1. CD4/CD8 ratio

In vitro models of T cell ageing indicate differences in the apoptotic susceptibility of CD4 and CD8 T cell populations. Pawelec et al. (1996) reported that CD4 T cell clones showed increased apoptosis to CD95 ligation as the cells underwent increasing population doublings. However, long term cultures of senescent CD8 cells show an increased resistance to a range of apoptotic stimuli, although expressing CD95 on all cells (Spaulding et al., 1999). The heightened apoptotic susceptibility of aged CD4 T cells would suggest impaired recognition of TAA's upon antigen presentation and therefore an overall suppressed immune surveillance mechanism.

The implications of senescent CD8 T cells' apoptotic resistance are unclear within the context of the aged immune system. These aged cells are functionally competent (Spaulding et al., 1999), and increased perforin expression has been reported in clonally expanded CD8 cells (Chamberlain et al., 2000) suggesting that they would be cytotoxic in immune-surveillance, particularly so since they would be resistant to the tumours' immune protection through immunocyte apoptosis. However, the general observation of increased tumour load in the elderly would counter this view and it may be that other regulatory aspects affect the CD8 tumour response.

7.2. Costimulators

CD28 is a key receptor that affects both T cell activation and susceptibility to apoptosis. Costimulation via CD28 in superantigen-challenged cultures leads to resistance to CD95-induced apoptosis (McLeod et al., 1998), which may be due to increased levels of the anti-apoptotic intracellular regulators c-FLIPs and $Bclx_L$ (Boise and Thompson, 1997; Kirchhoff et al., 2000a,b). However, in some situations CD28 may lead to an increased responsiveness to CD95 (Boussiotis et al., 1997). In aged cultures and also in elderly patient-derived cells expression of CD28 is dramatically reduced on both CD4 and CD8 T cells (Effros et al., 1994; Fangoni et al., 1996; Weyand et al., 1998). In support of the protective role of CD28 signalling, CD28[low] cells, aged in culture, have been found to be more susceptible to apoptosis (Dennett et al., 2002).

CD25, the IL-2 α-chain receptor, is also a key costimulator in T cells. As mentioned previously, it is thought that, with age, T cells become more dependent on exogenous IL-2 for survival due to reduced IL-2 secretion. In addition, we, and others, have found that expression of CD25 is reduced in both aged T cells and T cell clones (Haynes et al., 2000; Dennett et al., 2002). The decreased presence of CD25 correlates with increased apoptotic susceptibility in cultures with exogenous supply of IL-2 (Dennett et al., 2002), suggesting a role for CD25 signalling in apoptotic regulation. IL-2 has been reported to play a role in T cell AICD through inhibition of FLIP generation (Algeciras-Schimnich et al., 1999). However, we and others, have not been able to observe changes in FLIP levels with age (unpublished observations), although, due to the association of FLIP levels with cell cycle (Algeciras-Schimnich et al., 1999), unsynchronised cultures may negate changes being observed.

The decreased expression of CD28 results in a diminished proliferative response (Adibzadeh et al., 1995), although this does not appear to affect cytokine production (Weyand et al., 1998). The ability of the cell to remain operational even with reduced CD28 receptor expression also suggests the possibility of a predominant role for CTLA4. CTLA4 is upregulated through activation and, upon CD80/86 ligation, generates a strong negative signal to the T cell (Chambers et al., 1996). This, together with the reduced expression of surface CD28, would suggest the balance of T cell signals tending away from activation. Indeed, studies in mice (Wakikawa et al., 1997) and human T cell clones (unpublished data) show an increased expression of CTLA4 in aged T cells. Further work is required to elucidate the role of CTLA-4 in aged T cells. A new member to the CD28/CTLA-4 receptor family is ICOS (Hutloff et al., 1999). ICOS is expressed on activated T cells and stimulates secretion of IL-10, suggesting an immunosuppressive role, although recent studies suggest an effect on Th1 : Th2 balance (Greenwald et al., 2002). Little work has been undertaken on the role of ICOS in aged T cells, although studies have shown increased expression on aged T cell clones (Pawelec et al., 2002). Ligation of ICOS has been reported to protect cells from apoptosis (Beier et al., 2000) and, since it can be up-regulated on CD28 negative cells, it may play an important role in apoptotic regulation in aged T cells.

There are myriad receptors that either directly costimulate or provide additional support for costimulation that may have roles in apoptotic regulation particularly in the absence of CD28. However, interestingly, in some cases the lack of CD28 negates signalling through some receptors. Borthwick et al. (2000) found that in CD8 CD28$^+$ cells costimulation through Cd11a and CD18 reduced AICD, although CD28$^-$ cells remained susceptible to apoptosis. Therefore, overall, aged T cells respond differently to the costimulatory environment they find themselves in due to the age-related changes in receptor expression and signalling.

In the context of immune surveillance the lack of CD28 and CD25 on aged T cells would seriously compromise the attacking T cells. Although the CD28$^-$ cells do not show dysfunction in all functional aspects they do have a reduced proliferative capacity. Therefore, the recognition, and subsequent immune response, of TAA's is compromised. In addition, at the T cell : tumour interface there would be heightened apoptosis of the infiltrating cells through tumour-derived CD95-L, although any AICD involvement would be reduced.

7.3. P-glycoprotein

P-glycoprotein (Pgp) is a 170–180 kDa protein encoded by the multi drug resistance-1 (MDR-1) human gene responsible for cross-resistance of mammalian cells to a number of chemotherapeutic agents. It functions as an ATP-dependent efflux pump by intercepting the drug as it moves through the lipid membrane and flips it from the inner to the outer leaflet and into the extracellular media (Higgins and Gottesman, 1992). In addition to tumour cells, Pgp has been found on barrier systems including the gut and blood-brain barrier. Of particular relevance here, Pgp is also expressed in haemapoetic stem cells, NK cells, B and T cells (Cordon-Cardo et al., 1990; Chaudhary and Roninson, 1991; Gupta et al., 1992). It has been found that CD8 cells express higher levels than CD4 cells (Gupta et al., 1992; Klimecki et al., 1994) and therefore has an impact on tumour attacking capacity. Pgp influences T cell function through affecting cytokine e.g. IL-2 transportation and release (Drach et al., 1996; Raghu et al., 1996). In aged mice and humans an increase in the proportion of Pgp^{hi} cells has been found (Gupta and Gollapudi, 1993; Witkowski and Miller, 1993) and, in the mice studies, this was associated with dysfunctional status (Witkowski and Miller, 1993). The failure of Pgp^{hi}, but not Pgp^{lo}, aged T cells to respond to TCR stimulation appears to be associated with impaired TCR signalling (Witowski and Miller, 1990) and may account for the decline with age of IL-2 responsiveness. Of particular interest are the findings that Pgp may function as an inhibitor of caspase activity in human T cells (Smyth et al., 1998) and therefore act as an apoptotic regulator. The changes in apoptotic capability seen in aged T cells may be associated with alterations in Pgp function due to its influence on the caspases and IL-2, although the latter two factors have been found not to be related (Gollapudi and Gupta, 2001). In particular the predominantly high expression on CD8 cells may correlate with inhibition of the caspase pathway resulting in age-related apoptotic resistance. CD4 cells, however, also show high expression of Pgp with age (Gollapudi and Gupta, 2001) and this finding, together with their apoptotic susceptibility, would not support a role for Pgp in caspase modulation. However, other regulatory factors undoubtedly impact on overall apoptotic capability in peripheral T cells, since we have found that in in vitro studies, of predominantly CD4 cells, Pgp function does not correlate with apoptotic susceptibility, although it remains able to affect caspase activity (unpublished observations). In the context of aged immune : tumour interactions Pgp would appear to have a role, primarily in protection of attacking T cells from chemotherapeutic intervention but also in the apoptotic susceptibility of T cells from tumour counter attack.

7.4. Cytokines

Cytokines play a vital role in both differentiation of T cells and functional regulation. Upon ageing there is a skewing of aged CD4 cells upon antigenic stimulation towards generating type 2 cytokines (IL-5, IL-6 and IL-10) due to a diminution in the ability of aged cells to generate type 1 (IL-2, IFNγ and TNFα) cytokines

(Shearer, 1997). Apoptosis has been reported to be affected by the T cells cytokine profile. There appears to be general agreement that Th2 cells are resistant to CD95-induced apoptosis, although a CD95-independent pathway equally affects both cell subsets (Ramsdell et al., 1994; Varadhachary et al., 1997; Zhang et al., 1997). The type 2 cytokine IL-6 is particularly associated to age-related changes in apoptotic capability. Increases in IL-6 levels occur with age and correlate with several age-related clinical problems (Ershler, 1993) and indeed IL-6 has been purported to be a biomarker of ageing (Pawelec et al., 2002). The increase in IL-6 levels will affect the apoptotic capability of T cells by reducing AICD through inhibition of CD95:CD95-L interactions (Ayroldi et al., 1998) thus leading to apoptotic resistance. However, clonally expanded CD4 and CD8 type 1 T cells remain susceptible to AICD (Ledru et al., 1998). The onset of type 1 T cell characteristics correlates kinetically with susceptibility, although it appears that cytokine production itself is not responsible for the unequal susceptibility to AICD (Zhang et al., 1997). Therefore, the apoptotic capability of T cells would appear to be influenced by the cytokine profile of the cell.

In the context of immune surveillance, IL-6 and TGFβ are particularly relevant. IL-6 is key in tumourgenesis and can also be generated from vascular smooth muscle cells in response to CD95 ligation (O'Connell et al., 2001). As discussed above, the IL-6 environment surrounding the tumour would be suggested to inhibit AICD at the immune:tumour interface compounding the age-associated type 2-related apoptotic resistance of infiltrating T cells. However, a cocktail of cytokines influence the immune:tumour interface with TGFβ being of particular importance. TGFβ is classically known as an immunosuppressor and a key effector-cytokine of regulatory T cells (Gorelik and Flavell, 2001), however it has recently been acknowledged as having stimulatory effects on tumour progression and invasion (Akhurst and Derynck, 2001). The suppressive effects are particularly relevant to understanding immune-privilege at the tumour site. TGFβ affects T cells through inhibiting IL-2 production and differentiation (Gorelik and Flavell, 2001), therefore suppressing proliferation of attacking T cells and cytotoxic T cell production. TGFβ may also affect ageing T cells since preliminary results from our group suggest that TGFβ is generated by aged T cell clones and influences function (unpublished observations). In addition, TGFβ generated from multiple myeloma (MM) cell lines (Cook et al., 1999) and patient-derived bone marrow (unpublished observations) suppresses T cell activity through IL-2 signalling and reduction in CD25 expression. The latter effect has also been reported on tumour infiltrating lymphocytes from breast carcinoma, although this was not correlated with TGFβ effects (Lopez et al., 1998). The effect of TGFβ on IL-2 signalling appears not to affect apoptotic capability of T cells (Cook et al., 1999), however, preliminary results utilising T cells cocultured with MM patient-derived bone marrow show increased CD95-L-induced apoptosis (unpublished observations). Additionally, the presence of TGFβ at an immune-privileged site influences the pro-inflammatory response by inhibiting neutrophil activation (Chen et al., 2001). As mentioned previously, this effect is particularly important in affecting the response generated through CD95-L counter attack (Green and Ferguson, 2001) and tipping the balance towards an anti-inflammatory response.

Overall, TGFβ supports tumour progression and suppresses immune attack and indeed, in an aged system, promotes many of the classic T cell biomarkers of ageing, namely reduced IL-2 production, signalling and changes in apoptotic capability.

8. Summary

In summary, aged T cells show classic signs of dysfunction that may detract from immune surveillance of tumours tipping the balance in the tumour's favour. Certainly, aged CD4 T cells would have a poor response to presented TAA's and may be more susceptible to apoptotic stimuli. However, CD8 cells at the tumour site could be argued to be able to generate effective responses to TAA's, being resistant to tumour counter-attack. However, the cytokine environment will play a role in the responsiveness of aged T cells. The tumour-derived generation of TGFβ will not favour differentiation of CD8 to cytotoxic T cells and would also reduce IL-2 signalling. In addition, the lack of aged T cell costimulation through CD28 would also support an immunosuppressed state sustaining tumour progression. Clearly the functional properties of an ageing immune system directly impact on cancer prognosis in the elderly and a greater understanding of the biomarkers associated with ageing in T cells is needed to provide possible interventionist strategies.

References

Abidzadeh, M., Pohla, H., Rehbein, A., Pawelec, G., 1996. The T cell in the ageing individual. Mech. Ageing Dev. 91, 145–154.

Adibzadeh, M., Pohla, H., Rehbein, A., Pawelec, G., 1995. Long-term culture of monoclonal human T lymphocytes: models for immunosenescence? Mech. Ageing Dev. 83, 171–183.

Aggarwal, S., Gupta, S., 1998. Increased apoptosis of T cell subsets in aging humans: altered expression of Fas (CD95), Bcl-2 and Bax. J. Immunol. 160, 1627–1637.

Aggarwal, S., Gupta, S., 1999. Increased activity of caspase 3 and caspase 8 in anti-Fas-induced apoptosis in lymphocytes from ageing humans. Clin. Exp. Immunol. 117, 285–290.

Akhurst, R.J., Derynck, R., 2001. TGFb signalling in cancer – a double-edged sword. Trends Cell Biol. 11, S44–S51.

Alderson, M.R., Tough, T., Davis-Smith, T., Braddy, S., Falk, B., Schooley, K.A., Goodwin, R.G., Smith, C.A., Ramsdell, F., Lynch, D.H., 1995. Fas ligand mediates activation-induced cell death in human T lymphocytes. J. Exp. Med. 181, 71.

Algeciras-Schimnich, A., Griffith, T.S., Lynch, D.H., Paya, C.V., 1999. Cell cycle-dependent regulation of FLIP levels and susceptibility to Fas-mediated apoptosis. J. Immunol. 162, 5205–5211.

Arai, H., Chan, S.Y., Bishop, D.K., Nabel, G.J., 1997. Inhibition of the alloantibody response by CD95 ligand. Nat. Med. 3, 843–848.

Ayroldi, E., Zollo, O., Cannarile, L., D'Adamio, F., Grohmann, U., Delfino, D.V., Riccardi, C., 1998. Interleukin-6 (IL-6) prevents activation-induced cell death: IL-2-independent inhibition of Fas/FasL expression and cell death. Blood 92, 4212–4219.

Beier, K.C., Hutloff, A., Dittrich, A.M., Heuck, C., Rauch, A., Buchner, K., Ludewig, B., Ochs, H.D., Mages, H.W., Kroczek, R.A., 2000. Induction, binding specificity and function of human ICOS. Eur. J. Immunol. 30, 3707–3717.

Boise, L.H., Thompson, C.B., 1997. Bcl-XL can inhibit apoptosis in cells that have undergone Fas-induced protease activation. PNAS 94, 3759.

Borthwick, N.J., Lowdell, M., Salmon, M., Akbar, A.N., 2000. Loss of CD28 expression on CD8(+) T cells is induced by IL-2 receptor gamma chain signalling cytokines and type I IFN, and increases susceptibility to activation-induced apoptosis. Int. Immunol. 12, 1005–1013.

Boucher, N., Dufeu-Duchesne, T., Vicaut, E., Farge, D., Effros, R.B., Schachter, F., 1998. CD28 expression in T cell aging and human longevity. Exp. Gerontol. 33, 267–282.

Boussiotis, V.A., Lee, B.J., Freeman, G.J., Gribben, J.G., Nadler, L.M., 1997. Induction of T cell clonal anergy results in resistance, whereas CD28– mediated costimulation primes for susceptibility to Fas- and Bax-mediated programmed cell death. J. Immunol. 159, 3156–3167.

Chamberlain, W.D., Falta, M.T., Kotzin, B.L., 2000. Functional subsets within clonally expanded CD8 + memory T cells in elderly humans. Clin. Immunol. 94, 160–172.

Chambers, C.A., Krummel, M.F., Boitel, B., Hurwitz, A., Sullivan, T.J., Fournier, S., Cassell, D., Brunner, M., Allison, J.P., 1996. The role of CTLA-4 in the regulation and initiation of T cell responses. Immunol. Rev. 153, 27–46.

Chao, D.T., Korsmeyer, S.J., 1998. BCL-2 family: regulators of cell death. Annu. Rev. Immunol. 16, 394–419.

Chaudhary, P.M., Roninson, I.B., 1991. Expression and activity of P-glycoprotein, a multidrug efflux pump, in human hematopoietic stem cells. Cell 66, 85–94.

Chen, W., Frank, M.E., Jin, W., Wahl, S.M., 2001. TGF-beta released by apoptotic T cells contributes to an immunosuppressive milieu. Immunity 14, 715–725.

Cohen, J.J., Duke, R.C., Fadok, V.A., Sellins, K.S., 1992. Apoptosis and programmed cell-death in immunity. Annu. Rev. Immunol. 10, 267–293.

Cook, G., Campbell, J.D., Carr, C.E., Boyd, K.S., Franklin, I.M., 1999. Transforming growth factor beta from multiple myeloma cells inhibits proliferation and IL-2 responsiveness in T lymphocytes. J. Leukoc. Biol. 66, 981–988.

Cordon-Cardo, C., O'Brien, J.P., Boccia, J., Casals, D., Bertino, J.R., Melamed, M.R., 1990. Expression of the multidrug resistance gene product (P-glycoprotein) in human normal and tumor tissues. J. Histochem. Cytochem. 38, 1277–1287.

Dennett, N.S., Barcia, R.N., McLeod, J.D., 2002. Age associated decline in CD25 and CD28 expression correlate with an increased susceptibility to CD95 mediated apoptosis in T cells. Exp. Gerontol. 37, 271–283.

Dhein, J., Walczak, H., Baumler, C., Debatin, K.M., Krammer, P.H., 1995. Autocrine T-cell suicide mediated by APO-1/(Fas/CD95). Nature 373, 438–441.

Drach, J., Gsur, A., Hamilton, G., Zhao, S., Angeler, J., Fiegl, M., Zojer, N., Radere, M., Habrel, M., Huber, H., 1996. Involvement of P-glycoprotein in the transmembrane transport of interleukin-2, IL-4 and interferon-gamma in normal T lymphocytes. Blood 88, 1747–1754.

Effros, R.B., Boucher, N., Porter, V., Zhu, X., Spaulding, C., Walford, R.L., Kronenberg, M., Cohen, D., Schachter, F., 1994. Decline in CD28+ T cells in centenarians and in long-term T cell cultures: a possible cause for both in vivo and in vitro immunosenescence. Exp. Gerontol. 29, 601–609.

Ershler, W.B., 1993. Interleukin-6: a cytokine for gerontologists. J. Am. Geriatr. Soc. 41, 176–181.

Fangoni, F.F., Vescovini, R., Mazzola, M., Bologna, G., Nigro, E., Lavagetto, G., Franceschi, C., Passeri, G., Sansoni, P., 1996. Expansion of cytotoxic CD8(+)CD28(−) T cells in healthy ageing people, including centenarians. Immunology 88, 501–507.

Gollapudi, S., Gupta, S., 2001. Anti-P-glycoprotein antibody-induced apoptosis of activated peripheral blood lymphocytes: a possible role of P-glycoprotein in lymphocyte survival. J. Clin. Immunol. 21, 420–430.

Gorelik, L., Flavell, R.A., 2001. Immune-mediated eradication of tumors through the blockade of transforming growth factor-beta signaling in T cells. Nat. Med. 7, 1118–1122.

Green, D.R., Ferguson, T.A., 2001. The role of Fas ligand in immune privilege. Nat. Rev. Mol. Cell. Biol. 2, 917–924.

Greenwald, R.J., Latchman, Y.E., Sharpe, A.H., 2002. Negative co-receptors on lymphocytes. Curr. Opin. Immunol. 14, 391–396.

Griffith, T.S., Yu, X., Herndon, J.M., Green, D.R., Ferguson, T.A., 1996. CD95-induced apoptosis of lymphocytes in an immune privileged site induces immunological tolerance. Immunity 5, 7–16.

Gupta, S., Gollapudi, S., 1993. P-glycoprotein (MDR 1 gene product) in cells of the immune system: Its possible role and alterations in aging and human immunodeficiency virus-1 infection. J. Clin. Immunol. 13, 289–301.

Gupta, S., Kim, C.H., Tsuruo, T., Gollapudi, S., 1992. Preferential expression of multidrug resistance gene 1 product (P-glycoprotein), a functionally active efflux pump in human CD8+ T cells. A role in cytotoxic effector function. J. Clin. Immunol. 12, 451–458.

Hahne, M., Rimoldi, D., Schroter, M., Romero, P., Schreier, M., French, L.E., Schneider, P., Bornand, T., Fontana, A., Lienard, D., Cerottini, J., Tschopp, J., 1996. Melanoma cell expression of Fas(Apo-1/CD95) ligand: implications for tumor immune escape. Science 274, 1363–1366.

Haynes, L., Eaton, S.M., Swain, S.L., 2000. The defects in effector generation associated with aging can be reversed by addition of IL-2 but not other related gamma(c)-receptor binding cytokines. Vaccine 18, 1649–1653.

Higgins, C.F., Gottesman, M.M., 1992. Is the multidrug transporter a flippase? Trends Biochem. Sci. 17, 18–21.

Hill, L.L., Shreedhar, V.K., Kripke, M.L., Owen-Schaub, L.B., 1999. A critical role for Fas ligand in the active suppression of systemic immune responses by ultraviolet radiation. J. Exp. Med. 189, 1285–1294.

Hohlbaum, A.M., Moe, S., Marshak-Rothstein, A., 2000. Opposing effects of the transmembrane and soluble Fas ligand expression on inflammation and tumor survival. J. Exp. Med. 191, 1209–1219.

Hutloff, A., Dittrich, A.M., Beier, K.C., Eljaschewitsch, B., Kraft, R., Anagnostopoulos, I., Kroczek, R.A., 1999. ICOS is an inducible T-cell co-stimulator structurally and functionally related to CD28. Nature 397, 263–266.

Kelliher, M.A., Grimm, S., Ishida, Y., Kuo, F., Stanger, B.Z., Leder, P., 1998. The death domain kinase RIP mediates the TNF-induced NF-kappaB signal. Immunity 8, 297–303.

Kirchhoff, S., Muller, W.W., Krueger, A., Schmitz, I., Krammer, P.H., 2000a. TCR-mediated up-regulation of c-FLIPshort correlates with resistance toward CD95-mediated apoptosis by blocking death-inducing signaling complex activity. J. Immunol. 165, 6293–6300.

Kirchhoff, S., Muller, W.W., Li-Weber, M., Krammer, P.H., 2000b. Up-regulation of c-FLIPshort and reduction of activation-induced cell death in CD28-costimulated human T cells. Eur. J. Immunol. 30, 2765–2774.

Klimecki, W.T., Futscher, B.W., Grogan, T.M., Dalton, W.S., 1994. P-glycoprotein expression and function in circulating blood cells from normal volunteers. Blood 83, 2451–2458.

Kroemer, G., Zamzami, N., Susin, S.A., 1997. Mitochondrial control of apoptosis. Immunol. Today 18, 44–51.

Ledru, E., Lecoeur, H., Garcia, S., Debord, T., Gougeon, M.-L., 1998. Differential susceptibility to activation-induced apoptosis among peripheral Th1 subsets:correlation with Bcl-2 expression and consequences for AIDS pathogenesis. J. Immunol. 160, 3194–3206.

Lin, Y., Devin, A., Rodriguez, Y., Liu, Z.G., 1999. Cleavage of the death domain kinase RIP by caspase-8 prompts TNF- induced apoptosis. Genes Dev. 13, 2514–2526.

Lopez, C.B., Rao, T.D., Feiner, H., Shapiro, R., Marks, J.R., Frey, A.B., 1998. Repression of interleukin-2 mRNA translation in primary human breast carcinoma tumor-infiltrating lymphocytes. Cell. Immunol. 190, 141–155.

Manukata, S., Enomot, T., Tsujimoto, M., Otsuki, Y., Miwa, H., Kanno, H., Aozasa, K., 2000. Expression of Fas ligand and other apoptosis-related genes and their prognostic significance in epithelial ovarian neoplasms. Br. J. Cancer 82, 1446–1452.

Martinon, F., Holler, N., Richard, C., Tschopp, J., 2000. Activation of a pro-apoptotic amplification loop through inhibition of NF-kappaB-dependent survival signals by caspase-mediated inactivation of RIP. FEBS Lett. 468, 134–136.

McLeod, J.D., Walker, L.S., Patel, Y.I., Boulougouris, G., Sansom, D.M., 1998. Activation of human T cells with superantigen (staphylococcal enterotoxin B) and CD28 confers resistance to apoptosis via CD95. J. Immunol. 160, 2072–2079.

Miyawaki, T., Uehara, T., Nibu, R., Tsuji, T., Yonehara, S., Taniguchi, N., 1992. Differential expression of apoptosis-related Fas antigen on lymphocyte subpopulations in human peripheral blood. J. Immunol. 149, 3753.

Murtaza, A., Kuchroo, V.K., Freeman, G.J., 1999. Changes in the strength of co-stimulation through the B7/CD28 pathway alter functional T cell responses to altered peptide ligands. Int. Immunol. 11, 407–416.

O'Connell, J., Bennett, M.W., O'Sullivan, G.C., Collins, J.K., Shanahan, F., 1999. The Fas counterattack: cancer as a site of immune privilege. Immunol. Today 20, 46–52.

O'Connell, J., Houston, A., Bennett, M.W., O'Sullivan, G.C., Shanahan, F., 2001. Immune privilege or inflammation? Insights into the Fas ligand enigma. Nat. Med. 7, 271–274.

O'Connell, J., O'Sullivan, G.C., Collins, J.K., Shanahan, F., 1996. The Fas counter attack: Fas-mediated T cell killing by colon cancer cells expressing Fas ligand. J. Exp. Med. 184, 1075–1082.

Ottonello, L., Tortolina, G., Amelotti, M., Dallegri, F., 1999. Soluble Fas ligand is chemotactic for human neutrophilic polymorphonuclear leukocytes. J. Immunol. 162, 3601–3606.

Pawelec, G., Barnett, Y., Forsey, R., Frasca, D., Globerson, A., McLeod, J., Caruso, C., Franceschi, C., Fulop, T., Gupta, S., Mariani, E., Mocchegiani, E., Solana, R., 2002. T cells and aging, January 2002 update. Front. Biosci. 7, d1056–d1183.

Pawelec, G., Effros, R.B., Caruso, C., Remarque, E., Barnett, Y., Solana, R., 1999a. T cells and aging (update February 1999). Front. Biosci. 4, D216–D269.

Pawelec, G., Muller, R., Rehbein, A., Hahnel, K., Ziegler, B.L., 1999b. Finite lifespans of T cell clones derived from CD34+ human haematopoietic stem cells in vitro. Exp. Gerontol. 34, 69–77.

Pawelec, G., Sansom, D., Rehbein, A., Adibzadeh, M., Beckman, I., 1996. Decreased proliferative capacity and increased susceptibility to activation-induced cell death in late-passage human CD4+ TCR2+ cultured T cell clones. Exp. Gerontol. 31, 655–668.

Raghu, G., Park, S.W., Robinson, I.B., Mechetner, E.B., 1996. Monoclonal antibodies against p-glycoprotein, an MDR1 gene product, inhibits interleukin-2 release from PHA activated lymphocytes. Exp. Haematol. 24, 1258–1264.

Ramsdell, F., Seaman, M.S., Miller, R.E., Picha, K.S., Kennedy, M.K., Lynch, D.H., 1994. Differential ability of Th1 and Th2 T cells to express Fas ligand and to undergo activation-induced cell death. Int. Immunol. 6, 1545.

Restifo, N.P., 2000. Not so Fas: re-evaluating the mechanisms of immune privilege and tumor escape. Nat. Med. 6, 493–495.

Scaffidi, C., Fulda, S., Srinivasan, A., Friesen, C., Li, F., Tomaselli, K.J., Debatin, K.M., Krammer, P.H., Peter, M.E., 1998. Two CD95 (APO-1/Fas) signaling pathways. EMBO J. 17, 1675–1687.

Seliger, B., Maeurer, M.J., Ferrone, S., 1997. TAP off – tumors on. Immunol. Today 18, 292–299.

Shearer, G.M., 1997. Th1/Th2 changes in aging. Mech. Ageing Dev. 94, 1–5.

Shimizu, S., Narita, M., Tsujimoto, Y., 1999. Bcl-2 family of proteins regulate the release of apoptogenic cytochrome c by the mitochondrial channel VDAC. Nature 399, 483–487.

Smith, C.A., Farrah, T., Goodwin, R.G., 1994. The TNF receptor superfamily of cellular and viral proteins: activation, costimulation and death. Cell 76, 959–962.

Smyth, M.J., Krasovskis, E., Sutton, V.R., Johnstone, R.W., 1998. The drug efflux protein, P-glycoprotein, additionally protects drug-resistant tumor cells from multiple forms of caspase-dependent apoptosis. Proc. Natl. Acad. Sci. USA 95, 7024–7029.

Spaulding, C., Guo, W., Effros, R.B., 1999. Resistance to apoptosis in human CD8+ T cells that reach replicative senescence after multiple rounds of antigen-specific proliferation. Exp. Gerontol. 34, 633–644.

Suda, T., Takahashi, T., Golstein, P., Nagata, S., 1994. Molecular cloning and expression of the Fas ligand, a novel member of the tumour necrosis factor family. Cell 75, 1169.

Suzuki, I., Fink, P.J., 2000. The dual functions of fas ligand in the regulation of peripheral CD8+ and CD4+ T cells. Proc. Natl. Acad. Sci. USA 97, 1707–1712.

Tanaka, M., Itai, T., Adachi, M., Nagata, S., 1998. Downregulation of Fas ligand by shedding. Nat. Med. 4, 31–36.

Ting, A.T., Pimentel-Muinos, F.X., Seed, B., 1996. RIP mediates tumor necrosis factor receptor 1 activation of NF-kappaB but not Fas/APO-1-initiated apoptosis. EMBO J. 15, 6189–6196.

Tortorella, C., Loria, M.P., Piazzolla, G., Schulzekoops, H., Lipsky, P.E., Jirillo, E., Antonaci, S., 1997. Age related impairment of T cell proliferative responses related to the decline of CD28+ T cell subsets. Arch. Gereontol. Geriatr. 26, 55–70.

Varadhachary, A.S., Perdow, S.N., Hu, C., Ramanarayanan, M., Selgame, P., 1997. Differential ability of T cell subsets to undergo activation-induced cell death. PNAS 94, 5778.

Vaziri, H., Schachter, F., Uchida, I., Wei, L., Zhu, X., Effros, R.B., Cohen, D., Harley, C.B., 1993. Loss of telomeric DNA during aging of normal and trisomy 21 human lymphocytes. Am. J. Hum. Genet. 52, 661–667.

Wakikawa, A., Utsuyama, M., Hirokawa, K., 1997. Altered expression of various receptors on T cells in young and old mice after mitogenic stimulation: a flow cytometric analysis. Mech. Ageing Dev. 94, 113–122.

Walker, L.S., McLeod, J.D., Boulougouris, G., Patel, Y.I., Ellwood, C.N., Hall, N.D., Sansom, D.M., 1999. Lack of activation induced cell death in human T blasts despite CD95L up-regulation: protection from apoptosis by MEK signalling. Immunology 98, 569–575.

Wesselborg, S., Janssen, O., Kabelitz, D., 1993. Induction of activation-induced cell death (apoptosis) in activated but not resting peripheral-blood T-cells. J. Immunol. 150, 4338.

Weyand, C.M., Brandes, J.C., Schmidt, D., Fulbright, J.W., Goronzy, J.J., 1998. Functional properties of CD4+ CD28− T cells in the aging immune system. Mech. Ageing Dev. 102, 131–147.

Witkowski, J.M., Miller, R.A., 1993. Increased function of P-glycoprotein in T lymphocyte subsets of aging mice. J. Immunol. 150, 1296–1306.

Witowski, J.M., Miller, R.A., 1990. Calcium signal abnormalities in murine T lymphocytes that express the multidrug transporter P-glycoprotein. Mech. Ageing Dev. 107, 165–180.

Zhang, J., Ma, B., Marshak-Rothstein, A., Fine, A., 1999. Characterisation of a novel cis-element that regulates Fas lignd expression in corneal endothelial cells. J. Biol. Chem. 274, 26,537–26,542.

Zhang, X., Brunner, T., Carter, L., Dutton, R.W., Rogers, P., Bradley, L., Sato, T., Reed, J.C., Green, D., Swain, S.L., 1997. Unequal death in T helper cell (Th)1 and Th2 effectors: Th1, but not Th2, effectors undergo rapid Fas/Fas-l-mediated apoptosis. J. Exp. Med. 185, 1837–1849.

Advances in
Cell Aging and
Gerontology

A road to ruins: an insight into immunosenescence

Sudhir Gupta

*Division of Basic and Clinical Immunology, Medical Sciences I, C-240, University of California,
Irvine, CA 92697, USA*
Correspondence address: Tel.: + 1-949-824-5818; fax: + 1-949-824-4362. E-mail: sgupta@uci.edu

Contents

1. Introduction

Ageing is a stochastic process. It is genetically determined (because genetic constitution determines its course) and may appear to be programmed because it encompasses stereotypic biochemical responses to particular cellular states. Ageing is accompanied by lymphopenia, a steady decline in immune functions, and increased frequency of infections, cancer, and autoimmune phenomena. Immunological abnormalities in ageing include a defect in signaling pathways, a decreased T cell response to mitogens and antigens, altered cytokine expression and increase in memory T cell phenotype (reviewed in Pawelec et al., 2002). Lymphopenia is shared by both CD4 + and CD8 + T cells (Fagiola et al., 1993; Gupta, 2002). Fagiola et al. (1993) have also shown increased TNF-α production in aged humans.

2. Apoptosis

Apoptosis or programmed cell death plays an important role in embryogenesis, metamorphosis, cellular homeostasis, and tissue atrophy and tumor regression.

Advances in Cell Aging and Gerontology, vol. 13, 173–189

SIGNALING PATHWAYS OF APOPTOSIS

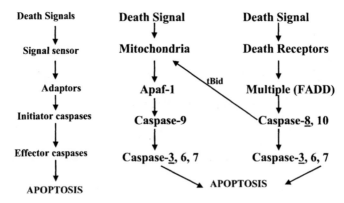

Fig. 1. Signaling steps of two major pathways of apoptosis. FADD, Fas-associated death domain.

In the immune system, apoptosis appears to play a crucial role in selection of the T cell repertoire, deletion of self-reactive T and B lymphocytes, deletion of peripheral effector T cells following termination of an immune response, regulation of immunological memory and in the cytotoxicity of target cells by CTL and NK cells (Gupta, 2000c, 2001b; Krammer, 2000). There are two major signaling pathways of apoptosis (Fig. 1), the death receptor pathway and the mitochondrial pathway (Ashkanazi and Dixit, 1998; Hengartner, 2002; Green and Reed, 1998). The mitochondrial pathway of apoptosis in ageing has not been explored. In the death receptor signaling pathway the signal is provided by the interaction between the ligand and death receptor. The death receptors belong to the tumor necrosis factor receptor (TNFR) superfamily and the ligands belong to the tumor necrosis factor (TNF)/ nerve growth factor family. Members of the death receptor family contain one to five cysteine-rich repeats in their extracellular domain and a death domain (DD) in their cytoplasmic tail. The DD is essential for initiating apoptotic signals. Following ligation of the death receptor by ligand, the death receptors oligomerize and recruit adapter molecules and initiator caspases to a complex, the death-inducing signaling complex (DISC), which results in the activation of the initiator caspases. The initiator caspases then cleave effector caspases (caspase-3, -6, and -7) to generate active effector caspases, which in turn cleave a number of target substrates resulting in morphologic and biochemical characteristics of apoptosis. The process of apoptosis is tightly regulated by a number of gene products that promote or block apoptosis at different stages. The most extensively studied and perhaps most important are the Bcl-2 family proteins (Reed, 1997). Bcl-2 and Bcl-X_L are anti-apoptotic, whereas Bax is pro-apoptotic. These molecules either homodimerize or heterodimerize with molecules of opposing function (e.g. Bax may heterodimerize with Bcl-2 or Bcl-x_L). The net influence on apoptosis appears to be dependent upon the relative concentrations of these molecules in heterodimers and the ratio between pro- and anti-apoptotic molecules. Though the regulatory role of Bcl-2 family proteins in the mitochondrial pathway of apoptosis is well established (Green and Reed, 1998; Hengartner, 2002), their role in the regulation of death receptor signaling is controversial (Scaffidi et al., 1998) and may be cell type specific. Though the two pathways appear to be distinct, death receptor signaling may mediate apoptosis through the mitochondrial pathway via tBid (Li et al., 1998).

3. Activation-induced cell death

Once a cellular immune response has taken place most of the cells are removed by activation-induced cell death (AICD), which appears to be primarily mediated by Fas (CD95)–Fas ligand (CD95L) interactions (Krammer, 2000). Other molecules may also be involved in this process; the TNF/TNFR system has been suggested to play a preferential role in apoptosis of CD8 + T cells (Lenardo, 1997), although more recent data suggest that TNF–TNFR is involved in apoptosis of both CD4 + and CD8 + T cells (Aggarwal et al., 1999; Gupta, 2000b, 2002). On the practical side, there are basic differences in in vitro-induced apoptosis during AICD, CD95–CD95L, and TNF–TNFR signaling. In AICD, pre-activated T cells are re-activated with the same stimulus, whereas in the CD95–CD95L system, activated cells are stimulated with anti-CD95 monoclonal antibodies or soluble CD95L, and in the TNF–TNFR system, activated T cells are stimulated with TNF-α. Within the AICD system, resistance or sensitivity is dependent upon proper engagement of TCRs by specific antigen bound to MHC molecules, antigen concentration, and co-stimulatory signals (Di Somma et al., 1999). It has been suggested that AICD could be inhibited in memory T cells activated in vivo by a foreign antigen, but may become operative when the antigen has been cleared. These observations are particularly important when we discuss apoptosis in ageing and to explain some observations that appear to be contradictory. Another fact we should keep in mind is that there are differences between in vitro replicative senescence and in vivo senescence. In in vitro replicative senescence, during repeated stimulation in culture, there is a possibility of selecting an apoptosis-resistant population, which may be different from that present in vivo during ageing. Finally, we should also consider that apoptosis might be different among various lymphoid compartments (i.e. peripheral blood vs spleen, vs lymph nodes). Finally, in humans consideration should be given to other factors, including nutritional and socio-economic status of the aged and the use of medications and nutritional and vitamin supplements.

4. CD95/CD95L pathway of apoptosis

CD95, a 45–52 kDa glycoprotein, is a type I transmembrane receptor that is widely expressed and constitutively present on lymphocytes; however, CD95 ligand (CD95L), a type II transmembrane protein, displays a more restricted expression. It is lacking from resting lymphocytes and is induced upon activation of lymphocytes. CD95L can be cleaved from the cell surface by metalloproteases and therefore CD95L may be found in the soluble form in vivo and can trigger apoptosis

CD95-Mediated Signaling Pathway of Apoptosis

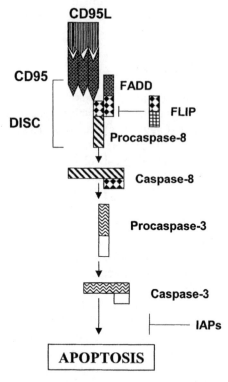

Fig. 2. CD95-mediated apoptosis signaling. DISC, death-inducing signaling complex; FLIP, flice inhibitory protein; IAPs, inhibitor of apoptosis proteins.

(Tanaka et al., 1995). The steps of the CD95-mediated apoptosis signaling pathway are shown in Fig. 2 (Gupta, 2000a). Upon ligation with CD95L, or anti-CD95 monoclonal antibodies, CD95 undergoes trimerization. Since cytoplasmic domains do not have intrinsic enzymatic activity, they recruit an adapter protein, the fas-associated death domain (FADD), which interacts with the DD of CD95 via its own DD. FADD also contain a death effector domain (DED) and through homologous interaction, it recruits procaspse-8 (Flice) to form a DISC, which serves as a platform to initiate enzymatic activation of the apoptotic pathway. Procaspase-8 is auto-proteolytically cleaved to generate active caspase-8, which is released from the DISC into the cytoplasm where it serves as an enzyme for effector caspases. Active caspase-8 cleaves a number of effector procaspases including procaspases-3 to generate active effector caspases, which cleave and activate a number of substrates, resulting in characteristic apoptosis. These substrates include structural proteins (e.g. lamin, fodrin, actin, gelsolin), enzymes including poly (ADP-ribose) polymerase (PARP), DNA-PKcs, PKC θ, and ICAD, and transcription factors like NF-κB.

This classical pathway occurs in so-called type I cells (Scaffidi et al., 1998). In type II cells, procaspases-8 levels are very low and thus hardly any DISC is formed; therefore the caspase cascade has to be amplified via the mitochondrial pathway (Scaffidi et al., 1998). Caspase-8 cleaves Bid, a Bcl-2 family member; truncated Bid (tBid) translocates to and activates mitochondria resulting in the release of cytochrome *c*, activation of caspase-9 and finally of effector caspases, predominantly caspase-3 resulting in apoptosis (Li et al., 1998).

Apoptosis mediated by the CD95–CD95L interaction is regulated by other DED-containing molecules including FLIP (Flice-inhibitory protein). These proteins contain two DEDs. Cellular FLIP (cFLIP) is present in two alternatively spliced isoforms, the long (FLIP$_L$) and short (FLIP$_S$) forms. FLIP$_L$ contains a COOH-terminal domain beyond two DEDs and resembles caspase-8 and -10. However, FLIP$_L$ is devoid of protease activity (Irmler et al., 1997). Stable over-expression of FLIP$_L$ or FLIP$_S$ results in resistance to receptor-mediated apoptosis. More recently, it has been reported that FLIP also promotes activation of NF-κB and Erk signaling pathways by recruiting adapter proteins (including RIP, TRAF-1, 2, 3 and Raf) involved in this signaling (Kataoka et al., 2000). Therefore, FLIP is not simply an inhibitor of death receptor-induced apoptosis but also mediates the activation of NF-κB. The DED of FLIPs binds to CD95–FADD complexes and inhibits the recruitment and activation of procaspases-8 and therefore acts as an anti-apoptotic molecule (Hu et al., 1997; Irmler et al., 1997; Thome et al., 1997; Yeh et al., 2000). Since FADD is also used as an adaptor molecule for TNF-induced apoptosis, FLIP is an anti-apoptotic for both CD95 and TNFR-mediated apoptosis.

5. TNFR/TNF pathway of apoptosis

Tumor necrosis factor (TNF-α) is a pleiotropic cytokine that exerts its biological activities via two distinct cell surface receptors, i.e. TNFR-I and TNFR-II (Rothe et al., 1992; Tartaglia and Goeddel, 1992; Vandenabeele et al., 1995a; Screaton and Xu, 2002; Gupta, 2001a,b). Both TNF receptors have significant homology in their extracellular domains; however, they differ structurally in their cytoplasmic domains. TNFR-I contains a DD, whereas TNFR-II lacks DD. TNFR-I mediates signaling for both activation (via NF-κB) and for cell death, whereas TNFR-II mediates an activation signal. Furthermore, the expression of these two receptors is differentially regulated (Thomas et al., 1990; Ware et al., 1991). TNF receptors do not exhibit enzymatic function, and therefore, have to depend on recruitment of other proteins for signaling (Yuan, 1997; Rath and Aggarwal, 1999).

Steps representing the events in TNF-R-induced death are shown in Fig. 3. Upon ligation with TNF, TNF-RI undergo trimerization and induce association of the receptors' DD and subsequent recruitment of an adapter protein containing homologous DD; the TRADD (TNF receptor-associated death-domain). TRADD functions as a platform adapter that recruits several signaling molecules to the activated receptor, including FADD, TRAF-2 (TNF-R-associated factor-2) and receptor

TNF-R Signaling Pathway of Apoptosis

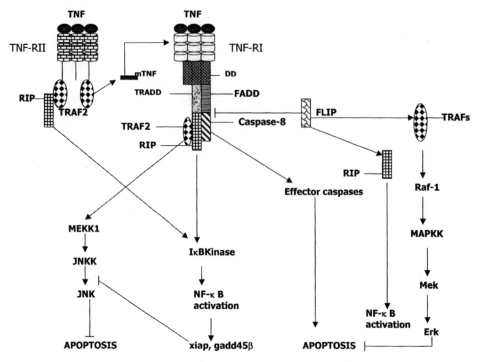

Fig. 3. Signaling pathway of TNF-induced apoptosis. DD, death domain; RIP, receptor interactive protein; TRADD, TNF receptor-associated death domain; TRAFs, TNF receptor-associated factors; JNK, Janus kinase; MAPKK, mitogens-activated protein kinase kinase.

interactive protein (RIP). TRAF-2 and RIP stimulate pathways leading to activation of NF-κB, whereas FADD mediates activation of apoptosis (Hsu et al., 1996a,b). Therefore, FADD is a common conduit for both CD95- and TNF-RI-mediated apoptosis. Recruitment of TRADD and FADD following ligation of TNFR-I with TNF-α results in the recruitment (to DISC) and autocatalytic activation of procaspase-8 to mature active caspase-8. Active caspase-8 cleaves effector procaspase 3, 6, and 7 to generate mature active effector caspase-3, -6, and -7. Mature effector caspases cleave a number of substrates mentioned above culminating in the morphological and biochemical changes of apoptosis.

Unlike TNF-RI, TNF-RII lacks a cytoplasmic DD, interaction between TNF-α and TNF-RII results in binding of TRAF1 and TRAF2 to the cytoplasmic portion of TNF-RII. However, it has been reported that TNF-RII may play an important role in the regulation of apoptosis through TNF-RI. Several investigators have reported that TNF-RII potentiates TNF-R-I-mediated apoptosis (Declercz et al., 1998; Grell et al., 1999; Haridas et al., 1998; Kelliher et al., 1998; Pimentel-Muinos and Seed, 1999; Tartaglia et al., 1993; Vandenabeele et al., 1995b; Weiss et al., 1998). Pimentel-Muinos and Seed (1999) have shown that RIP, a DD kinase that is required for NF-κB activation by TNF-RI, can dramatically affect signaling through TNF-RII. They showed that in the presence of RIP, TNF-RII triggers apoptosis, whereas in the absence of RIP, TNF-RII activates NF-κB. However, RIP KO mice show increased sensitivity to TNF-α-induced apoptosis (Kelliher et al., 1998). We have not been able to demonstrate that TNFR-II potentiates TNFR-I-mediated apoptosis (Gupta, unpublished observation). Therefore, the role of TNF-RII in enhancing TNF-induced apoptosis remains unclear.

NF-κB is one of the transcription factors that play an important role in the regulation of immune response genes (Baeuerle and Henkel, 1994; Ghosh et al., 1998). NF-κB exists as either heterodimers or homodimers of a subfamily of the Rel family of proteins. The predominant form of NF-κB is a heterodimer comprising p50 (NF-κB1) and p65 (RelA). Other forms contain p52 (NF-κB2), RelB, and c-Rel subunits (Ghosh et al., 1998). A number of genes, including cytokines, chemokines, cell surface receptors and adhesion molecules, are targets of NF-κB. In unstimulated cells, NF-κB is kept in the cytoplasm through interaction with the inhibitory protein termed IκB (inhibitor κB) (Baldwin, 1996). When cells are exposed to inducers of NF-κB, such as TNF, IκB is phosphorylated at two specific serine residues (Brown et al., 1995; Zandi et al., 1998; Delahase et al., 1999; Franzoso and Siebenlist, 1995). This phosphorylation is a signal for ubiquitination and degradation of IκB by the 26S proteosome (Pahl, 1999). Free NF-κB dimmers are released and translocated to the nucleus, where they activate transcription of target genes. The protein kinase complex that phosphorylates IκB in response to proinflammatory signals (e.g. TNF-α) contains two catalytic subunits, IKKα and IKKβ (or IKK1 and IKK2) and a regulatory subunit, IKKγ or NEMO (NF-κB essential modulator) (Zandi et al., 1998; Karin, 1999). The predominant form of IKK in mammalian cells is an IKKα:IKKβ heterodimer associated with IKKγ. Delahase et al. (1999) demonstrated that phosphorylation of two sites at the activation loop of IKKβ is essential for the activation of IKK by TNFα. They showed that IKKβ and not the IKKα is the target for TNFα and IL-1. Once activated, IKKβ is autophosphorylated at a carboxyl terminal serine cluster resulting in decreased IKK activity and may thus prevent prolonged cellular activation. Furthermore, it has been demonstrated that IKKβ is essential for protecting cells from apoptosis (Li et al., 1999), including T cells from TNF-α-induced apoptosis (Senftleben et al., 2001).

NF-κB has been shown to inhibit apoptosis, especially that triggered by TNF-α (Antwerp et al., 1996; Beg and Baltimore, 1996; Liu et al., 1996). It has been demonstrated that a number of anti-apoptotic genes are upregulated by NF-κB activation, including cIAP1 and cIAP2 (Chu et al., 1997; You et al., 1997). It was not until recently that it was shown how NF-κB mediates its anti-apoptotic effect. De Smaele et al. (2001), Tang et al. (2001) have demonstrated that NF-κB activated by TNF induces two genes, the *gadd45β* and *xiap*, where the proteins encoded by these genes inhibit apoptosis mediated by JNK activation. We are currently studying the effect of age on the expression and regulation of these two genes. A detailed signaling pathway of TNF-α-induced NF-κB activation and mechanism of anti-apoptotic effect of NF-κB is shown in Fig. 4.

NF-κB Activation and mechanism of anti-apoptotic effect of NF-κB

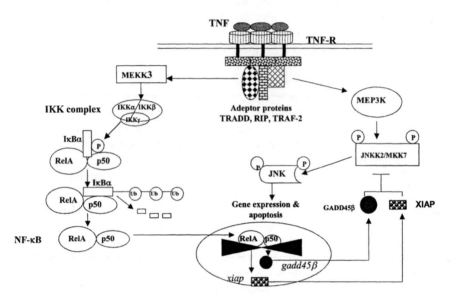

Fig. 4. TNF-induced activation of NF-κB and mechanism of anti-apoptotic effect of NF-κB. Ub, ubiquitination; P, phosphorylation.

6. Apoptosis of T cells in aged mice

There is a controversy regarding the proportions and numbers of CD4+ and/or CD8+ T cells in aged mice, especially with regard to CD4+ T cells. Previous studies describe either a decreased proportion of CD4+ T cells in lymph nodes or peripheral blood from aged mice (Boersma et al., 1985; Calahan et al., 1993; Miller, 1997; Grossmann et al., 1993); no change with age in CD4+ T cells in spleen or lymph nodes (Kirschmann and Murasko, 1992; Komuro et al., 1990; Sidman et al., 1987); or increases in CD4+ T cells with age (Engwerda et al., 1996). Similarly, CD8+ T cells in aged mice are increased (Gonzalez-Quintial and Theofilopolous, 1992; Linton et al., 1996), decreased (Komuro et al., 1990; Utsuyama and Hirokawa, 1987), or unchanged (Miller, 1997) depending on the study. Hsu et al. (2001) observed increased CD8+ and decreased CD4+ T cells resulting in reversal of CD4+/CD8+ T cell ratio. The reason for such diverse observations is unclear. There appears to be an agreement, however, that memory CD4+ and CD8+ T cell subsets are increased in aged mice.

In mice, apoptosis has been examined predominantly for AICD and no studies have been published of TNF-TNFR apoptosis. Zhou et al. (1995) generated *fas* transgenic mice and compared immunological status of young and old transgenic mice with wild-type littermates. They found that Fas expression and Fas-induced apoptosis was decreased in old wild type mice, but not in old transgenic mice. In old wild type mice, there was an increase in CD44+ Fas− cells, decreased IL-2 production and increased IL-4 and IL-10 production. In the transgenic mice these changes were not observed. Furthermore, age-related thymus involution was prevented in the Fas-transgenic mice. However, the life-span of the transgenic mice was not increased. Spaulding et al. (1997) showed that T cell apoptosis induced by irradiation, heat shock or CD3-stimulation was reduced in old compared to young mice. Polyak et al. (1997) also reported higher levels of in vivo and in vitro lymphocyte apoptosis after irradiation in young compared to old mice. More recently, Hsu et al. (2001) have reported decreased apoptosis in activated CD8+ cells as compared to CD4+ T cells from aged mice; however this difference was observed only at 96 h of activation and no such difference was observed at 48 h. In contrast, Chrest et al. (1995), using cells from the lymph nodes of Balb/C mice, have shown an increased T cell apoptosis upon anti-CD3 engagement. Pahlavani and Vargas (2000) also reported increased activation-induced apoptosis in old Fischer 344 rats. Ex vivo T cells from very old mice have also been reported to be more susceptible to TCR-mediated AICD than those from young or old mice (Provinciali et al., 1998). Furthermore, the expression of CD95 in CD4 cells of old mice infected with *M. tuberculosis* is not decreased (Roberts and Orme, 1998). Telford and Miller (1999) also observed increased AICD in aged mice and not unexpectedly decreased apoptosis in CD8+ T cells after withdrawal of an agonist. The reason for these contradictory results could be any, or a combination, of several factors outlined above.

7. T cell apoptosis in aged humans

In humans during ageing there is a progressive decline in immunological functions (Pawelec et al., 2002; Gupta, 2000c). Although there is also a controversy regarding lymphopenia in humans, recently significant decreases in total lymphocytes and both CD4+ and CD8+ T cells have been reported (Fagiola et al., 1993; Gupta, 2002). Our data comprised analysis of 40 young and 40 old subjects. Aged subjects were of middle class socio-economic status and living an active and independent life. They were not taking any vitamins or nutritional supplements at least 1-week prior to study. Similar to mice, there appear to be increased proportions of memory T cells in ageing. In vivo, $CD4^+$ cells in old donors showed significantly decreased CD45RB expression (Kudlacek et al., 1995). Salmon et al. (1994) studied the effect of in vitro replicative senescence on the expression of CD45RB isoforms. They observed that CD45RA expression (marker for naïve cells) was lost rapidly after activation of naïve cells in vitro, whereas loss of RB expression was gradual and occurred over many cell cycles, and a reciprocal increase in CD45RO expression (marker for memory cells) was demonstrated. The progressive shift from RB^{hi+} to RB^{lo+} was paralleled by gradual loss of Bcl-2 and acquisition of CD95, as well as a gradual loss of the ability to secrete sufficient IL-2 to maintain autocrine proliferation (whereas IL-4 secretion remains intact). This eventually results in cells becoming dependent on exogenous IL-2 not only for growth but also for their survival, because without sufficient IL-2; cells undergo apoptosis. Longevity extension for T cell clones might therefore be achieved by upregulation of Bcl-2 or Bcl-2 family-member Bcl-x$_L$. It may be that Bcl-x$_L$ is critical because protection from apoptosis in resting T cells cultured on fibroblast monolayers is associated with maintenance of Bcl-x$_L$ but not Bcl-2 expression (Gombert et al., 1996). It should be emphasized; however, that any of these molecules might provide an anti-apoptotic function. Aggarwal and Gupta (1998) reported decreased Bcl-2 and increased Bax expression both at the protein and mRNA levels but no increase in Bcl-x$_L$ expression in lymphocytes from aged humans. In contrast, Effros (1998) observed increased Bcl-2 expression in in vitro replicative senescent cultures of CD8 cells, which are resistant to apoptosis and show reduced levels of caspase-3. In contrast, several experiments suggest that under certain circumstances CD3/CD28-stimulated CD8 cells may be susceptible to apoptosis despite their expression of Bcl-x$_L$ (Laux et al., 2000). Borthwick et al. (2000) also demonstrated that CD8 + CD28 + T cells when stimulated with cytokines with common γ chain (IL-2, IL-7 and IL-15) lost CD28 expression and became susceptible to AICD.

Replicative senescence studies of AICD and CD95-mediated cell death in aged humans have been performed either in vivo or in vitro. The results of these two types of studies are contrasting. In vivo studies show increased apoptosis in human T cell subsets, whereas in vitro replicative senescence studies show decreased apoptosis, especially in CD8 + T cells. However, more studies are needed to resolve the issue of resistance to apoptosis of CD8 + CD28− T cells.

The proportion of CD95 + CD4 + and CD8 + T cells is increased with age in vivo (Aggarwal and Gupta, 1998; Gupta, 2000c; Miyawaki et al., 1992; Phelouzat et al., 1997; Potestio et al., 1998; Shinohara et al., 1995). Aggarwal and Gupta (1998) reported increased expression of CD95 in both CD45RA + naïve and CD45RO + memory subpopulations of CD4 + and CD8 + T cells from aged subjects. Shinohara et al. (1995) also observed increased CD95 + memory CD4 + and CD8 + T cells in aged humans; however, they did not analyze the expression of CD95 in naïve T cell subsets. Aggarwal and Gupta (1998) demonstrated that CD95 over-expression was at both the protein and mRNA level. In contrast, decreased proportions of CD95 + T cell subsets have been reported in very advanced (> 90 years of age) ageing (Aspinall et al., 1998; Potestio et al., 1999; McNerlan et al., 1999). This discrepancy cannot be explained on the basis of different anti-CD95 monoclonal antibodies used in different studies because Aggarwal and Gupta (1998) demonstrated that upregulation of CD95 was also present at the gene level. The amount of soluble Fas in the blood of elderly donors is significantly increased as compared to young donors (Seishima et al., 1996). Aggarwal et al. (1997) have also demonstrated that neonatal T cell subsets express less CD95 as compared to young subjects. Therefore, it is likely that there is an age-dependent increase in the expression of CD95 and very advanced aged humans (> 90 years or centenarians) are immunological privileged due to down regulation of CD95 (and perhaps CD95L and less apoptosis). Then one can argue whether the study of centenarian is a true reflection of the effect of ageing on the immune system and for that matter on other systems as well. It has been demonstrated that CD95-mediated apoptosis correlates better with the expression of CD95L rather than with the expression of CD95 (Gupta et al., 1998; Yel et al., 1999). Aggarwal and Gupta (1998) reported increased expression of CD95L in both CD4+ and CD8+ T cells from aged humans. A number of investigators have observed increased AICD and anti-CD95-induced apoptosis in ageing (Aggarwal and Gupta, 1998; Lechner et al., 1996; Potestio et al., 1998; Drouet et al., 1999). Aggarwal and Gupta (1998) reported increased anti-CD95-induced apoptosis in both CD4+ and CD8+ T cells and their naïve and memory subsets. However, significantly more apoptosis was observed in CD45RO+ memory subsets as compared to CD45RA+ naïve T cells in both young and ageing subjects. Therefore, it appears that increased apoptosis of T cells and T cell subsets in aged humans is not simply due to an accumulation of memory T cells. More recently, it has been demonstrated that increased memory T cells in ageing are not due to prolonged activation in vivo and differentiation from naïve cells; rather, memory cells are produced in large numbers in the ageing bone marrow environment (Timm and Thoman, 1999). We also observed no difference, using several markers; including CD25, CD38, and CD69, in activation markers of CD4+ and CD8+ T cells and their naïve and memory subsets in aged humans as compared to young controls. Increased apoptosis was associated with increased expression and early and greater activation of both caspase-8 and -3 in aged subjects as compared to young subjects (Aggarwal and Gupta, 1999). Furthermore, Bcl-2 (anti-apoptotic) expression was decreased whereas Bax expression (pro-apoptotic) was increased in ageing subjects. Herndon et al. (1997) reported increased apoptosis in CD3+CD45RO− (naïve) T cells in ageing as compared to young following in vitro culture of lymphocytes stimulated with PHA and cultured with IL-2 for up to 6 days. Similarly, have shown increased susceptibility to apoptosis during clonal expansion of T cells from aged humans as compared to young subjects. Phelouzat et al. (1996, 1997) also reported increased AICD in lymphocytes from aged subjects, using suprapharmacologic concentrations of PMA and ionomycin. In contrast, an age-related decrease in CD95-negative cells and a reciprocal increase in CD28-negative cells have been reported (Fagnoni et al., 2000). Spaulding et al. (1997) observed resistance to apoptosis in human CD8+ T cells that reached replicative senescence after multiple rounds of antigen-specific proliferation. This is associated with upregulation of Bcl-2. It is also suggested that CD8+CD28− T cells that, in aged subjects are expanded, are resistant to AICD. It is likely that in replicative senescence apoptosis-sensitive cells are deleted during early repetitive stimulation leaving behind the apoptosis-resistant population of cells. Furthermore, there might be a differential sensitivity of this subset to different apoptotic stimuli.

We have made extensive studies of TNF-α-induced apoptosis in human ageing (Aggarwal et al., 1999; Gupta, 2000b, 2002). We showed that CD4 and CD8 cells from the elderly were also more susceptible to TNF-α-induced apoptosis. Furthermore, we have demonstrated that increased TNF-α-induced apoptosis was present both in CD45RA+ and CD45RO+ T cells, suggesting that increased apoptosis in ageing was not merely a consequence of increased accumulation of memory type cells. Increased apoptosis in ageing lymphocytes was associated with both increased expression of TNF-RI and decreased expression of TNF-RII, both at the mRNA and protein level, in both CD4+ and CD8+ T cell subsets. Furthermore, increased expression of TRADD and FADD was observed in lymphocytes from aged subjects both at the level of mRNA and protein. A decreased expression of TRAF-2 and TNF-RII and early and increased activation of caspases-8 and -3 were also observed in aged humans. Furthermore, we have observed increased sensitivity of T cell subsets from aged humans to TNF-α-induced apoptosis that is associated with increased and early activation of caspases. The sensitivity of T cells to TNF-α-induced apoptosis appears to be age-dependent; cord blood lymphocytes are least sensitive (Aggarwal et al., 2000) whereas aged T cells are most sensitive to TNF-α-induced apoptosis (Aggarwal et al., 1999). Because FADD is a common conduit for both CD95- and TNFR-mediated apoptosis and FADD expression is increased in ageing, we examined the role of FADD in increased sensitivity of aged lymphocytes to TNF-induced apoptosis. A significant decrease in TNF-induced apoptosis was observed when a dominant negative FADD was expressed in aged T cells (unpublished observation). This would suggest that over-expression of FADD may in part be responsible for increased sensitivity of aged T cells to TNF-induced and perhaps CD95-mediated apoptosis.

In ageing, studies of NF-κB pathways are few and limited. Pahlavani and Harris (1996) reported decreased NF-κB DNA binding activity in nuclear extracts of concanavalin A-stimulated splenic lymphocytes from old Fischer rats as compared to young rats. We have also observed decreased TNF-induced DNA-binding activity of NF-κB in lymphocytes from aged humans as determined by a recently developed ELISA assay (unpublished data). Whisler et al. (1996) reported decreased levels of NF-κB in unstimulated and PHA, PMA and anti-CD3-stimulated T cells from five aged humans as compared to five young subjects. Trebilcock and Ponnappan (1996) demonstrated decreased induction of NF-κB in response to PMA and TNF. These authors further suggested that decreased induction of NF-κB could be due to decreased proteosome-mediated degradation of IκB (Ponnappan et al., 1999). None of these investigators studied IKK activity or investigated a role of decreased NF-κB activation in apoptosis. We have also observed decreased phosphorylation of Iκ-B and activation of NF-κB in TNF-treated lymphocytes from aged humans (manuscript in preparation). We have observed that IKKβ expression (at the protein level) is comparable between young and aged T cells. However, over-expression of IKKβ by transfection resulted in an inhibition of apoptosis to a greater extent in T cells than in T cells from aged humans (unpublished observation). This may indicate that even though the levels of IKKβ are comparable in young and aged, the activity of IKKβ in aged T cells may be lower than that in T cells from young subjects.

In summary, lymphocytes from aged humans appear to be increasingly sensitive to CD95- and TNF-induced apoptosis; however cells, especially CD8+CD28−, from replicative senescence appear to be resistant to AICD. In mice, data appear to be contradictory. Studies are in progress to delineate mitochondrial pathway of apoptosis in aged humans.

Acknowledgements

The work cited is in part supported by a grant from UPHS RO1AG-18313.

References

Aggarwal, S., Gupta, A., Nagata, S., Gupta, S., 1997. Programmed cell death (apoptosis) in cord blood lymphocytes. J. Clin. Immunol. 17, 63–73.

Aggarwal, S., Gupta, S., 1998. Increased apoptosis of T cell subsets in aging humans: altered expression of Fas (CD95), Bcl-2 and Bax. J. Immunol. 160: 1627–1637, 1998.

Aggarwal, S., Gupta, S., 1999. Increased activity of caspase-3 and caspase-8 during Fas-mediated apoptosis in lymphocytes from ageing humans. Clin. Exp. Immunol. 117, 285–290.

Aggarwal, S., Gollapudi, S., Gupta, S., 1999. Increased TNF-α-induced apoptosis in lymphocytes from aged humans: changes in TNF-α receptor expression and activation of caspases. J. Immunol. 162, 2154–2161.

Aggarwal, S., Gollapudi, S., Yel, L., Gupta, A.S., Gupta, S., 2000. TNF-α-induced apoptosis in neonatal lymphocytes: TNFRp55 expression and downstream pathways of apoptosis. Gene Immun. 1, 271–279.

Antwerp, D.J.V., Martin, S.J., Kafri, T., Green, D.R., Verma, I.M., 1996. Suppression of TNF-α-induced apoptosis by NF-κB. Science 274, 787–789.

Ashkanazi, A., Dixit, V.M., 1998. Death receptors: signaling and modulation. Science 281, 1305–1308.

Aspinall, R., Carroll, J., Jiang, S.S., 1998. Age-related changes in the absolute number of CD95 positive cells in T cell subsets in the blood. Exp. Gerontol. 33, 581–591.

Baldwin, A.S., 1996. The NF-κB and IκB proteins: new discoveries and insights. Annu. Rev. Immunol. 14, 649–681.

Baeuerle, P.A., Henkel, T., 1994. Function and activation of NF-κB in the immune system. Annu. Rev. Immunol. 12, 141–179.

Beg, A.A., Baltimore, D., 1996. An essential role for NF-κB in preventing TNF-α-induced cell death. Science 274, 782–784.

Boersma, W.J.A., Steinmeier, F.A., Haaijman, J.J., 1985. Age-related changes in the relative numbers of Thy-1 and Lyt-2-bearing peripheral blood lymphocytes in mice: a longitudinal approach. Cell Immunol. 93, 417–430.

Brown, K., Gerstberger, S., Carlson, L., Franzoso, G., Siebenlist, U., 1995. Control of IκB-α proteolysis by site-specific, signal induced phosphorylation. Science 281, 1360–1363.

Borthwick, N.J., Lowdell, M., Salmon, M., Akbar, A.N., 2000. Loss of CD28 expression on CD8 + T cells is induced by IL-2 receptor γ chain signalling cytokines and type I IFN, and increases susceptibility to activation-induced apoptosis. Int. Immunol. 12, 1005–1013.

Calahan, J.E., Kappler, J.W., Marrack, P., 1993. Unexpected expansion of CD8 bearing cells in old mice. J. Immunol. 151, 665–669.

Chrest, F.J., Buchholz, M.A., Kim, Y.H., Kwon, T.-K., Nordin, A.A., 1995. Anti-CD3-induced apoptosis in T cells from young and old mice. Cytometry 20, 33–42.

Chu, Z.L., McKinsey, T.A., Liu, L., Gentry, J.J., Malim, M.H., Ballard, D.W., 1997. Suppression of tumor necrosis-factor-induced cell death by inhibitor of apoptosis c-IAP2 is under NF-kB control. Proc. Natl. Acad. Sci. USA 94, 10,057–10,062.

Declercz, W., Denecker, G., Fiers, W., Vandenabeele, P., 1998. Cooperation of both TNF receptors in inducing apoptosis: involvement of the TNF receptor-associated factor binding domain of the TNF receptor 75. J. Immunol. 161, 390–399.

Delahase, M., Hayakawa, M., Chen, Y., Karin, M., 1999. Positive and negative regulation of IκB kinase activity through IKKβ subunit phosphorylation. Science 284, 309–313.

De Smaele, E., Zazzeroni, F., Papa, S., Nguyen, D.U., Jin, R., Cong, R., Franzoso, G., 2001. Induction of gadd45β by NF-kB downregulates pro-apoptotic JNK signaling. Nature 414, 308–313.

Di Somma, M.M., Somma, F., Montani, M.S.G., Mangiacasale, R., Cundari, E., Piccolella, E., 1999. TCR engagement regulates differential responsiveness of human memory T cells to Fas (CD9)-mediated apoptosis. J. Immunol. 162, 3851–3858.

Drouet, M., Lauthier, F., Charmes, J.P., Sauvage, P., Ratinaud, M.H., 1999. Age-associated changes in mitochondrial parameters on peripheral human lymphocytes. Exp. Gerontol. 34, 843–852.

Effros, R.B., 1998. Replicative senescence in the immune system: impact of the hayflick limit on T cell functions in elderly. Am. Hum. Genet. 62, 1003–1007.

Engwerda, C.R., Fox, B.S., Handwerger, B.S., 1996. Cytokine production by T lymphocytes from young and aged mice. J. Immunol. 156, 3621–3630.

Fagnoni, F.F., Vescovini, R., Passeri, G., Bologna, G., Pedrazzoni, M., Lavagetto, G., Casti, A., Franceschi, C., Passeri, M., Sansoni, P., 2000. Shortage of circulating naïve CD8(+) T cells provide new insights on immunodeficiency in aging. Blood 95, 2860–2868.

Fagiola, U., Cossarizza, A., Scala, E., Fanales-Belasio, E., Ortolani, C., Cozzi, E., Monti, D., Franceschi, C., Paganelli, R., 1993. Increased cytokine production in mononuclear cells of healthy elderly people. Eur. J. Immunol. 23, 2375–2378.

Franzoso, G., Siebenlist, U., 1995. Control of IκB-α proteolysis by site-specific, signal-induced phosphorylation. Science 267, 1485–1490.

Ghosh, S., May, M.J., Kopp, E.B., 1998. NF-κB and rel proteins: evolutionarily conserved mediators of immune responses. Annu. Rev. Immunol. 16, 225–260.

Gombert, W., Borthwick, N.J., Wallace, D.L., Viner, N., Hyde, H., Boffil, M., Pilling, D., Beverely, P.C.L., Janossy, G., Salmon, M., Akbar, A.N., 1996. Fibroblasts prevent apoptosis of IL-2-deprived T cells without inducing proliferation: a selective effect on Bcl-x(L) expression. Immunology 34, 397–404.

Gonzalez-Quintial, R., Theofilopolous, A.N., 1992. V beta gene repertoire in aging mice. J. Immunol. 149, 230–236.

Green, D., Reed, J.C., 1998. Mitochondria and apoptosis. Science 28, 1309–1312.

Grell, M., Zimmermann, G., Gottfried, E., Chen, C.M., Grunwald, U., Huang, D.C., Wu Lee, Y.H., Durkop, H., Englemann, H., Scheurich, P., 1999. Induction of cell death by tumor necrosis factor (TNF) receptor 2, CD40, and CD30: a role of TNFR1 activation by endogenous membrane-anchored TNF. EMBO J. 18, 3034–3043.

Grossmann, A., Maggio-Price, L., Jinneman, J.C., Rabinowitch, P.S., 1993. Influence of aging on free intracellular calcium and proliferation of mouse T cell subsets from various lymphoid organs. Cell Immunol. 135, 118–131.

Gupta, S., 2000a. Suicidal journey in the Fas (t) track. Recent Res. Dev. Immunol. 2, 11–19.

Gupta, S., 2000b. Molecular and biochemical pathways of apoptosis in lymphocytes from aged humans. Vaccine 18, 1596–1601.

Gupta, S., 2000c. Molecular steps of cell suicide: an insight into immune senescence. J. Clin. Immunol. 20, 229–239.

Gupta, S., 2001a. Molecular steps of TNF receptor-mediated apoptosis. Curr. Mol. Med. 1, 299–306.

Gupta, S., 2001b. Molecular steps of death receptor and mitocondrial pathways of apoptosis. Life Sci. 69, 2957–2964.

Gupta, S., 2002. Tumor necrosis factor-α-induced apoptosis in T cell subsets from aged humans. Receptor expression and downstream signaling events. Exp. Gerontol. 37, 293–299.

Gupta, S., Aggarwal, S., Nguyen, T., 1998. Increased spontaneous apoptosis of T lymphocytes in DiGeorge anomaly. Clin. Exp. Immunol. 113, 65–71.

Haridas, V., Darnay, B.G., Natrajan, K., Helle, R., Aggarwal, B.B., 1998. Overexpression of the p80 TNFR leads to TNF-dependent apoptosis, nuclear factor-kappa B activation. J. Immunol. 160, 3152–3162.

Hengartner, M.O., 2002. The biochemistry of apoptosis. Nature 407, 770–776.

Herndon, F.J., Hsu, H.C., Mountz, J.D., 1997. Increased apoptosis of CD45RO— T cells with aging. Mech. Ageing Dev. 94, 123–134.

Hsu, H., Shu, H.B., Pan, M.G., Goeddel, D.V., 1996a. TRADD-TRAF2 and TRADD-FADD interactions define two distinct TNF receptor 1 signal transduction pathways. Cell 84, 299–308.

Hsu, H., Huang, J., Shu, H.B., Baichwal, V., Goeddel, D.V., 1996b. TNF-dependent recruitment of the protein kinase RIP to the TNF receptor-1 signaling complex. Immunity 4, 387–396.

Hsu, C., Shi, J., Yang, P., Xu, X., Dodd, C., Matsuki, Y., Zhang, H.-G., Mountz, J.D., 2001. Activated CD8 + T cells from aged mice exhibit decreased activation-induced cell death. Mech. Ageing Dev. 122, 1663–1684.

Hu, S., Vincenz, C., Ni, J., Gentz, R., Dixit, V.M., 1997. I-Flice, a novel inhibitor of tumor necrosis factor receptor-1 and CD95-induced apoptosis. J. Biol. Chem. 272, 17,255–17,257.

Irmler, M., Thome, M., Hahne, M., Schneider, P., Hofmann, K., Steiner, V., Bodmer, J.-L., Schriter, M., Burns, K., Mattmann, C., Rimoldi, D., French, L.E., Tschopp, J., 1997. Inhibition of death receptor signals by cellular FLIP. Nature 388, 190–195.

Karin, M., 1999. How NF-kB is activated: the role of the IkB kinase (IKK) complex. Oncogene 18, 6867–6874.

Kataoka, T., Budd, R.C., Holler, N., Thome, M., Martinon, F., Irmler, M., Burns, K., Hahne, M., Kennedy, N., Kovascovics, M., Tschopp, J., 2000. The caspase-8 inhibitor FLIP promotes activation of NF-κB and Erk signaling pathways. Curr. Biol. 10, 640–648.

Kelliher, M.A., Grimm, S., Ishida, Y., Kuo, F., Stranger, B.Z., Leder, P., 1998. The death domain kinase RIP mediates the TNF-induced NF-kappa B signal. Immunity 8, 297–303.

Kirschmann, D.A., Murasko, D.M., 1992. Splenic and inguinal lymph node T cells from aged mice respond differently to polyclonal and antigenic stimuli. Cell. Immunol. 139, 42–437.

Komuro, T., Sano, K., Asano, Y., Tada, T., 1990. Analysis of age-related degeneracy of T cell repertoire: localized functional failure in CD8 + T cells. Scand. J. Immunol. 32, 545–553.

Krammer, P.H., 2000. CD95's deadly mission in the immune system. Nature 407, 789–795.

Kudlacek, S., Jahanideh-Kazempour, S., Graninger, W., Willvonseder, R., Pietschmann, P., 1995. Differential expression of various T cell surface markers in young and elderly subjects. Immunobiology 192, 198–204.

Laux, I., Koshnan, A., Tindell, C., Bae, D., Zhu, X., June, C.H., Effros, R.B., Nel, A., 2000. Response differences between CD4 + and CD8 + T cells during CD28 costimulation: implication for immune cell-based therapies and studies related to the expansion of double-positive T cells during aging. Clin. Immunol. 96, 187–197.

Lechner, H., Amort, M., Steger, M.M., Maczek, C., Grubeck-Loebenstein, B., 1996. Regulation of CD95 (Apo-1) expression and the induction of apoptosis in human T cells: changes in old age. Int. Arch. Allergy Immunol. 110, 238–243.

Lenardo, M.J., 1997. The molecular regulation of lymphocyte apoptosis. Semin. Immunol. 9, 1–5.

Li, H., Zhu, H., Xu, C., Yuan, J., 1998. Cleavage of Bid by caspase-8 mediates mitochondrial damage in the Fas pathway of apoptosis. Cell 94, 491–501.

Li, Z.W., Chu, W.M., Hu, Y.L., Delhase, M., Deerinck, T., Ellisman, M., Johnson, R., Karin, M., 1999. The IKKβ subunit of IκB kinase (IKK) is essential for nuclear factor-κB activation and prevention of apoptosis. J. Exp. Med. 189, 1839–1845.

Linton, P.J., Haynes, L., Klinman, N.R., Swain, S.L., 1996. Antigen-independent changes in naïve CD4 T cells with aging. J. Exp. Med. 184, 1891–1900.

Liu, Z.G., Hsu, H., Goeddel, D.V., Karin, M., 1996. Dissection of TNF receptor 1 effector functions: JNK activation is not linked to apoptosis while NF-kappa-B activation prevents cell death. Cell 87, 565–576.

McNerlan, S.E., Alexander, H.D., Rea, I.M., 1999. Age-related reference intervals for lymphocyte subsets in whole blood of healthy individuals. Scand. J. Clin. Lab. Invest. 59, 89–92.

Miyawaki, T., Uehara, T., Nibu, R., Tsuji, T., Yachie, A., Yonehara, S., Taniguchi, N., 1992. Differential expression of apoptosis-related Fas antigen on lymphocyte subpopulations in human peripheral blood. J. Immunol. 149, 3753–3758.

Miller, R.A., 1997. Age-related changes in T cell surface markers: a longitudinal analysis in genetically heterogeneous mice. Mech. Ageing Dev. 96, 181–196.

Pahl, H.L., 1999. Activators and target genes of Rel/NF-kB transcription factors. Oncogene 18, 6855–6866.

Pahlavani, M., Harris, M.D., 1996. The age-related changes in DNA binding activity of AP-1, NF-κB, and Oct-1 transcription factors in lymphocytes from rats. Age 19, 45–54.

Pahlavani, M.A., Vargas, D.A., 2000. Activation-induced apoptosis in T cells from young and old Fisher 344 rats. Int. Arch. Allergy Immunol. 122, 182–189.

Pawelec, G., Barnett, Y., Effros, R., Forsey, R., Frasca, D., Globerson, A., Mariani, E., McLeod, J., Caruso, C., Franceshi, C., Fulop, T., Gupta, S., Mocchegiani, E., Solana, R., 2002. T cells and aging. Front. Biosci. 7, d1056–d1183, May 1.

Phelouzat, M.A., Arbogast, A., Laforge, T., Quadri, R.A., Proust, J.J., 1996. Excessive apoptosis of mature T lymphocytes is a characteristic feature of human immune senescence. Mech. Ageing Dev. 88, 25–38.

Phelouzat, M.A., Laforge, T., Arbogast, A., Quadri, R.A., Boutet, S., Proust, J.J., 1997. Susceptibility to apoptosis of T lymphocytes from elderly humans is associated with increased in vivo expression of functional Fas receptors. Mech. Ageing Dev. 96, 35–46.

Pimentel-Muinos, F.X., Seed, B., 1999. Regulated commitment of TNF receptor signaling: a molecular switch for death or activation. Immunity 11, 783–793.

Polyak, K., Wu, T.T., Hamilton, S.R., Kinzler, K.W., Vogelstein, B., 1997. Less death in dying. Cell Death Differ. 4, 242–246.

Ponnappan, U., Zhong, M., Trebilcock, G.U., 1999. Decreased proteosome-mediated degradation in T cells from the elderly: a role in immune senescence. Cell. Immunol. 192, 167–174.

Potestio, M., Caruso, C., Gervasi, F., Scialabba, G., D'Anna, C., DiLorenzo, G., Balistreri, C.R., Candore, G., Romano, C.C., 1998. Apoptosis and ageing. Mech. Ageing Dev. 102, 221–237.

Potestio, M., Pawelec, G., DiLorenzo, G., Candore, G., D'Anna, C., Gervasi, F., Lio, D., Tranchida, G., Caruso, C., Romano, C.C., 1999. Age-related changes in the expression of CD95 (APO-1/Fas) on blood lymphocytes. Exp. Gerontol. 34, 659–673.

Provinciali, M., DiStefano, G., Stronati, S., 1998. Flow cytometric analysis of CD3/TCR complex, zinc, and glucocorticoid-mediated regulation of apoptosis and cell cycle distribution in thymocytes from old mice. Cytometry 32, 1–8.

Rath, P.C., Aggarwal, B.B., 1999. TNF-induced signaling in apoptosis. J. Clin. Immunol. 19, 350–364.

Reed, J.C., 1997. Double identity for protein of Bcl-2 family. Nature 387, 773–778.

Roberts, W.G., Orme, I.M., 1998. CD95 expression in aged mice infected with tuberculosis. Infect. Immun. 66, 5036–5040.

Rothe, J., Gehr, G., Loetcher, H., Lesslauer, W., 1992. Tumor necrosis factor receptor-structure and function. Immunol. Res. 11, 81–90.

Salmon, M., Pilling, D., Borthwick, N.J., Viner, N., Janissy, G., Bacon, P.A., Akbar, A.N., 1994. The progressive differentiation of primed T cells is associated with an increased susceptibility to apoptosis. Eur. J. Immunol. 24, 892–899.

Scaffidi, C., Fulda, S., Srinivasan, A., Friesen, C., Li, F., Tomaselli, K.J., Debatin, K.M., Krammer, P.H., Peter, M.E., 1998. Two CD95 (Apo-1/Fas) signaling pathways. EMBO J. 17, 1675–1687.

Screaton, G., Xu, X.-N., 2002. T cell life and death signaling via TNF-receptor family members. Curr. Opin. Immunol. 12, 316–322.

Seishima, M., Takemura, M., Saito, K., Sano, H., Minatoguchi, S., Fujiwara, H., Hachiya, T., Noma, A., 1996. Highly sensitive ELISA for soluble fas in serum: increased soluble fas in the elderly. Clin. Chem. 42, 1911–1914.

Senftleben, U., Li, Z.-W., Baud, V., Karin, M., 2001. IKKβ is essential for protecting T cells from TNFα-induced apoptosis. Cell 14, 217–230.

Shinohara, S., Sawada, T., Nishioka, Y., Tohma, S., Kisaki, T., Inoue, T., Ando, K., Ikeda, M., Fuji, H., Ito, K., 1995. Differential expression of Fas antigen and Bcl-2 protein on CD4+ T cells, CD8+ T cells and monocytes. Cell. Immunol. 163, 303–338.

Sidman, C.L., Luther, E.A., Marshall, J.D., Nguyen, K.A., Roopenian, D.C., Worthen, S.M., 1987. Increased expression of major histocompatibility complex antigens on lymphocytes from aged mice. Proc. Natl. Acad. Sci. USA 84, 7624–7628.

Spaulding, C.C., Walford, R.L., Efros, R.B., 1997. The accumulation of non-replicative, non-functional, senescent T cells with age is avoided in colorically restricted mice by an enhancement of T cell apoptosis. Mech. Ageing Dev. 93, 25–33.

Tang, G., Minemoto, Y., Dibling, B., Purcell, N.H., Li, Z., Karin, M., Lin, A., 2001. Inhibition of JNK activation through NF-κB target genes. Nature 414, 313–317.

Tanaka, M., Suda, T., Takahashi, T., Nagata, S., 1995. Expression of the functional fas ligand in activated lymphocytes. EMBO J. 14, 223–239.

Tartaglia, L.A., Goeddel, D.V., 1992. Two TNF receptors. Immunol. Today 13, 151–153.

Tartaglia, L., Pennica, D., Goddel, D.V., 1993. Ligand passing: the 75-kDa tumor necrosis factor (TNF) receptor recruits TNF for signaling by the p55-kDa TNF receptor. J. Biol. Chem. 268, 18,542–18,548.

Telford, W.G., Miller, R.A., 1999. Aging increases CD8 T cell apoptosis induced by hyperstimulation but decreases apoptosis induced by agonist withdrawal in young. Cell. Immunol. 191, 131–138.

Thome, M., Schneider, P., Hofmann, C., Fickenscher, H., Meini, E., Neipel, C., Mattmann, C., Burns, K., Bodmer, J., Schroter, M., Scaffidi, C., Krammer, P., Peter, M., Tschopp, J., 1997. Viral Flice-inhibitory proteins (FLIPs) prevent apoptosis induced by death receptors. Nature 386, 517–521.

Thomas, B., Grell, M., Pfizenmaier, K., Scheurich, P., 1990. Identification of a 60-kDa tumor necrosis factor (TNF) receptor as the major signal transducing component in TNF responses. J. Exp. Med. 172, 1019–1023.

Timm, J.A., Thoman, M.L., 1999. Maturation of CD4+ lymphocytes in the aged microenvironment results in a memory-enriched population. J. Immunol. 162, 711–717.

Trebilcock, G.U., Ponnappan, U., 1996. Evidence for lowered induction of nuclear factor kappa B in activated human T lymphocytes during aging. Gerontology 42, 137–146.

Utsuyama, M., Hirokawa, K., 1987. Age-related changes of splenic T cells in mice – a flow cytometric analysis. Mech. Ageing Dev. 40, 89–102.

Vandenabeele, P., Declercq, W., Beyaert, R., Fiers, W., 1995a. Two tumor necrosis factor receptors: structure and function. Trends Cell Biol. 5, 392–399.

Vandenabeele, P., Declercq, W., Vanhaesebroeck, B., Grooten, J., Fiers, W., 1995b. Both TNF receptors are required for TNF-mediated induction of apoptosis in PC60 cells. J. Immunol. 154, 2904–2913.

Ware, C.F., Crowe, P.D., Van Arsdale, T.L., Andrews, J.L., Grayson, M.H., Jerzy, R., Smith, C.A., Goodwin, R.G., 1991. Tumor necrosis factor (TNF) receptor expression in T lymphocytes: differential regulation of the type 1 receptor during activation of resting and effector T cells. J. Immunol. 147, 4229–4238.

Weiss, T., Grell, M., Siekienski, K., Muhlenbeck, F., Durkop, H., Pfizenmaier, K., Scheurich, P., Wajant, H., 1998. TNFR80-dependent enhancement of TNFR60-induced cell death is mediated by TNFR-associated factor 2 and is specific for TNFR60. J. Immunol. 161, 3136–3142.

Whisler, R.L., Beiqing, L., Chen, M., 1996. Age-related decreases in IL-2 production by human T cells are associated with impaired activation of nuclear transcriptional factors AP-1 and NF-AT. Cell. Immunol. 169, 185–195.

Yeh, W.C., Itie, A., Elia, A.J., Ng, M., Shu, H.B., Wakeham, A., Mirtsos, C., Suzuki, N., Bonnard, M., Goeddel, D.V., Mak, T.W., 2000. Requirement of casper (c-FLIP) in regulation of death receptor-induced apoptosis and embryonic development. Immunity 12, 633–642.

Yel, L., Aggarwal, S., Gupta, S., 1999. Cartilage-hair hypoplasia syndrome: increased apoptosis of T lymphocytes is associated with altered expression of Fas (CD95), FasL (CD95L), IAP, Bax and Bcl-2. J. Clin. Immunol. 19, 428–435.

Yuan, J., 1997. Transducing signals of life and death. Curr. Opin. Cell Biol. 9, 247–251.

You, M., Ku, P., Hrdickova, R., Bose, H.R., 1997. ch-IAP, a member of the inhibitor of apoptosis protein family is a mediator of the antiapototic activity of the v-Rel oncoprotein. Mol. Cell. Biol. 17, 7328–7341.

Zandi, E., Chen, Y.I., Karin, M., 1998. Direct phosphorylation of IκB by IKKα and IKKβ: discrimination between free and NF-κB-bound substrate. Science 281, 1360–1363.

Zhou, T., Edwards, C.K., 3rd, Mountz, J.D., 1995. Prevention of age-related T cell apoptosis defect in CD2-fas-transgenic mice. J. Exp. Med. 182, 129–137.

Advances in
Cell Aging and
Gerontology

Genetic damage and ageing T cells

Owen A. Ross[a,b], Martin D. Curran[b], Derek Middleton[a,b],
Brian P. McIlhatton[b], Paul Hyland[a], Orla Duggan[a],
Kathryn Annett[a], Christopher Barnett[a] and Yvonne Barnett[a]

[a]*School of Biomedical Sciences, University of Ulster, Coleraine, Northern Ireland, BT52 1SA, UK*
Correspondence address: Tel.: +44-28-7032-4627; fax: +44-28-7032-4965.
E-mail: ya.barnett@ulster.ac.uk (Y.B.)
[b]*Northern Ireland Regional Histocompatibility and Immunogenetics Laboratory, City Hospital, Belfast,
Northern Ireland, BT9 7TS, UK*

Contents

1. Introduction – T cells and ageing

The deterioration of the immune system with age (termed "immunosenescence") is thought to contribute to morbidity and mortality in the aged due to decreased resistance to infection and, possibly, certain cancers. T cell function is significantly altered in vivo and in vitro in elderly compared to young individuals (Pawelec et al., 2002). Several early studies have already suggested a positive association between

Advances in Cell Aging and Gerontology, vol. 13, 191–215

good T cell function in vitro and individual longevity (Roberts-Thomson et al., 1974; Murasko et al., 1988; Wayne et al., 1990), and between absolute lymphocyte counts and longevity (Bender et al., 1986). Improved understanding of the immunosenescence process and its causes may lead to significant enhancement of the human health span and the quality of later life, and reductions in the cost of medical care in old age.

2. Free radicals, T cells and ageing

The Free Radical Theory of Ageing, first proposed in 1956 by Denham Harman, postulates that reactive oxygen species (ROS), produced mainly via the aerobic metabolic pathway in the mitochondria, are the major source of intrinsic oxidative damage for the cell (Harman, 1956; Beckman and Ames, 1998; Cadenas and Davies, 2000; Sastre et al., 2000). Table 1 illustrates many of the other sources and pathways that can also result in free radical production. Harman proposed his hypothesis after observing the parallels between the affects of ageing and those of ionising radiation, including mutagenesis, cancer and gross cellular damage. Harman postulated that the free radicals cause random tissue damage, at the cellular level, which impairs the function and proliferative capacity of the cell and that ageing is a result of the failure of the protective mechanisms to combat the ROS induced damaged. The damaging effects of ROS, and the defence mechanisms established to combat the oxidative damage caused by them, has placed the Free Radical Theory of Ageing firmly in the centre of the ageing research field.

Numerous reports exist which document an age-related accumulation of damage to cellular biomolecules (lipids, proteins, carbohydrates and nucleic acids) in various cells of the body (reviewed in Barnett, 1994). Such damage may alter the structure and thus function of the affected biomolecules. Changes in structure may have detrimental consequences leading to alterations in cellular function which may affect tissue/organ function and which may result in the development of a number of pathologies. Table 2 summarises major age-related alterations in biomolecule structure and the resultant physiological consequences of such structural changes.

Highly reactive aldehydic products, produced as by-products of lipid peroxidation, have the potential to be an extremely deleterious endogenous source of free radical stress and are known to react with proteins and DNA causing significant damage (Choe et al., 1995). These lipid peroxidation by-products have long half-lives and longer ranges relative to free radicals.

Various cytosolic oxidases and dehydrogenases are known to produce free radicals (Yu and Yang, 1996). For example, constitutive/inducible nitric oxidase synthases (NOS) that create nitric oxide free radicals (NO•) are now recognised as widely distributed cytosolic enzymes that play many key physiological roles (Schmidt and Walter, 1994). The induction of NOS can become unregulated with increasing age, and NOS have the potential to be major contributors to age related oxidative stress (Yu and Yang, 1996).

Certain free radical stresses may also be occupational hazards for T cells. At sites of inflammation T cells are exposed to high concentrations of ROS and reactive

Table 1
Sources of free radicals

Endogenous sources	Exogenous sources
Plasma membrane	*Radiation*
Lipoxygenase	Ionising
Cycloxygenase	Ultraviolet
NADPH oxidase	
Mitochondria	*Drug oxidation*
Electron transport	Acetaminophen
Ubiquinone	Carbontetrachloride
NADH dehydrogenase	Cocaine
Microsomes	*Oxidising gases*
Electron transport	Oxygen
Cytochrome p450	Ozone
Cytochrome b5	Nitrogen dioxide
Peroxisomes	*Xenobiotic elements*
Oxidases	As, Pb, Hg, Cd
Flavoproteins	
Phagocytic cells	*Redox cycling substances*
Neutrophils	Paraquat
Macrophytes	Diquat
Eosinophils	Alloxan
Endothelial cells	Doxorubicin
Autoxidation reactions	*Heat shock*
Metal catalysed reactions	
Other	Cigarette smoke and
	combustion products
Haemoglobin, flavins, xanthine oxidase, monoamine oxidase, galactose oxidase, indolamine oxidase, tryptophan dooxygensae	

Adapted and modified from Halliwell (1987) and Freeman (1984).

nitrogen intermediates, produced by activated neutrophils, macrophages and T cells, as a normal part of the immune response. These reactive species have the capacity to cause damage to the cells involved in the immune response (Metzger et al., 1980; Gregory et al., 1993) and any functional impairment incurred may have long-term consequences.

2.1. Mitochondria, T cells and ageing

Harman, in 1972, was the first to postulate that the intracellular organelle the mitochondrion, may contribute to the mechanisms of ageing. Since then, the mito-chondria have been continually implicated in playing a pivotal role in the ageing process (Harman, 1972; Brierley et al., 1997; Lee and Wei, 1997; Papa and Skulachev, 1997; Miquel, 1998; Brand, 2000; Kowald and Kirkwood, 2000;

Table 2
Major age-related alterations in biomolecule structure and the resultant physiological consequences of such structural changes

Biomolecule	Alteration	Physiological consequence
Lipids	Lipid peroxidation	Oxidised membranes become rigid, lose selective permeability and integrity. Cell death may occur. Peroxidation products can act as cross-linking agents and may play a role in protein aggregation, the generation of DNA damage and mutations, and the age-related pigment lipofuscin
Proteins	Racemisation, deamination, oxidation, carbamylation	Alterations to long-lived proteins may contribute to ageing and/or pathologies. Crosslinking and formation of advance glycosylation end products (AGEs) which can severely affect protein structure and function. Effects on the maintenance of cellular homeostasis
Carbohydrates	Fragmentation, depolymerisation, glucose autooxidation	Alters physical properties of connective tissue. Glycosylation of proteins in vivo with subsequent alteration of biological function
Nucleic acids	Strand breaks, base adducts, loss of 5-methyl cytosine from DNA	Damage could be expected to interfere with the processes of transcription, translation and DNA replication. Such interference may reduce a cell's capacity to synthesise vital polypeptides/proteins. In such circumstances cell death may occur. The accumulation of a number of hits in critical cellular genes associated with the control of cell growth and division has been shown to result in the process of carcinogenesis
		Dedifferentiation of cells (5-methylcytosine plays an important role in switching off genes as part of gene regulation). If viable, such dedifferentiated cells may have altered physiology and may contribute to altered tissue/organ function

Melov, 2000; Rustin et al., 2000; Berdanier and Everts, 2001; Salvioli et al., 2001). The mitochondrion is distinct as an organelle, being the only extra-nuclear site of DNA in the cell. For this reason the mitochondrial genome (mtDNA) is often referred to as the "other human genome".

It has been reported that the number of mitochondria decreases with age but those remaining increase in size, especially after 60 years of age (Ozawa, 1997). No gross derangement has been observed histologically but shortened cristae, matrix vacuolisation and loss of dense granules have been reported (Shigenaga et al., 1994). The activity of the mitochondrial enzymes involved in the electron transport chain has been shown to progressively decline with age and each individual complex displays differing rates of decline (Lee and Wei, 1997). There are 80 polypeptide subunits incorporated into the five complexes of the electron transport chain and 67 of these are encoded for by nuclear DNA. The enzymes encoded by nuclear DNA have been reported to express no significant change with age (Ozawa, 1997). The complexes that show deterioration with age all involve subunits that are encoded for in the mtDNA.

The mitochondria are a major source of ROS generated in vivo. It has been reported that 1–2% of the total electron flow through the electron transport chain results in the generation of superoxide radicals $O_2^{\bullet}$ and 1–5% of all oxygen consumption during respiration ends up in ROS (Richter et al., 1995).

These ROS can have deleterious effects within the mitochondria. Trounce et al. (1989), Yen et al. (1989), first demonstrated that the respiratory function of the mitochondria gradually declines with age in human tissues. Hayakawa et al. (1993), demonstrated that mtDNA accumulates damage with age; levels of 8-hydroxy-2-deoxyguanosine were shown to increase with age and correlated with the rate of accumulation of mtDNA with deletions.

Many authors have hypothesised a "vicious cycle" involving mtDNA abnormalities and defects in bioenergetic production (Wei et al., 1998; Ozawa, 1999; Lee and Wei, 2001). It is postulated that both mutations and energy defects accumulate during senescence in a mutually exacerbating cycle of molecular damage and mitochondrial dysfunction, fuelled by oxidative stress (Nagley and Wei, 1998).

There is still relatively little known about modifications in the mitochondria of immune system cells with ageing. However, as mitochondria play a central role in apoptosis, a major process in T cell death, and in ATP production, which may be compromised by oxidative damage, age-associated defects in mitochondrial function are postulated to be important in T cell function and age-related immunodeficiency (Shigenaga et al., 1994; Brenner et al., 1998; Drouet et al., 1999).

2.2. Mitochondrial mutation

The mtDNA is a covalently closed circular double-stranded genome of 16,569 bp and the complete sequence was published in 1981 (Anderson et al., 1981). The mtDNA genome encodes for 22 tRNAs, two rRNAs and 13 polypeptides and it is estimated that mtDNA undergoes mutations at 5–10 times faster than nuclear DNA. There are at least two factors implicated in promoting this high rate of mutation: (i) there are no histone proteins to protect mtDNA from damage; and (ii) there is a very inefficient repair mechanism for damaged mtDNA (Bohr and Anson, 1999; Bohr and Dianov, 1999).

2.2.1. Common deletion mtDNA4977

The established decline in mitochondrial bioenergetic function during human ageing in various tissues has been attributed to the occurrence and accumulation of mutations (deletions, base substitutions and frame-shifts) in the mtDNA (Brierley et al., 1998; Liu et al., 1998a; Graff et al., 1999; Michikawa et al., 1999; Fernandez-Moreno et al., 2000; Kopsidas et al., 2000; Wang et al., 2001). Deletions of mtDNA are the most common mutations associated with human diseases and the ageing process (Kovalenko et al., 1997; Cormio et al., 2000; Cottrell et al., 2000; Raha and Robinson, 2000). Deletions are epitomised by the "common deletion", a

4977 bp mtDNA deletion (mtDNA4977) that occurs, between two 13 bp direct repeats, at nucleotide positions mt8470–13447. The 4977 bp deleted region encodes for seven polypeptides essential for the enzyme complexes of the OXPHOS pathway, four for complex I (ND3, ND4, ND4L and ND5), one for complex IV (CO II), two for complex V (ATP6 and ATP8) and five tRNA genes (Porteous et al., 1998).

The mtDNA4977 deletion has been reported to accumulate in a variety of human tissues with increasing age including; heart, brain, lung, skeletal muscle (post-mitotic), skin and liver (non-post-mitotic) (Kao et al., 1997; Liu et al., 1998b; Bogliolo et al., 1999; Lu et al., 1999; Cormio et al., 2000; Kim et al., 2000). A number of previous investigations have failed to detect the deleted form in the blood of healthy individuals, although increased levels of mtDNA4977 have been consistently reported in the blood of diseased patients (e.g. Kearns–Sayre syndrome, Pearson's pancreas syndrome and mitochondrial myopathies – Biagini et al., 1998; Von Wurmb et al., 1998). We have suggested that PCR amplification conditions are critical for its detection after investigations in our laboratory showed the mtDNA4977 deletion in both ex vivo DNA samples and a library of human CD4+ T cell clones (Ross et al., 2002).

2.2.2. Point mutations

As well as mtDNA4977, single nucleotide point mutations within the mtDNA accumulate with age (Michikawa et al., 1999; Wang et al., 2001). Less than 10% of mtDNA is non-coding and approximately 90% of this non-coding DNA is located in an area termed the control region. An age-dependent large accumulation of point mutations in the control region for mtDNA replication of human skin fibroblasts was observed by Michikawa et al. (1999) and this was recently followed by the reported accumulation of muscle-specific mtDNA mutations with age within the control region (Wang et al., 2001). The control region is the most variable region of the human mtDNA genome and is the location of the displacement loop (D-loop). The 1122 bp D-loop of the mtDNA is a non-coding region that is located between mt16024–576 and is known to contain a number of polymorphic variants.

Mutational analysis of a 444 bp section of the D-loop (mt15978–16422) was performed in our laboratory on two groups of ex vivo lymphocyte DNA samples from 19 to 45- and 85 to 90-year-old cohorts (Ross et al., 2002) using a novel heteroduplex methodology (reference strand conformational analysis, RSCA) which identifies sequence variations through differing band migration patterns obtained via non-denaturing polyacrylamide gel electrophoresis (Arguello et al., 1998). Although no accumulation of damaged mtDNA was observed in our study, the high resolution of RSCA identified a total of 16 different mtDNA species (RSCA profiles). Examination of the 16 various RSCA profiles observed within the two age groups revealed no statistically significant associations. However, a level of polyplasmy was observed in both age groups although no significant difference with age was found (Ross et al., 2002).

The low level of mtDNA damage with age that was observed in our studies would suggest that the mitochondrion might not be as prone to DNA damage as previously believed. Recently, it has been reported that mitochondrial endogenous oxidative damage may have been overestimated in the past (Anson et al., 2000) and although mtDNA does not encode DNA repair proteins (Bohr and Anson, 1999), repair of damaged mtDNA does take place, compensating for oxidative damage (Bohr and Anson, 1999; Bohr and Dianov, 1999).

Antioxidant defences and DNA repair mechanisms have been shown to affect the level of DNA damage, and age-related changes in these parameters have also been reported in lymphocytes (Mendoza-Nunez et al., 1999; Doria and Frasca, 2001; Hyland et al., 2002). The absence of an age-related increase in the level of mtDNA damage in lymphocytes may in part be due to the high turnover rate of blood cells which does not permit the damaged mtDNA to accumulate in any significant number (Meissner et al., 2000). It is also worth noting that mtDNA is actively turned over in both mitotic and post-mitotic tissue every 7–31 days, depending on the cell type and tissue therefore ridding the mitochondrial population of many of the mutant forms (Kopsidas et al., 2000).

2.2.3. Genomic polymorphisms

Through the heteroduplex work performed in our laboratory (Ross et al., 2002) it was observed that a variety of mtDNA species exist within the cells of the immune system in vivo. This probably reflects the polymorphic nature of the mtDNA genome rather than any accumulation of mtDNA damage with age. Certainly, the observation of similar polyplasmic frequencies in both age groups does not support the hypothesis that an age-related accumulation of genetic abnormalities of the mtDNA plays a significant role in the dysfunction of the immune system with age.

The possibility that certain inherited mtDNA polymorphisms, rather than accumulated damage, may predispose certain individuals to be more resistant to age-related diseases, providing superior immune function and promoting longevity is an attractive concept, and is supported by recent studies in French, Japanese, Italian and Irish populations (Ivanova et al., 1998; Tanaka et al., 1998; De Benedictis et al., 1999; Ross et al., 2001).

Tanaka et al. (1998) reported a mtDNA polymorphism at mt5178, a C to A transversion, to be at a significantly increased frequency within a sample of Japanese centenarians compared to a control group aged < 46 years old. The frequency was reported at 62% in 37 centenarians and 45% in the 252 control samples. This polymorphism, mt5178A, was not observed within the Irish population. The Irish population is a relatively closed population and therefore does not display the genotypical diversity that is expressed globally. This polymorphism, mt5178A, is linked to the Asian haplogroup M and results in the absence of an Alu I restriction site and has not as yet been reported in the European population (Torroni et al., 1996). It is therefore understandable that a polymorphism associated with the Japanese population is not found within the Irish population.

Ivanova et al. (1998) observed the mt9055A polymorphism at a frequency of 13.2% in the French study of 248 centenarians and at 6.6% in a control group of 285 adults. Ross et al. (2001) observed mt9055A at a frequency of 5% in the control group (*n* = 100, 19–45 years old) and 9% in the aged group (*n* = 129, 80–97 years old) of the Irish population. These results concur with the finding of the French group, although the values did not attain statistical significance, suggesting that mt9055A may be associated with successful ageing. However, it is not a major factor in the ageing process and would therefore not be a good predictive marker for successful ageing.

It is now well established that the numerous mtDNA sequence variants/haplotypes, defined by restriction enzyme analysis, that exist within the European Caucasian population can be subsumed within nine broad mtDNA haplogroups; H, I, J, K, T, U, V, W, and X, differentiated on the basis of ancestral polymorphisms (Torroni et al., 1996).

De Benedictis et al. (1999) reported a statistically significant increase in the frequency of the broad J haplogroup, defined by the polymorphic restriction site mt16065G, 23% in male centenarians from the north of Italy compared to 2% in younger males from the same region. In contrast to these findings a study in our laboratory (Ross et al., 2001) failed to show any J haplogroup association in the Irish population, with either age or gender. However, through phylogenetic analysis the Irish study identified two distinct branches forming the J haplogroup. The first branch is defined by the presence of the Hinf I restriction site mt16389G and displays a significantly increased frequency in the Irish aged population, while the second branch is defined by the absence of the site mt16000G and has a significantly decreased frequency in the Irish aged population. Unfortunately, in the Italian study they only tested for the defining restriction site for the J haplogroup, mt16065G, and it is therefore not possible to ascertain which branch their individuals assigned to the J haplogroup predominantly belonged (De Benedictis et al., 1999). It is tempting to speculate that the Italian male centenarians used in their study fall in to the first branch, given the increased prevalence of this in the Irish aged subjects.

De Benedictis et al. (2000) postulate that their finding of an increased frequency of the J haplogroup in centenarians indicates that the J-specific genetic background can affect longevity, possibly by conferring increased resistance to the deleterious effects of the accumulation of mtDNA mutations with age. In contrast, the same haplogroup is reported to have an elevated frequency in some complex diseases (Rose et al., 2001), e.g. Leber Hereditary Optic Neuropathy, a rare neurodegenerative disease of young adults that results in blindness due to optic nerve degeneration (Torroni et al., 1997). Rose et al. (2001) have proposed that the complex interactions between mtDNA and nuclear genes may play a role in an individual's susceptibility to both disease and prolonged longevity, according to the specific genetic background. The presence of two distinct branches of the J haplogroup may hold the key as to why the haplogroup as a whole could be not only associated with successful ageing but also a number of complex diseases. Similarly, it is important to point out that any specific nucleotide or haplogroup defining mtDNA polymorphism associated with ageing or disease may not be directly involved but simply acting as a marker for other tightly linked polymorphisms occurring elsewhere in the mtDNA genome that directly affect longevity. Certainly, if the significant association of the J haplogroup with ageing is confirmed in future studies, it will be necessary to carry out large scale sequencing studies to pinpoint the exact nucleotide sequence differences responsible, and determine their functional impact on the ageing process.

3. T cell genetic damage in vivo and in vitro

Several studies have linked an intact, competent immune response with longevity (Roberts-Thomson et al., 1974; Murasko ct al., 1988; Wayne et al., 1990; Bender et al., 1986). This is consistent with the observation that decreased T cell proliferative potential in the elderly may not be evident in the extremely old (Sansoni et al., 1997; Franceschi et al., 1995), i.e. those individuals surviving longest are those with the highest proliferative responses. This is supported by the findings in mice, that the longest-lived strains are those with the highest proliferative responses (discussed in Pawelec et al., 2002).

3.1. Genetic damage in vivo

Previous work in our laboratory has shown that T cells in vivo accumulate DNA damage and mutation over a lifetime (King et al., 1994; Barnett and King, 1995; King et al., 1997). Healthy, free living subjects from four age groups were examined; 35–39, 50–54, 65–69, and 75–80 years. Basal levels of DNA damage, as measured by the alkaline comet assay, were found to increase with age in lymphocytes sampled from the first three age groups. In addition, there was an increased frequency of chromosomal aberrations in the 65–69 year group, compared to the first two age groups. However, levels of DNA damage in the 75–80 years age group (all healthy, successfully aged) were shown to be very similar to those within the 35–39 years age group.

Mutant frequency at the *HPRT* gene locus in lymphocyte samples increased as a function of age at a rate of approximately 1.33% per year to 69 years. In contrast, the mean *HPRT* mutant frequency of the 75–80-year-old group was not significantly different to that of the 35–39 years group.

King et al. (1997) demonstrated an age-related decrease in DNA repair capacity in lymphocytes in the 35–39, 50–54, and 65–69 years age groups, however, the 75–80 years old subjects displayed DNA repair capacities similar to the 35–39 years old group. The subjects of the NONA Immune study investigated by Hyland et al. (2002) represent a population-based sample of the oldest elderly individuals drawn from the municipality of Jönköping, located in south-central Sweden, and have reached extreme old age (86–94 years old). The aims of the investigation were to assess the magnitude of in vivo antioxidant capacity, and to assess the levels and types of DNA damage in mononuclear cells from these aged subjects. Cells from aged NONA subjects displayed levels of DNA single strand breaks and oxidatively damaged sites equivalent to those in cell samples from a middle-aged control group (40–60 years old). This data supported the importance of maintenance of genomic stability in longevity and was in agreement with the findings of our previous work (King et al., 1994, 1997; Barnett and King, 1995, 1997).

Taken together, these findings on genetic damage levels and DNA repair capacity in vivo could be explained by donor selection pressures resulting in an association of healthspan and longevity with maintenance of genetic stability due, at least in part, to the retention of DNA repair capacity.

3.2. T cell genetic damage in vitro

In addition to in vivo studies, further insights into the relationship between genetic damage and the ageing of T cells have been gained from the in vitro human peripheral blood derived CD4+ T cell clone model of immunosenescence.

The propagation of T cell clones in vitro provides a good model for investigating age-associated changes to T cell function and replicative senescence. Human T cell clones can be maintained for extended periods in tissue culture, and are known to have finite lifespans (Duquesnoy and Zeevi, 1983; Effros et al., 1994; Effros and Walford, 1984; Grubeck-Loebenstein et al., 1994; Mariani et al., 1990; McCarron et al., 1987; Perillo et al., 1989, 1993).

Many important similarities have been found between the T cell clone model of immunosenescence and the ageing of T cells in vivo, including decreased expression of CD28 and IL-2R, decreased IL-2 production, and decreases in telomere length and telomerase induction (reviewed in Pawelec et al., 2002). Previous studies have also demonstrated the presence of T cells in vivo with characteristics of senescent cultured cells, including shortened telomeres (reviewed in Effros and Pawelec, 1997), in Down's syndrome (a condition of accelerated ageing) and in centenarians (Franceschi et al., 1991, 1995).

In line with the in vivo evidence, Hyland et al. (2000, 2001) have demonstrated an increase in oxidative DNA damage with increasing age using the in vitro model of T cell ageing. Levels and types of DNA damage were determined in independent T cell clones as a function of their in vitro lifespan.

Alkaline comet assays revealed low levels of DNA damage as the clones progressed through their in vitro lifespan, with a significant increase in DNA damage in the majority of the clones prior to the end of their lifespans. The results of modified comet assays for the detection of oxidised purines and pyrimidines revealed an age-related increase in the oxidative DNA damage in the T cell clones as they aged (Fig. 1).

It is not yet known if the age-related increase in DNA damage within T cells in vivo is sufficient to result in T cell replicative arrest or senescence, which would have major implications for T cells required to take part in an immune response. However, results indicate that T cells containing mutations in genes coding for normal cellular metabolism may be selected against in vivo. T cells containing such mutations might have a reduced proliferative capacity, lowered response to proliferative stimuli, or may become non-viable. For example, Inamizu et al. (1986) demonstrated that the phytohaemagglutinin-induced mitogenic response of mouse splenic T cells was inversely related to the frequency of HPRT-negative mutant T cells, and Podlutsky et al. (1996) demonstrated that human T lymphocytes containing mutations in the *HPRT* gene have reduced proliferation rates in vitro.

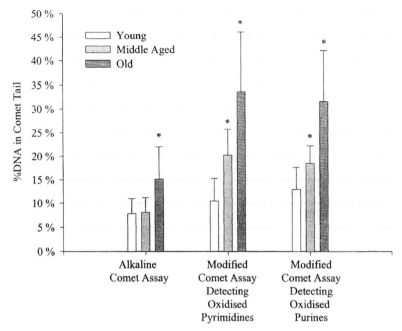

Fig. 1. Levels and types of DNA damage accumulating in vitro with age in human peripheral blood derived CD4 + T cell clones (data summarised from Hyland et al., 2001). DNA damage was assessed using the alkaline comet assay, and modified versions to detect oxidatively damaged nucleotides, and is expressed as percentage DNA in comet tail. Bars represent mean ± SD from 11 clones throughout their in vitro lifespans. *Indicates that a value is statistically significantly higher than those from younger samples, $P < 0.05$.

The work of Dempsey et al. (1983) to quantify the number of lymphocytes containing *HPRT* gene mutations in carriers of the Lesch–Nyhan mutation, found that only 1–9% of lymphocytes carried a mutation. This percentage is much lower than would be predicted (50%) for such carriers, suggesting that HPRT-negative mutant lymphocytes are selected against in vivo.

3.3. Oxidative DNA damage to T cells and longevity in vitro

There is experimental data to support the hypothesis that oxidative damage contributes to replicative senescence. When cultured at reduced oxygen tension levels, normal human diploid fibroblasts can achieve more population doublings (PD) than those cultured under standard oxygen tension levels before reaching senescence (Chen et al., 1995a; Packer and Fuehr, 1977). It is logical then, that compounds with antioxidant capacity may have the ability to extend the lifespan of cultured cells (Chen et al., 1995a; McFarland and Holliday, 1994, 1999).

Recent results from our laboratories showed that phenyl-*tert*-butylnitrone (PBN) extended the in vitro lifespan of T cell clones (Table 3). PBN is one of the most widely used spin-traps in free radical research. It selectively reacts with (traps) free

Table 3
Effect of supplementation with 250 μM PBN on in vitro longevity of human
peripheral blood derived CD4+ T cell clones

Clone	Population doublings achieved	
	PBN	Controls
385–2	64.7	62.0
385–7	87.8	74.0
399–35	70.5	65.4
399–37	72.7	63.2
400–23	76.3	68.9
400–60	Not analysed	56.5
Mean ± SD	74.4 ± 8.6*	65.0 ± 6.0

*Statistically significant difference compared to control group, $P < 0.05$.

radicals forming a stable spin-adduct. Reports have demonstrated that PBN protects against oxidative damage caused by ischemia-reperfusion (Phillis and Clough-Helfman, 1990; Bradamante et al., 1992), oxidation of low density lipoproteins (Thomas et al., 1994), and can reverse the accumulation of high levels of oxidatively damaged proteins in the brains of old gerbils (Carney et al., 1991). PBN has also been reported to significantly extend the lifespan of normal (Saito et al., 1998) and senescence accelerated mice (Edamatsu et al., 1995), and human diploid fibroblast cells (Chen et al., 1995a), and to increase proliferative capacity of murine haematopoetic stem cells (Kashiwakura et al., 1997).

Recent work on fibroblasts has suggested that N-hydroxylamine is the active group, and N-hydroxylamines have been observed to decrease the endogenous production of oxidants, reverse the acceleration of senescence induced by hydrogen peroxide (H_2O_2), and significantly decrease the level of apurinic/apyrimidinic sites in DNA (Atamna et al., 2000).

Figure 2 highlights the antioxidant properties of PBN, significantly lowering the amounts of DNA damage detected throughout the clones' extended lifespans. Lower levels of DNA single strand breaks, and of oxidative lesions, were found at all time points analysed. PBN potentiates a strong reduction in the oxidative stress on the T cell clones in culture, and it appears plausible that the observed extension of lifespan in the clones is due to amelioration of the levels of oxidative DNA damage. There is still discussion about the mechanisms of PBN's antioxidative properties (Barclay and Vinqvist, 2000; Reinke et al., 2000), however, it is clear that it can modulate the effects of oxidative stress both in vivo and in vitro.

4. Defences against free radical induced genetic damage

An elaborate network of defence mechanisms have evolved in vivo to try and minimise damage from excessive levels of free radicals. These defence mechanisms include the enzymatic antioxidants, non-enzymatic antioxidants and stress proteins.

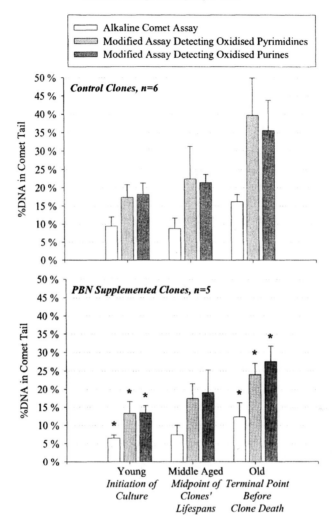

Fig. 2. Effects of 250 μM PBN supplementation on levels of DNA damage within T cell clones as a function of age in vitro. Bars represent the mean ± SD for clones within a group, for the basic comet assay or enzyme modified comet assays. *Denotes statistically significant difference compared to control, $P = 0.05$.

4.1. Enzymatic antioxidants

Protective enzymes have been discovered that act by converting highly reactive ROS into less reactive/inactive moieties. Such enzymatic antioxidants include super-oxide dismutase (SOD; EC 1.15.1.1), glutathione peroxidase (GPx; EC 1.11.1.9), catalase (CAT; EC 1.11.1.6) and caeruloplasmin (CPL; EC 1.16.3.1). SOD is an enzymatic scavenger of the superoxide anion ($O_2^{\bullet}$), which is a highly oxidative species. The superoxide anion is converted to H_2O_2, a reactive compound which must be further dealt with. Hydrogen peroxide is subsequently converted to water and oxygen by CAT. Another enzymatic antioxidant defence is provided by the selenium-dependent enzyme GPx. GPx catalyses the reduction of various organic hydroperoxides and H_2O_2 (Ozawa, 1995) with reduced glutathione to form oxidised glutathione, an alcohol group and water.

4.2. Non-enzymatic antioxidants

Non-enzymatic antioxidants are generally ubiquitous molecules widely spread in vivo. There are a number of proteins that specifically capture various biological forms of iron and copper (e.g. albumin, bilirubin, transferrin, metallothioneins) that, in their unbound states, can catalyse free radical formation. However, the most studied defences against damage by ROS are provided by small non-enzymatic antioxidants such as the hydrophilic vitamin C and glutathione, and lipophilic vitamin E and carotenoids in membranes and lipoproteins (discussed in Beckman and Ames, 1998).

The quinones are natural lipophilic, hydrophobic antioxidants that include coenzyme Q. Coenzyme Q has gained considerable attention over the past few years as it is located in the mitochondrial electron transport chain and links complex I to complex III of the respiratory chain (Beyer, 1992). Coenzyme Q, in its reduced state, participates in cellular defence against oxidative damage by preventing lipid peroxidation in the biological membranes and low-density lipoproteins within the cell.

4.3. Stress proteins

Heat shock proteins (HSP) are a group of highly conserved proteins that are induced by elevated temperatures or a variety of cellular stresses. The heat shock response is a defence reaction that involves a number of programmed changes in gene transcription and translation and is exhibited by essentially all cells when placed under stress. The cells reduce their overall rates of gene transcription and translation for a short period of time and produce HSP. HSP act as molecular chaperones by binding to unfolded or misfolded proteins and by promoting either correct refolding or proteolytic degradation. Oxidative damage is connected with conditions such as ischemia, inflammation and ageing (Plumier et al., 1995; Jacquier-Sarlin et al., 1994). Inactive or altered proteins, e.g. abnormal enzymes, are reported to accumulate within humans as they age (Stadtman, 1992). The exact molecular targets for HSP protection from oxidative stress is still unresolved. Cell membranes, DNA and proteins are all thought to be protected by HSP and it has been suggested that heat shock may increase the expression or the activity of the antioxidant enzymes catalase and SOD (Wheeler et al., 1995). It has been proposed that following heat shock the partially altered proteins expose hydrophobic regions on their surface and it is these sites that the majority of HSP recognise and bind to, facilitating refolding and preventing irreversible aggregation. The ability of HSP to act as molecular chaperones may also explain the cytoprotective effects that are observed during tissue damage (Jindal, 1996). Different model systems have demonstrated that the induction of HSP, particularly HSP 70, declines in response to heat shock with age, rendering the organism more vulnerable to stress damage (Favatier et al., 1997).

Members of the 70 kDa HSP family (HSP 70) are particularly implicated in the role of stress protection by acting as molecular chaperones and are consistently associated with defences against conditions involving ROS. HSP 70 family members are located to three cellular regions; the cytoplasm (HSP 72–73), the endoplasmic reticulum (HSP 78) and the mitochondria (HSP 70, 75). Mitochondria may be selectively protected from oxidative injury by HSP (Polla et al., 1996), and some HSP (HSP 65) are essentially mitochondrial proteins (Feige and Polla, 1994).

Three genes coding for members of the HSP 70 family are located within the major histocompatibility locus (MHC), a duplicated locus of two intronless genes, HSP 70-1 and -2, encoding two identical protein products and a third intronless gene, HSP 70-HOM, are part of the HLA class III region (Milner and Campbell, 1990).

HSP play a role in immune responses, and are associated with a role in antigen presentation (Arnold et al., 1995). It has been postulated that HSP are a central link in the innate and adaptive immune responses (Anderson and Srivastava, 2000).

4.4. In vivo antioxidant status

In vivo antioxidants are important for the maintenance of genomic stability through their ability to scavenge/remove potentially damaging free radicals. Previous studies have shown the levels of antioxidant enzymes GPx, CAT, and CPL to be significantly increased with advanced age (King et al., 1997; Mecocci et al., 2000), but that levels of non-enzymatic antioxidants decrease with age (King et al., 1997; Mecocci et al., 2000). It is thought that this is due to age-related increased free radical levels depleting the non-enzymatic antioxidants and inducing increased levels of antioxidant enzymes. However, in considering centenarians it has recently been shown that they maintain their levels of non-enzymatic antioxidants (ascorbic acid and α-tocopherol) and adopt a particular antioxidant profile which may contribute to their healthy ageing (Mecocci et al., 2000).

Hyland et al. (2002) used the Ferric Reducing Ability of Plasma (FRAP) assay to examine the total antioxidant capacity of plasma from control (middle-aged) and NONA subjects. The FRAP assay detects the activity of the major species of non-enzymatic plasma antioxidants; uric acid, ascorbic acid, α-tocopherol, bilirubin, and proteins (albumin), which have been estimated to contribute 60, 15, 5, 5, and 10% of the detected activity, respectively (Benzie and Strain, 1996). The results of this study showed that the successfully aged NONA subjects had maintained a high level of protective antioxidant capacity in their plasma, which was significantly greater than the middle-aged controls (Fig. 3).

Thus, it appears that those individuals successfully reaching extreme old age, such as centenarians and participants in the longitudinal NONA Immune study, maintain their antioxidant defence systems resulting in a lower basal load of oxidative DNA

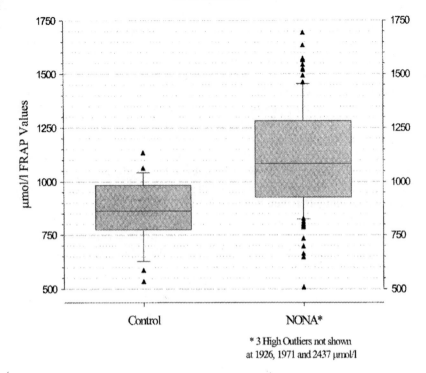

Fig. 3. Distribution of FRAP assay values for control and NONA samples (adapted from Hyland et al., 2002). FRAP assay measures antioxidant ability of plasma. Control (40–60 years old) and NONA (86–94 years old) subjects were participating in the longditudinal Swedish NONA Immune Study. The graph plots the median, 10th, 25th, 75th, and 90th percentiles as vertical boxes with error bars.

damage within the cells of the immune system with age. In vivo antioxidant capacity would therefore appear to be anti-immunosenescent.

4.5. DNA repair processes

The defences against free radical damage described above are imperfect, and in the case of DNA damage, if it is not repaired, it may become fixed by the process of DNA replication to form mutations. DNA is the only biomolecule for which a true repair system exists. Once DNA damage is recognised by cells it is subject to removal by any one of a range of lesion specific repair enzymes. The resultant gap is appropriately filled and ligation restores the continuity of the sugar-phosphate backbone (Lindahl and Wood, 1999; Mol et al., 1999). However, DNA repair systems are imperfect and there is accumulating evidence for increased DNA damage and mutation both in vivo and in vitro as a function of age.

In humans, genetic polymorphisms exist in DNA repair genes. Some of these may be associated with altered DNA repair efficiency, contributing to inter-individual differences in DNA repair capacity (Duell et al., 2000). There are also age related decreases in the fidelity of DNA polymerase and $3'-5'$ exonuclease activity

(Taguchi et al., 1998). Microsatellite stability in human cells is primarily reliant on the mismatch repair system, and has been suggested as a marker of the integrity of the mismatch repair system. However, microsatellite instability has been observed to increase with age in human leukocytes (BenYehuda et al., 2000). The mitochondrial genome is also subject to repair mechanisms, however, mtDNA repair is a very inefficient process (Bohr and Anson, 1999; Bohr and Dianov, 1999).

5. Genetic damage and T cell mediated immune responses

T cells accumulate DNA damage and mutations in vivo and in vitro as a function of age. This may compromise the proliferative capacity of lymphocytes and when T cells containing critical levels of genetic damage are required to undergo rapid clonal expansion in the presence of the appropriate antigen, insufficient numbers may be produced rendering the immune response sub-optimal. If DNA damage is detected by dividing cells a series of molecular events, mediated by cyclin dependent kinases and their inhibitors, can result in cell cycle arrest to allow time for repair of the damage, or to enable deletion of the affected cells by apoptosis.

In in vitro cellular senescence studies, expression of the cyclin-dependent kinase inhibitor (CKI) protein $p16^{INK4a/CDKN2a}$ seems to increase gradually during the lifespan and persist at high levels during senescence (Li et al., 1994; Rogan et al., 1995; Alcorta et al., 1996; Hara et al., 1996; Reznikoff et al., 1996; Zindy et al., 1997) and during quiescence in human T cells (Erickson et al., 1998). Mitra et al. (1999) recently reported that short term increase of p16 expression led to a reversible cell cycle arrest, whereas following sustained increase for 3 days a significant fraction of cells were unable to resume growth and displayed features associated with cellular senescence long after the return of p16 levels to normal. This supports the suggestion of Stein et al. (1999) that p16 up-regulation may be part of a terminal differentiation program that is activated in senescent cells.

Our group has recently determined changes in the levels of CKI with age in T cells in vitro (Hyland et al., unpublished data). CKI protein levels for p16 and p21 were determined by western blot analysis at three time points during the in vitro lifespan of peripheral blood derived CD4+ T cell clones (Table 4). The oldest T cell clone samples displayed the highest levels of p16 and p21.

The Cip/Kip family of CDK inhibitors (p21, p27, and $p57^{Kip2}$) displays a broad range of specificity and can inhibit all of the G1 cyclin–CDK complexes (Polyak et al., 1994; Lee et al., 1995; Matsuoka et al., 1995; Xiong et al., 1993; Harper et al., 1995). The expression of p21 is inducible by direct transcriptional activation by the DNA damage-inducible p53 tumour suppressor gene product (El-Deiry et al., 1993; Dulic et al., 1994) and by ROS (Qiu et al., 1996).

It is not yet known if the age-related increase in DNA damage within T cells, in vitro and in vivo, is sufficient to result in T cell replicative arrest through the action of CKI. Such arrest would have major implications for T cells required to undergo rapid clonal expansion as part of an immune response. If the proliferative capacity of T cells is compromised the immune response could be sub-optimal.

Table 4
Levels of CKIs p16 and p21 in four peripheral blood derived CD4 + T cell clones throughout their in vitro lifespans

Clone	Protein	Relative age of clone		
		Young	Midlife	Old
x385–7	p16	n.d.	63.61	47.28
	p21	40.87	91.57	45.66
x399–37	p16	14.04	36.61	59.74
	p21	n.d.	24.71	82.91
x400–23	p16	40.08	168.06	303.99
	p21	n.d.	80.39	151.52
x400–60	p16	16.21	59.57	55.46
	p21	150.00	198.96	240.26

Results are expressed as relative integrated optical densities (arbitrary units). n.d., not determined.

Also, if such an arrest leads to apoptosis then the immune systems memory is compromised.

6. Conclusions

Many reports support a relationship between longevity and genomic stability. This may be underpinned by antioxidant defences which help to reduce free radical induced damage and furthered by DNA repair processes which help to control the level of mutation to the mitochondrial and nuclear genomes, thus helping to maintain cellular function. Thus, as summarised by Franceschi et al. (1995), Steinmann and Hartwig (1995), for maintenance of healthy immune function into extreme old age, there is a requirement for a high level of lymphocyte genomic stability (low spontaneous breaks, maintenance of antioxidant levels, repair of damage etc.) which otherwise are known to increase with age in average, non-centenarian, donors (King et al., 1994, 1997).

Acknowledgements

The authors wish to acknowledge; the European Union for funding support under the aegis of European Union Concerted Action on the Molecular Biology of Immunosenescence (EUCAMBIS; Biomed 1 contract CT94-1209); Immunology and Ageing in Europe (IMAGINE; QLK6-CT-1999-O2031); The South Dublin County Council, Ireland, for a studentship award to Paul Hyland; The Department of Employment and Learning, Northern Ireland, and the Northern Ireland Histocompatibility and Immunogenetics Research Fund for a studentship award to Owen A. Ross; Anders Wikby, Boo Johansson and Andrea Tompa of the University College of Health Sciences, Jönköping, Sweden, for their collaboration; and Graham Pawelec of the Medizinische Universitsklinik und Poliklinik, Tuebingen, Germany, for provision of T cell clones for our studies.

References

Alcorta, D.A., Xiong, Y., Phelps, D., Hannon, G., Beach, D., Barret, J.C., 1996. Involvement of the cyclin-dependent kinase inhibitor p16^{INK4a} in replicative senescence of normal human fibroblasts. Proc. Natl. Acad. Sci. USA 93, 13,742–13,747.

Anderson, S., Bankier, A.T., Barrell, B.G., De Bruijn, M.H., Coulson, A.R., Drouin, J., Eperon, I.C., Nierlich, D.P., Roe, B.A., Sanger, F., Schreier, P.H., Smith, A.J., Staden, R., Young, I.G., 1981. Sequence and Organisation of the human mitochondrial genome? Nature 290, 457–465.

Anderson, K.M., Srivastava, P.K., 2000. Heat, heat shock, heat shock proteins and death: a central link in innate and adaptive immune responses. Immunol. Lett. 74(1), 35–39.

Anson, R.M., Hudson, E., Bohr, V.A., 2000. Mitochondrial endogenous oxidative damage has been overestimated. FASEB J. 14, 355–360.

Arguello, J.R., Little, A.M., Pay, A.L., Gallardo, D., Rojas, I., Marsh, S.E.E., Goldman, J.M., Madrigal, J.A., 1998. Mutation detection and typing of polymorphic loci through double-strand conformation analysis. Nat. Genet. 18, 192–194.

Arnold, D., Faath, S., Rammensee, H.-G., Schild, H., 1995. Cross-priming of minor histocompatibility antigen-specific cytotoxic T-cells upon immunization with the heat shock protein gp96. J. Exp. Med. 182, 855–889.

Atamna, H., Paler-Martinez, A., Ames, B.N., 2000. *N*-*t*-butyl hydroxylamine, a hydrolysis product of alpha-phenyl-*N*-*t*-butyl nitrone, is more potent in delaying senescence in human lung fibroblasts. J. Biol. Chem. 275(10), 6741–6748.

Barclay, L.R.C., Vinqvist, M.R., 2000. Do spin traps also act as classical chain-breaking antioxidants? A quantitative kinetic study of phenyl-*tert*-butylnitrone (PBN) in solution and in liposomes. Free Radic. Biol. Med. 28(7), 1079–1090.

Barnett, Y.A., 1994. Nutrition and the ageing process. Br. J. Biomed. Sci. 51(3), 278–287.

Barnett, Y.A., King, C.M., 1995. An investigation of *in vivo* antioxidant status, mutation and DNA repair capacity in human lymphocytes as a function of age. Mut. Res. 338, 115–128.

Beckman, K.B., Ames, B.N., 1998. The free radical theory of aging matures. Physiol. Rev. 78, 547–581.

Bender, B.S., Nagel, J.E., Adler, W.H., Andres, R., 1986. Absolute peripheral blood lymphocyte count and subsequent mortality of elderly men. The Baltimore Longitudinal Study of Aging. J. Am. Geriatr. Soc. 34, 649–654.

BenYehuda, A., Globerson, A., Krichevsky, S., On, H.B., Kidron, M., Friedlander, Y., Friedman, G., BenYehuda, D., 2000. Ageing and the mismatch repair system. Mech. Ageing Dev. 121, 173–179.

Benzie, I.F.F., Strain, J.J., 1996. The ferric reducing ability of plasma (FRAP) as a measure of "antioxidant power": The FRAP Assay. Anal. Biochem. 239, 70–76.

Berdanier, C.D., Everts, H.B., 2001. Mitochondrial DNA in aging and degenerative disease. Mut. Res. 475, 169–184.

Beyer, R.E., 1992. An analysis of the role of coenzyme Q in free radical generation and as an antioxidant. Biochem. Cell Biol. 70, 390–403.

Biagini, G., Pallotti, F., Carraro, S., Sgarbi, G., Pich, M.M., Lenaz, G., Anzivino, F., Gualandi, G., Xin, D., 1998. Mitochondrial DNA in platelets from aged subjects. Mech. Ageing Dev. 101, 269–275.

Bogliolo, M., Izzoti, A., De Flora, S., Carli, C., Abbondandolo, A., Degan, P., 1999. Detection of the '4977bp' mitochondrial DNA deletion in human atherosclerotic lesions. Mutagenesis 14(1), 77–82.

Bohr, V.A., Anson, R.M., 1999. Mitochondrial DNA repair pathways. J. Bioenerg. Biomembr. 31, 391–398.

Bohr, V.A., Dianov, G.L., 1999. Oxidative DNA damage processing in nuclear and mitochondrial DNA. Biochimie 81, 155–160.

Bradamante, S., Monti, E., Paracchini, L., Lazzarini, E., Piccinini, F., 1992. Protective activity of the spin trap *tert*-butyl-alpha-phenyl nitrone (PBN) in reperfused rat-heart. J. Mol. Cell. Cardiol. 24(4), 375–386.

Brand, M.D., 2000. Uncoupling to survive? The role of mitochondrial inefficiency in ageing. Exp. Gerontol. 35, 811–820.

Brenner, C., Marzo, I., Kromer, G., 1998. A revolution in apoptosis: from a nucleocentric to a mitochondriocentric perspective. Exp. Gerontol. 33, 543–553.

Brierley, E.J., Johnson, M.A., James, O.F.W., Turnbull, D.M., 1997. Mitochondrial involvement in the ageing process. Mol. Cell. Biochem. 174, 325–328.

Brierley, E.J., Johnson, M.A., Lightowlers, R.N., James, O.F.W., Turnbull, D.M., 1998. Role of mitochondrial DNA mutations in human aging: implications for the central nervous system and muscle. Ann. Neurol. 43, 217–223.

Cadenas, E., Davies, K.J.A., 2000. Mitochondrial free radical generation, oxidative stress, and aging. Free Radic. Biol. Med. 29(3–4), 222–230.

Carney, J.M., Starke-Reed, P.E., Oliver, C.N., Landum, R.W., Cheng, M.S., Wu, J.F., Floyd, R.A., 1991. Reversal of age-related increase in brain protein oxidation, decrease in enzyme activity, and loss in temporal and spatial memory by chronic administration of the spin-trapping compound *n-tert*-butyl-alpha-phenylnitrone. Proc. Natl. Acad. Sci. USA 88, 3633–3636.

Chen, Q., Fischer, A., Reagan, J.D., Yan, L.J., Ames, B.N., 1995a. Oxidative DNA damage and senescence of human diploid fibroblast cells. Proc. Natl. Acad. Sci. USA 92, 4337–4341.

Choe, M., Jackson, C., Yu, B.P., 1995. Lipid peroxidation contributes to age-related membrane rigidity. Free Radic. Biol. Med. 18, 977–984.

Cormio, A., Lezza, A.M.S., Vecchiet, J., Felzani, G., Marangi, L., Guglielmi, F.W., Francavilla, A., Cantatore, P., Gadaleta, M.N., 2000. mtDNA deletions in aging and in non-mitochondrial pathologies. Ann. N.Y. Acad. Sci. 908, 299–301.

Cottrell, D.A., Blakely, E.L., Borthwick, G.M., Johnson, M.A., Taylor, G.A., Brierley, E.J., Ince, P., Turnbull, D.M., 2000. Role of mitochondrial DNA mutations in disease and aging. Ann. N.Y. Acad. Sci. 908, 199–207.

De Benedictis, G., Rose, G., Carrieri, G., De Luca, M., Falcone, E., Passarino, G., Bonafe, M., Monti, D., Baggio, G., Bertolini, S., Mari, D., Mattace, R., Franceschi, C., 1999. Mitochondrial DNA inherited variants are associated with successful aging and longevity in humans. FASEB J. 13, 1532–1536.

De Benedictis, Carrieri, G., Varcasia, O., Bonafe, M., Franceschi, C., 2000. Inherited variability of the mitochondrial genome and successful aging in humans. Ann. N.Y. Acad. Sci. 908, 208–218.

Dempsey, J.L., Morley, A.A., Seshadri, R.S., Emmerson, B.T., Gordon, R., Bhagat, C.I., 1983. Detection of the carrier state for an X-linked disorder, the Lesch–Nyhan syndrome, by the use of lymphocyte cloning. Hum. Genet. 64, 288–290.

Doria, G., Frasca, D., 2001. Age-related changes of DNA damage recognition and repair capacity in cells of the immune system. Mech. Ageing Dev. 122, 985–998.

Drouet, M., Lauthier, F., Charmes, J.P., Sauvage, P., Ratinaud, M.H., 1999. Age-associated changes in mitochondrial parameters on peripheral human lymphocytes. Exp. Gerontol. 34, 843–852.

Duell, E.J., Wiencke, J.K., Cheng, T.J., Varkonyi, A., Zuo, Z.F., Ashok, T.D.S., Mark, E.J., Wain, J.C., Christiani, D.C., Kelsey, K.T., 2000. Polymorphisms in the DNA repair genes XRCC1 and ERCC2 and biomarkers of DNA damage in human blood mononuclear cells. Carcinogenesis 21, 965–971.

Dulic, V., Kaufmann, W.K., Wilson, S.J., Tlsty, T.D., Lees, E., Harper, J.W., Elledge, S.J., Reed, S.I., 1994. p53 dependent inhibition of cyclin-dependent kinase activities in human fibroblasts during radiation induced G1 arrest. Cell 76, 1013–1023.

Duquesnoy, R., Zeevi, A., 1983. Immunogenetic analysis of the HLA complex with alloreactive T cell clones. Hum. Immunol. 8, 17–23.

Edamatsu, R., Mori, A., Packer, L., 1995. The spin-trap *N-tert*-alpha-phenyl-butylnitrone prolongs the life span of the senescence accelerated mouse. Biochem. Biophys. Res. Commun. 211(3), 847–849.

Effros, R.B., Walford, R.L., 1984. T cell cultures and the Hayflick limit. Hum. Immunol. 9, 49–65.

Effros, R.B., Boucher, N., Porter, V., Zhu, X.M., Spaulding, C., Walford, R.L., Kronenberg, M., Cohen, D., Schachter, F., 1994. Decline in CD28 + T cells in centenarians and in long-term T cell cultures: a possible cause for both *in vivo* and *in vitro* immunosenescence. Exp. Gerontol. 29, 601–609.

Effros, R.B., Pawelec, G., 1997. Replicative senescence of T cells: does the Hayflick Limit lead to immune exhaustion? Immunol. Today 18(9), 450–454.

El-Deiry, W.S., Tokino, T., Velculescu, V.E., Levy, D.B., Parsons, R., Trent, J.M., Lin, D., Mercer, W.E., Kinzler, K.W., Vogelstein, B., 1993. WAF1, a potential mediator of p53 tumor suppression. Cell 75, 817–825.

Erickson, S., Sangfelt, O., Heyman, M., Castro, J., Einhorn, S., Grander, D., 1998. Involvement of the INK4 proteins p16 and p15 in T-lymphocyte senescence. Oncogene 17, 595–602.

Favatier, F., Bornman, L., Hightower, L.E., Eberhand, G., Polla, B.S., 1997. Variation in hsp gene expression and HSP polymorphism: do they contribute to differential disease susceptibilty and stress tolerance? Cell Stress Chaperones 2(3), 141–155.

Feige, U., Polla, B.S., 1994. HSP 70: a multi-gene, multi-structure, multi-function family with potential clinical applications. Experientia 50, 979–986.

Fernandez-Moreno, M.A., Bornstein, B., Petit, N., Garesse, R., 2000. The pathophysiology of mitochondrial biogenesis: towards four decades of mitochondrial DNA research. Mol. Gen. Metabol. 71, 481–495.

Franceschi, C., Monti, D., Cossarizza, A., Fagnoni, F., Passeri, G., Sansoni, P., 1991. Aging, longevity and cancer: studies in Down's syndrome and centenarians. Ann. N.Y. Acad. Sci. 621, 428–440.

Franceschi, C., Monti, D., Sansoni, P., Cossarizza, A., 1995. The immunology of exceptional individuals: the lesson of centenarians. Immunol. Today 16, 12–16.

Freeman, B.D., 1984. In: Armstrong, D., Sohal, R.S., Cutler, R.G., Slater, T.F. (Eds.), Free Radicals in Molecular Biology, Ageing and Disease. Raven Press, New York, 43 pp.

Graff, C., Clayton, D.A., Larsson, N.-G., 1999. Mitochondrial medicine – recent advances. J. Intern. Med. 246, 11–23.

Gregory, S.H., Wing, E.J., Hoffman, R.A., Simmons, R.L., 1993. Reactive nitrogen intermediates suppress the primary immunologic response to Listeria. J. Immunol. 150, 2901–2909.

Grubeck-Loebenstein, B., Lechner, H., Trieb, K., 1994. Long-term *in vitro* growth of human T cell clones: Can postmitotic 'senescent' cell populations be defined? Int. Arch. Allergy Immunol. 104, 232–239.

Halliwell, B., 1987. Free radicals, aging and disease. In: Cross, C.E. (Ed.), Oxygen Radicals and Human Disease. Ann. Internat. Med. 107, 528.

Hara, E., Smith, R., Parry, D., Tahara, H., Stone, S., Peters, G., 1996. Regulation of p16^{CDKN2} expression and its implications for cell immortalization and senescence. Mol. Cell. Biol. 16, 859–867.

Harman, D., 1956. Aging: a theory based on free radical and radiation chemistry. J. Gerontol. 11, 298–300.

Harman, D., 1972. The biological clock: the mitochondria? J. Am. Geriatr. Soc. 20, 145–147.

Harper, J.W., Elledge, S.J., Keyomarsi, K., Dynlacht, B., Tsai, L.-H., Zhang, P., Dobrowalski, S., Bai, C., Connell-Crowley, L., Swindell, E., Fox, M.P., Wei, N., 1995. Inhibition of cyclin dependent kinases by p21. Mol. Biol. Cell 6, 387–400.

Hayakawa, M., Sugiyoma, S., Hattori, K., Takasawa, M., Ozawa, T., 1993. Age-associated damage in mitochondrial DNA in human hearts. Mol. Cell. Biochem. 119, 95–103.

Hyland, P., Duggan, O., Hipkiss, A., Barnett, C., Barnett, Y., 2000. The effects of carnosine on oxidative DNA damage levels and *in vitro* lifespan in human peripheral derived CD4+ T cell clones. Mech. Ageing Dev. 121, 203–215.

Hyland, P., Barnett, C., Pawelec, G., Barnett, Y., 2001. Age-related accumulation of oxidative DNA damage and alterations in levels of p16$^{INK4a/CDKN2a}$, p21$^{WAF1/CIP1/SDI1}$ and p27^{KIP1} in human CD4+ T cell clones *in vitro*. Mech. Ageing Dev. 122, 1151–1167.

Hyland, P., Duggan, O., Turbitt, J., Coulter, J., Wikby, A., Johansson, B., Tompa, A., Barnett, C., Barnett, Y., 2002. Nonagenarians from the Swedish NONA Immune Study have increased plasma antioxidant capacity and similar levels of DNA damage in peripheral blood mononuclear cells compared to younger control subjects. Exp. Gerontol. 37, 465–473.

Inamizu, T., Kinohara, N., Chang, M.P., Makinodan, T., 1986. Frequency of 6-thioguanine-resistant T cells is inversely related to the declining T-cell activities in aging mice. Proc. Natl. Acad. Sci. USA 83, 2488–2491.

Ivanova, R., Lepage, V., Charron, D., Schachter, F., 1998. Mitochondrial genotype associated with French Caucasian centenarians. Gerontology 44, 349.

Jacquier-Sarlin, M.R., Fuller, K., Dinh-Xuan, A.T., Richard, M.J., Polla, B.S., 1994. Protective effects of hsp 70 in inflammation. Experientia 50, 1031–1038.

Jindal, S., 1996. Heat shock proteins: applications in health and disease. Trends Biotechnol. 14, 17–20.

Kao, S.-H., Liu, C.-S., Wang, S.-Y., Wei, Y.-H., 1997. Ageing-associated large-scale deletions of mitochondrial DNA in human hair follicles. Biochem. Mol. Biol. Int. 42(2), 285–298.

Kashiwakura, I., Kuwabara, M., Murakama, M., Hayase, Y., Takagi, Y., 1997. Effects of alpha-phenyl N-*tert*-butylnitrone, a spin trap reagent, on the proliferation of murine hematopoietic progenitor cells *in vitro*. Res. Commun. Mol. Pathol. Pharmacol. 98(1), 67–76.

Kim, U.-K., Kim, H.-S., Oh, B.-H., Lee, M.-M., Kim, S.-H., Chae, J.-J., Choi, H.-S., Choe, S.-C., Lee, C.-C., Park, Y.-B., 2000. Analysis of mitochondrial DNA deletions in four chambers of failing human heart: hemodynamic stress, age, and disease are important factors. Basic Res. Cardiol. 95, 163–171.

King, C.M., Gillespie, E.S., McKenna, P.G., Barnett, Y.A., 1994. An investigation of mutation as a function of age in humans. Mut. Res. 316, 79–90.

King, C.M., Bristow-Craig, H.E., Gillespie, E.S., Barnett, Y.A., 1997. *In vivo* antioxidant status, DNA damage, mutation and DNA repair capacity in cultured lymphocytes from healthy 75–80 year old humans. Mut. Res. 377, 137–147.

Kopsidas, G., Kovalenko, S.A., Heffernan, D.R., Yarovaya, N., Kramarova, L., Stojanoveki, D., Borg, J., Islam, M.M., Caragounis, A., Linnane, A.W., 2000. Tissue mitochondrial DNA changes. Ann. N. Y. Acad. Sci. 908, 226–243.

Kovalenko, S.A., Harms, P.J., Tanaka, M., Baumer, A., Kelso, J., Ozawa, T., Linnane, A.W., 1997. Method for in situ investigation of mitochondrial DNA deletions. Hum. Mut. 10, 489–495.

Kowald, A., Kirkwood, T.B., 2000. Accumulation of defective mitochondria through delayed degradation of damaged organelles and its possible role in the ageing of post-mitotic and dividing cells. J. Theor. Biol. 202(2), 145–160.

Lee, M.-H., Reynisdottir, I., Massague, J., 1995. Cloning of p57[Kip2] a cyclin dependent kinase inhibitor with unique domain structure and tissue distribution. Genes Dev. 9, 650–662.

Lee, H.C., Wei, Y.H., 1997. Role of mitochondria in human aging. Biomed. Sci. 4(6), 319–326.

Lee, H.C., Wei, Y.H., 2001. Mitochondrial alterations, cellular response to oxidative stress and defective degradation of proteins in aging. Biogerontology 2, 231–244.

Li, Y., Nichols, M.A., Shay, J.W., Xiong, Y., 1994. Transcriptional repression of the D-type cyclin dependent kinase inhibitor p16 by the retinoblastoma susceptibility gene product pRb. Cancer Res. 54, 6078–6082.

Lindahl, T., Wood, R.D., 1999. Quality control by DNA repair. Science 286(5446), 1897–1905.

Liu, V.W.S., Zhang, C., Nagley, P., 1998a. Mutations in mitochondrial DNA accumulate differentially in three different human tissues during ageing. Nucleic Acid Res. 26(5), 1268–1275.

Liu, V.W.S., Zhang, C., Pang, C.-Y., Lee, H.-C., Lu, C.-Y., Wei, Y.-H., Nagley, P., 1998b. Independent occurrence of somatic mutations in mitochondrial DNA of human skin from subjects of various ages. Hum. Mut. 11, 191–196.

Lu, C.-Y., Lee, H.-C., Fahn, H.-J., Wei, Y.-H., 1999. Oxidative damage elicited by imbalance of free radical scavenging enzymes is associated with large-scale mtDNA deletions in aging human skin. Mut. Res. 423, 11–21.

Mariani, E., Roda, P., Mariani, A.R., Vitale, M., Degrassi, A., Papa, S., Faccini, A., 1990. Age-associated changes in CD8+ and CD16+ cell reactivity: clonal analysis. Clin. Exp. Immunol. 81, 479–484.

Matsuoka, S., Edwards, M.C., Bai, C., Parker, S., Zhang, P., Baldini, A., Harper, J.W., Elledge, S.J., 1995. p57[Kip2], a structurally distinct member of the p21[Cip1] CDK inhibitor family, is a candidate tumor suppressor gene. Genes Dev. 9, 650–662.

McCarron, M., Osborne, Y., Story, C., Dempsey, J.L., Turner, R., Morley, A., 1987. Effect of age on lymphocyte proliferation. Mech. Ageing Dev. 41, 211–218.

McFarland, G.A., Holliday, R., 1994. Retardation of the senescence of cultured human diploid fibroblasts by carnosine. Exp. Cell Res. 212, 167–175.

McFarland, G.A., Holliday, R., 1999. Further evidence for the rejuvenating effects of the dipeptide l-carnosine on cultured human diploid fibroblasts. Exp. Gerontol. 34, 35–45.

Mecocci, P., Polidori, M.C., Troiano, L., Cherubini, A., Cecchetti, R., Pini, G., Straatman, M., Monti, D., Stahl, W., Sies, H., Franceschi, C., Senin, U., 2000. Plasma antioxidants and longevity: a study on healthy centenarians. Free Radic. Biol. Med. 288, 1243–1248.

Meissner, C., Mohamed, S.A., Klueter, H., Hamann, K., von Wurmb, N., Oehmichen, M., 2000. Quantification of mitochondrial DNA in human blood cells using an automated detection system. Forensic Sci. Int. 113, 109–112.

Melov, S., 2000. Mitochondrial oxidative stress. Physiologic consequences and potential for a role in aging. Ann. N.Y. Acad. Sci. 908, 219–225.

Mendoza-Nunez, V.M., Retana-Ugalde, R., Sanchez-Rodriguez, M., Altamirano-Lozano, M.A., 1999. DNA damage in lymphocytes of elderly patients in relation with total antioxidant levels. Mech. Ageing Dev. 108, 9–23.

Metzger, Z., Hoffeld, J.T., Oppenheim, J.J., 1980. Macrophage-mediated suppression: I Evidence for participation of both hydrogen peroxide and prostaglandins in suppression of murine lymphocyte proliferation. J. Immunol. 124, 983–988.

Michikawa, Y., Mazzucchelli, F., Bresolin, N., Scarlato, G., Attardi, G., 1999. Aging-dependent large accumulation of point mutations in the human mtDNA control region for replication. Science 286, 774–779.

Milner, C.M., Campbell, R.D., 1990. Structure and expression of the three MHC-linked HSP 70 genes. Immunogenetics 32, 242–251.

Miquel, J., 1998. An update on the oxygen stress-mitochondrial mutation theory of aging: genetic and evolutionary implications. Exp. Gerontol. 33(1–2), 113–126.

Mitra, J., Dai, C.Y., Somasundaram, K., El-Deiry, W., Satyamoorthy, K., Herlyn, M., Enders, G.H., 1999. Induction of p21$^{WAF1/CIP1}$ and inhibition of Cdk2 mediated by the tumor suppressor p16^{INK4a}. Mol. Cell. Biol. 19(5), 3916–3928.

Mol, C.D., Parikh, S.S., Putnam, C.D., Lo, T.P., Tainer, J.A., 1999. DNA repair mechanisms for the recognition and removal of damaged DNA bases. Annu. Rev. Biophys. Biomol. 28, 101–128.

Murasko, D.M., Weiner, P., Kaye, D., 1988. Association of lack of mitogen induced lymphocyte proliferation with increased mortality in the elderly. Aging Immunol. Infect. Dis. 1, 1–23.

Nagley, P., Wei, Y.-H., 1998. Ageing and mammalian mitochondrial genetics. TIG 14(12), 513–517.

Ozawa, T., 1995. Mechanism of somatic mitochondrial DNA mutations associated with age and diseases. Biochim. Biophys. Acta 1271, 177–189.

Ozawa, T., 1997. Genetic and functional changes in mitochondria associated with aging. Physiol. Rev. 77, 425–464.

Ozawa, T., 1999. Mitochondrial genome mutation in cell death and aging. J. Bioenerg. Biomembr. 31(4), 377–390.

Packer, L., Fuehr, K., 1977. Low oxygen concentration extends the lifespan of cultured human diploid cells. Nature 267, 423–425.

Papa, S., Skulachev, V.P., 1997. Reactive oxygen species, mitochondria, apoptosis and aging. Mol. Cell. Biochem. 174, 305–319.

Pawelec, G., Barnett, Y., Forsey, R., Frasca, D., Globerson, A., McLeod, J., Caruso, C., Franceschi, C., Fülöp, T., Gupta, S., Mariani, E., Mocchegiani, E., Solana, R., 2002. T cells and ageing. FIBS 7, 1056–1183.

Perillo, N.L., Walford, R.L., Newman, M.A., Effros, R.B., 1989. Human T lymphocytes possess a limited *in vitro* life span. Exp. Gerontol. 24, 177–187.

Perillo, N.L., Naeim, F., Walford, R.L., Effros, R.B., 1993. The *in vitro* senescence of human T-Lymphocytes – failure to divide is not associated with a loss of cytolytic activity or memory T-Cell phenotype. Mech. Ageing Dev. 67, 173–185.

Phillis, J.W., Clough-Helfman, C., 1990. Protection from cerebral ischemic-injury in gerbils with the spin trap agent *n-tert*-butyl-alpha-phenylnitrone (PBN). Neurosci. Lett. 116(3), 315–319.

Plumier, J.C.L., Ross, B.M., Currie, R.W., Angelidis, C.E., Kazlaris, H., Kollais, G., Pagoulatos, G.N., 1995. Transgenic mice expressing the human heat shock protein 70 have improved post ischemic myocardial recovery. J. Clin. Invest. 95, 1854–1860.

Podlutsky, A., Bastlova, T., Lambert, B., 1996. Reduced proliferation rate of hypoxanthine-phosphoribosyl transferase mutant human T-lymphocytes *in vitro*. Environ. Mol. Mutagen. 28, 13–18.

Polla, B.S., Kantengwa, S., François, D., Salvioli, S., Franceschi, C., Marsac, C., Cossarizza, A., 1996. Mitochondria are selective targets for the protective effects of heat shock against oxidative injury. Proc. Natl. Acad. Sci. USA 93, 6458–6463.

Porteous, W.K., James, A.M., Sheard, P.W., Porteous, C.M., Packer, M.A., Hyslop, S.J., Melton, J.V., Pang, C.-Y., Wei, Y.-H., Murphy, M.P., 1998. Bioenergetic consequences of accumulating the common 4977bp mitochondrial DNA deletion. Eur. J. Biochem. 257, 192–201.

Polyak, K., Kato, J.Y., Solomon, M.J., Scherr, C.J., Massague, J., Roberts, J.M., Koff, A., 1994. p27^{KIP1}, a cyclin-CDK inhibitor, links transforming growth factor-beta and contact inhibition to cell cycle arrest. Genes Dev. 8, 9–22.

Qiu, X., Forman, H.J., Schonthal, A.H., Cadenas, E., 1996. Induction of p21 mediated by reactive oxygen species formed during the metabolism of aziridinylbenzoquinones by HCT116 cells. J. Biol. Chem. 271, 31,915–31,921.

Raha, S., Robinson, B.H., 2000. Mitochondria, oxygen free radicals, disease and ageing. TIBS 25, 502–508.

Reinke, L.A., Moore, D.R., Sang, H., Janzen, E.G., Kotake, Y., 2000. Aromatic hydroxylation in PBN spin trapping by hydroxyl radicals and cytochrome p-450. Free Radic. Biol. Med. 28(3), 345–350.

Reznikoff, C.A., Yeager, T.R., Belair, C.D., Savelia, E., Puthenveetil, J.A., Stadler, W.M., 1996. Elevated p16 at senescence and loss of p16 at immortalization in human papilloma virus 16 E6, but not E7, transformed human uroepithelial cells. Cancer Res. 56, 2886–2890.

Richter, C., Gogvadze, V., Laffranchi, R., Schlapbach, R., Schnizer, M., Suter, M., Walter, P., Yaffee, M., 1995. Oxidants in mitochondria: from physiology to disease. Biochim. Biophys. Acta 1271, 67–74.

Roberts-Thomson, I.C., Whittingham, S., Youngchaiyud, U., Mackay, I.R., 1974. Ageing, immune response and mortality. Lancet 2, 368–370.

Rogan, E.M., Bryan, T.M., Hukku, B., Maclean, K., Chang, A.C.-M., Moy, E.L., Englezou, A., Warneford, S.G., Dalla-Pozza, L., Reddel, R.R., 1995. Alterations in p53 and p16^{INK4a} expression and telomere length during spontaneous immortalization of Li-Fraumeni syndrome fibroblasts. Mol. Cell. Biol. 15, 4745–4753.

Rose, G., Passarino, G., Carrieri, G., Altomare, K., Greco, V., Bertolini, S., Bonafe, M., Franceschi, C., De Benedictis, G., 2001. Paradoxes in longevity: sequence analysis of mtDNA haplogroup J in centenarians. Eur. J. Hum. Genet. 9, 701–707.

Ross, O.A., McCormack, R., Curran, M.D., Duguid, R.A., Barnett, Y.A., Rea, I.M., Middleton, D., 2001. Mitochondrial DNA polymorphism: its role in longevity of the Irish population. Exp. Gerontol. 36(7), 1161–1179.

Ross, O.A., Hyland, P., Curran, M.D., McIlhatton, B.P., Wikby, A., Johansson, B., Tompa, A., Pawelec, G., Barnett, C.R., Middleton, D., Barnett, Y.A., 2002. Mitochondrial DNA damage in lymphocytes: a role in immunosenescence? Exp. Gerontol. 37(2–3), 329–340.

Rustin, P., von Kleist-Retzow, J.-C., Vajo, Z., Rotig, A., Munnich, A., 2000. For debate: defective mitochondria, free radicals, cell death, aging-reality or myth-ochondria? Mech. Ageing Dev. 114, 201–206.

Saito, K., Yoshioka, H., Cutler, R.G., 1998. A spin trap, *N-tert*-butyl-alpha-phenylnitrone extends the life span of mice. Biosci. Biotechnol. Biochem. 62(4), 792–794.

Salvioli, S., Bonafe, B., Capri, M., Monti, D., Franceschi, C., 2001. Mitochondria, aging and longevity – a new perspective. FEBS Lett. 492, 9–13.

Sansoni, P., Fagnoni, F., Vescovini, R., Mazzola, M., Brianti, V., Bologna, G., Nigro, E., Lavagetto, G., Cossarizza, A., Monti, D., Franceschi, C., Passeri, M., 1997. T lymphocyte proliferative capability to defined stimuli and co-stimulatory CD28 pathway is not impaired in healthy centenarians. Mech. Ageing Dev. 96, 127–136.

Sastre, J., Pallardo, F.V., De la Asuncion, J.G., Vina, J., 2000. Mitochondria, oxidative stress and aging. Free Radic. Res. 32, 189–198.

Schmidt, H.H.H.W., Walter, U., 1994. NO at work. Cell 78, 919–925.

Shigenaga, M.K., Hagen, T.M., Ames, B.N., 1994. Oxidative damage and mitochondrial decay in aging. Proc. Natl. Acad. Sci. USA 91(23), 10,771–10,778.

Stadtman, E.R., 1992. Protein oxidation and aging. Science 257, 1220–1224.

Stein, G.H., Drullinger, L.F., Soulard, A., Dulic, V., 1999. Differential roles for cyclin-dependent kinase inhibitors p21 and p16 in the mechanisms of senescence and differentiation in human fibroblasts. Mol. Cell. Biol. 19(3), 2109–2117.

Steinmann, G., Hartwig, M., 1995. Immunology of centenarians. Immunol. Today 16, 549.

Taguchi, T., Fukuda, M., Toda, T., Ohashi, M., 1998. Age dependent decline in the 3' to 5' exonuclease activity involved in proofreading during DNA synthesis. Mech. Ageing Dev. 105, 75–87.

Tanaka, M., Gong, J.-S., Zhang, J., Yoneda, M., Yagi, K., 1998. Mitochondrial genotype associated with longevity. Lancet 35, 185–186.

Thomas, C.E., Ohlweiler, D.F., Kalyanaraman, B., 1994. Multiple mechanisms for inhibition of low-density-lipoprotein oxidation by novel cyclic nitrone spin traps. J. Biol. Chem. 269(45), 28,055–28,061.

Torroni, A., Huoponen, K., Francalacci, P., Petrozzi, M., Morrelli, L., Scozzari, R., Obinu, D., Savantaus, M.L., Wallace, D.C., 1996. Classification of European mtDNA from an analysis of three European populations. Genetics 144, 1835–1850.

Torroni, A., Petrozzi, M., D'Urbano, L., Sellitto, D., Zeviani, M., Carrara, F., Carducci, C., Leuzzi, V., Carelli, V., Barboni, P., De Negri, A., Scozzari, R., 1997. Haplotype and phylogenetic analyses suggest that one European-specific mtDNA background plays a role in the expression of Leber hereditary optic neuropathy by increasing the penetrance of the primary mutations 11778 and 14484. Am. J. Hum. Genet. 60(5), 1107–1121.

Trounce, I., Byrne, E., Marzuki, S., 1989. Decline in skeletal muscle mitochondrial respiratory chain function: possible factor in aging. Lancet 1, 637–639.

Von Wurmb, N., Oehmichen, M., Meissner, C., 1998. Demonstration of the 4977bp deletion in human mitochondrial DNA from intravital and postmortem blood. Mut. Res. 422, 247–254.

Wang, Y., Michikawa, Y., Mallidis, C., Bai, Y., Woodhouse, L., Yarasheski, K.E., Miller, C.A., Askanas, V., Engel, W.K., Bhasin, S., Attardi, G., 2001. Muscle-specific mutations accumulate with aging in critical human mtDNA control sites for replication. Proc. Natl. Acad. Sci. USA 98(7), 4022–4027.

Wayne, S.J., Rhyne, R.L., Garry, P.J., Goodwin, J.S., 1990. Cell-mediated immunity as a predictor of morbidity and mortality in subjects over 60. J. Gerontol. 45, 45–48.

Wheeler, J.C., Bieschke, E.T., Tower, J., 1995. Muscle-specific expression of Drosophila hsp 70 in response to aging and oxidative stress. Proc. Natl. Acad. Sci. USA 92, 10,408–10,412.

Wei, Y.-H., Pang, C.-Y., Lee, H.-C., Lu, C.-Y., 1998. Roles of mitochondrial DNA mutation and oxidative damage in human aging. Curr. Sci. 74, 887–893.

Xiong, Y., Hannon, G.J., Zhang, H., Casso, D., Kobayashi, R., Beach, D., 1993. p21 is a universal inhibitor of cyclin kinases. Nature 366, 701–704.

Yen, T.C., Chen, S.H., King, K.L., Wei, Y.-H., 1989. Liver mitochondrial respiratory functions decline with age. Biochem. Biophys. Res. Commun. 65, 994–1003.

Yu, P.B., Yang, R., 1996. Critical evaluation of the free radical theory of aging. Ann. N.Y. Acad. Sci. 786, 1–11.

Zindy, F., Quelle, D.E., Roussel, M.F., Sherr, C.J., 1997. Expression of the p16[INK4a] tumor suppressor versus other INK4 family members during mouse development and ageing. Oncogene 15, 203–211.

Advances in
Cell Aging and
Gerontology

Role of DNA-dependent protein kinase (DNA-PK), a protein with multiple intracellular functions, in cells of the ageing immune system

Daniela Frasca[a], Luisa Guidi[b] and Gino Doria[c]

[a]*Department of Microbiology and Immunology, School of Medicine, University of Miami,*
1600 NW 10th Ave, Miami, FL 33136, USA,
Correspondence address: Tel.: +1-305-243-6225; fax: +1-305-243-4623.
E-mail: dfrasca@med.miami.edu
[b]*Institute of Internal Medicine and Geriatrics, Catholic University of Rome, Rome, Italy,*
[c]*Department of Biology, University of Rome "Tor Vergata", Rome, Italy*

Contents

Abbreviations

A-T: ataxia-telangiectasia; Dbs: double strand break; DNA-PK: DNA-dependent protein kinase; DNA-PKcs: DNA-PK catalytic subunit; EMSA: electrophoretic mobility shift assay; HR: homologous recombination; MAPK: mitogen-activated protein kinase; NHEJ: non-homologous end joining; PBMC: peripheral blood mononuclear cells; PI3-K: phosphatidylinositol 3-kinase; TdT: terminal deoxynucleotidyl transferase.

Advances in Cell Aging and Gerontology, vol. 13, 217–230

1. Introduction

Double strand DNA breaks (dsb) are severe lesions that disrupt the genome integrity of the cell. Dsb are generated under physiological conditions, such as meiosis and recombination, and by chemical and radiation injuries. These DNA lesions must be rapidly recognized and repaired to prevent cell death or malignant transformation. Several studies have addressed the relationships between DNA damage, repair and ageing, and have suggested an age-dependent accumulation of DNA damage. Using cells from a variety of mammalian species, a positive correlation between life span and the DNA repair capacity has been demonstrated (Hart and Setlow, 1974). In particular, peripheral blood mononuclear cells (PBMC) (Lambert et al., 1979; Frasca et al., 1999) and epidermal cells (Nette et al., 1984) from elderly humans have revealed decreased DNA repair as compared to young controls.

2. Double strand DNA break recognition and repair mechanisms

In lower eukaryotes, the majority of dsb are repaired by homologous recombination (HR), a process whereby a homologous chromosome or a sister chromatid acts as a template to repair the break. In contrast, in mammalian cells the predominant dsb repair mechanism is the non-homologous end joining (NHEJ) pathway, which requires little or no sequence homology (Jackson, 1996; Kanaar and Hoeijmakers, 1997; Jeggo, 1998). Proteins operating in the HR process have been identified and include RecA in *Escherichia coli* and Rad 50–58 proteins in yeast (Friedberg, 1995). Human homologues of many of these proteins have also been identified, but their contribution to this repair mechanism is unclear (Ogawa et al., 1993; Muris et al., 1994; Kanaar et al., 1996).

Five proteins operating in the NHEJ have been identified and include three components of the DNA-PK complex such as ku 70, ku 80 and DNA-PKcs (now known as G22p1, Xrcc5 and Prkdc, respectively) as well as XRCC4 and DNA ligase IV (Jeggo, 1998). Mice carrying targeted disruption of the NHEJ pathway components share some common phenotypic features, including arrested lymphocyte development and increased sensitivity to ionizing radiation. Deficiencies in ku 70, ku 80, XRCC4 and DNA ligase IV, but not in DNA-PKcs, result in growth retardation and decreased proliferation in vitro. These anomalies are likely to result from an inability to repair DNA dsb occurring during normal DNA metabolism (Difilippantonio et al., 2000).

3. Role of the DNA-PK complex and associated factors in dsb recognition and repair and in telomere length maintenance

Eukaryotic cells contain many copies of the DNA-PK complex, which has been described to control several cellular functions (Jackson and Jeggo, 1995; Nussenzweig et al., 1996). The catalytic subunit of the complex, DNA-PKcs, is a member of the phosphatidylinositol 3-kinase (PI3-K) family and is involved in the phosphorylation of nuclear targets, thus affecting DNA repair, recombination, transcription and replication processes (Zhang and Yaneva, 1992; Troelstra and Jaspers, 1994; Roth et al., 1995; Weaver, 1995; Lieber et al., 1997; Yaneva et al., 1997). Beside DNA-PKcs, the best-characterized protein among the PI3-K family proteins is ATM, the protein deficient in Ataxia-Telangiectasia (A-T). A-T is a human disease characterized by immunodeficiency, neurological disorders, radiosensitivity and predisposition to cancer (Houldsworth and Lavin, 1980; Painter and Young, 1980; Kastan et al., 1992; Beamish and Lavin, 1994; Khanna et al., 1995). ATM has been linked to the detection and signalling of DNA damage (Rotman and Shiloh, 1998). DNA-PK, ATM, ATR (human A-T-related) (Cimprich et al., 1996), poly (ADP-ribose) polymerase (Boubnov and Weaver, 1995; Enoch and Norbury, 1995; Haffner and Oren, 1995; Hartley et al., 1995; Kupper et al., 1995; Lindahl et al., 1995; Savitsky et al., 1995; Schreiber et al., 1995; Weaver, 1995) are involved in DNA damage signalling and repair and the intracellular levels of these proteins seem to be the major rate-limiting factors in the repair of DNA breaks.

Ku 70−/− and ku 80−/− mice are very alike phenotypically, being hypersensitive to ionizing radiation and showing signs of premature senescence (Nussenzweig et al., 1996; Manis et al., 1998). These mice share many features with SCID and DNA-PKcs−/− mice, as they lack V(D)J recombination and exhibit severe combined immunodeficiency due to T and B lymphocyte arrest at early progenitor stages. Consistent with this growth defect, fibroblasts derived from ku 80−/− embryos show early loss of proliferating cells, prolonged doubling time, but intact cell-cycle checkpoints that prevent cells with damaged DNA from entering the cell-cycle, suggesting an important link between ku 80 and growth control. DNA-PKcs−/− mice exhibit normal growth, but they show severe immunodeficiency and radiation hypersensitivity. DNA-PKcs−/− mice are blocked in V(D)J coding but, unlike ku 70−/− or ku 80−/− mice, are not impeded in signal-end joint formation. Furthermore, these mice show hyperplasia and dysplasia of the intestinal mucosa and production of aberrant crypt foci, suggesting a novel role of DNA-PKcs in tumour suppression (Kurimasa et al., 1999).

The ku 70/80 heterodimer binds the ends of various types of DNA discontinuity and is involved in the repair of DNA breaks caused by V(D)J recombination, isotype switching, physiological oxidation reactions, ionizing radiation and some chemotherapeutic drugs (Troelstra and Jaspers, 1994; Lindahl et al., 1995; Weaver, 1995; Nussenzweig et al., 1996; Yaneva et al., 1997; Featherstone and Jackson, 1999). A role of ku in somatic mutation of rearranged immunoglobulin V region genes has also been suggested. The observation that somatic mutation events are accompanied by the creation of sites accessible to terminal deoxynucleotidyl transferase (TdT) in the V region and that these sites can lead to deletions and duplications in addition to the physiological single nucleotide substitutions could suggest a role for NHEJ in mutation processes. The ability of TdT to interact with ku could provide a mechanism for its recruitment at the sites of DNA breaks.

Recently, crystallization of DNA-bound ku has been obtained (Walker et al., 2001) and the structure of this compound determined. It appears that the two subunits of the ku heterodimer associate tightly and form a tetramer when bound to the two DNA ends of the break, as already postulated (Cary et al., 1997). Photo-cross-linking studies indicate that ku 70 is located proximal, and ku 80 distal, to the free DNA end. The ku heterodimer can translocate along DNA in an ATP-independent way (Devries, 1989; Paillard and Strauss, 1991), thus allowing several ku molecules to bind a single linear DNA fragment, leading to a cumulative binding of ku to DNA. The ku heterodimer has no sequence specificity when it binds to DNA ends, but it is supposed to load into a DNA end with a defined polarity in order to effectively organize the repair process. Its role is to stabilize broken DNA ends, bringing them together and preparing them for ligation (Ramsden and Gellert, 1998), as well as to prevent digestion of the broken ends by DNA exonucleases (Featherstone and Jackson, 1999). As ku has been described to activate mammalian DNA ligases in vitro (Ramsden and Gellert, 1998), it is possible that ku and a ligase may be sufficient to repair breaks without further processing before ligation. After binding to the DNA ends, the ku heterodimer recruits DNA-PKcs to the breaks and activates its kinase function (Gottlieb and Jackson, 1993; Jeggo et al., 1995). DNA-PKcs may then phosphorylate itself and/or other molecules that are required for DNA end rejoining. Alternatively, the DNA-PK complex may act as a scaffold to which other proteins required for DNA repair may be recruited. In addition to its role in DNA repair, DNA-PK may also act in the regulation of transcription by both RNA polymerase I and II (Ting et al., 1998).

The cell-cycle arrest induced by DNA damage may result in DNA repair and, therefore, in removal of the cell-cycle blockade, when the repair is completed. If, however, cells undergo apoptosis their capacity to repair damaged DNA is suppressed, also owing to the cleavage of repair enzymes by proteases (Candeias et al., 1997). DNA-PKcs has indeed been shown to be inactivated during programmed cell death, as it is specifically cleaved by caspase-3 or by a caspase-3-like protease with subsequent loss of its kinase potential (Casciola-Rosen et al., 1995; Han et al., 1996; Song et al., 1996). This DNA-PKcs inactivation blocks signalling and prevents repair of the fragmented genomic DNA produced during the apoptotic process.

Experiments performed with cell lines defective in XRCC4 have helped to clarify the role of this protein in dsb repair pathways. XRCC4 forms a tight and specific complex with DNA ligase IV (Critchlow et al., 1997; Grawunder et al., 1997; Leber et al., 1998) and stimulates its activity in vitro. XRCC4 might also act as an adaptor, bringing DNA ligase IV to the DNA-PK complex assembled on a DNA end, as suggested by the observation that XRCC4 interacts with DNA-PK in mammalian cell extracts.

NHEJ proteins (DNA-PKcs, ku and ATM) have also been described to be involved in transcriptional silencing and telomere length maintenance. They induce the formation of a heterochromatin-like state around the break (Tsukamoto et al., 1997; Boulton and Jackson, 1998), which in turn helps to prevent both transcription and replication processes, protects the break from non-specific nucleases and blocks undesired recombination reactions with other DNA molecules. DNA-PKcs−/− cells display an increased frequency of spontaneous telomeric fusions and anaphase bridges, suggesting a protective role of DNA-PKcs at the telomere site (Goytisolo et al., 2001). It is of note that the DNA-PK complex, which is primarily involved in dsb recognition and repair processes, also acts in combination with telomeric proteins in masking chromosome ends to prevent them from being recognized as dsb.

As normal functions of the immune system are dependent upon optimal cell division and clonal expansion of lymphocytes, it is of great importance to preserve the replicative capacity of lymphocytes. Deficiencies in telomere repair may result in accelerated senescence and ageing, as well as in genome instability, that leads to malignant transformation (Lansdorp, 2000).

3.1. Role of DNA-PK in dsb recognition and repair in cells of the ageing immune system

The DNA-PK complex is required for DNA dsb repair but its precise function is as yet unknown. The large size of the DNA-PK complex suggests that it may act as a scaffold to which other proteins are recruited. The catalytic subunit of the complex is unable to interact with DNA in the absence of ku, even in the presence of cross-linkers (Ting et al., 1998). Several lines of evidence have indeed shown that the interaction between DNA-PKcs and DNA is transient or weak in the absence of ku (Ting et al., 1998). Thus, ku is required to target the DNA-PKcs to the DNA ends.

The interaction of ku with DNA ends has been extensively studied. Using the Electrophoretic Mobility Shift Assay (EMSA) technique, we have demonstrated (Frasca et al., 1999) that ku forms multiple protein–DNA complexes, the number of which depends on the concentration of ku in the nuclear extract (Fig. 1). In the experiment in Fig. 1, PBMC (10^7/ml) from a young subject (25-year-old) were stimulated with phytohemagglutinin (PHA, 1.2%) for 1 h. After this time, nuclear extracts were prepared and tested by EMSA. All the bands due to the interaction of ku with DNA are supershifted using the polyclonal rabbit anti-human ku 70 and ku 80 antibodies, but not the monoclonal mouse anti-human IL-4 or the polyclonal rabbit anti-human DNA-PKcs antibodies (Fig. 2), as both anti-ku 70 and anti-ku 80 antibodies, upon incubation with the nuclear extract, were able to effectively prevent DNA-binding activity in the complex.

Mitogen stimulation of PBMC increases DNA-binding activity of ku, as compared to unstimulated controls (Fig. 3). Briefly, PBMC (10^7/ml) from a young subject (25-year-old) were left unstimulated or they were stimulated with different mitogens for 1 h, then nuclear extracts prepared and tested by EMSA. Results show that DNA-binding activity of ku can be detected, albeit at low levels, in unstimulated PBMC. Moreover, all the stimuli have comparable effects on the enhancement of DNA-binding activity of ku.

DNA-binding activity of ku was also investigated in human lymphocyte subsets. B cells were positively selected from the PBMC of a young subject (25-year-old) using anti-CD19 Microbeads, according to the MiniMacs protocol (Miltenyi). At the end of the purification procedure, cells were found to be almost exclusively CD19+ by cytofluorimetric analysis. CD45RA+ (naive) and CD45RO+ (memory) T cells were purified from the PBMC of the same young subject using

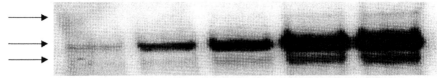

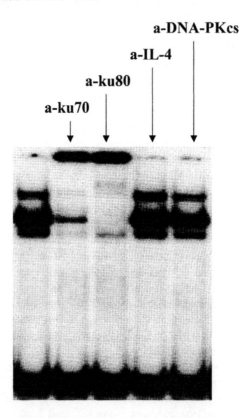

Fig. 1. Cumulative binding of ku to DNA, as evidenced by the use of the indicated amounts of nuclear extracts (from a young subject). Arrows indicate the different bands obtained when the protein concentration in the nuclear extract is varied.

Fig. 2. Supershift of the ku heterodimer by anti-ku 70 and anti-ku 80 antibodies, but not by anti-IL-4 or anti-DNA-PKcs antibodies. The nuclear extract is from a young subject.

anti-CD45RA or anti-CD45RO Microbeads. Positive CD45RA and CD45RO cells were additionally incubated with anti-CD45RO or anti-CD45RA Microbeads, respectively, in order to remove double positive cells. After this second incubation, cells were found to be almost exclusively ($>97\%$) CD45RA+ or CD45RO+ by cytofluorimetric analysis. After the purification procedure, cytoplasmic and nuclear extracts were prepared and tested by EMSA. DNA-binding activity of ku in nuclear

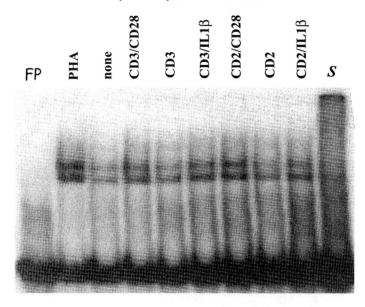

Fig. 3. Effect of PBMC stimulation on DNA-binding activity of ku. The nuclear extract is from a young subject.

and cytoplasmic extracts from PBMC, B cells, naive and memory T cells are shown in Fig. 4. DNA-binding of ku is evident only in the nuclear extracts of PBMC, B cells and naive, but not memory, T cells. Moreover, cytoplasmic ku is unable to bind DNA in any of the cell subsets. Again, the two bands observed reflect the cumulative binding of ku to DNA.

The relationship between the ability of a given organism to mount effective immune responses, and its ability to repair DNA damage, suggests that defects in the DNA damage recognition and repair proteins might lead to overlapping defects in both DNA repair and immune responsiveness. The link between the immunologic performance and life span, well supported by studies in mice (Smith and Walford, 1977; Puel et al., 1996; Doria and Frasca, 1997) and humans (Franceschi et al., 1995), led us to investigate age-related changes in the DNA repair capacity of cells of the human immune system. Briefly, PBMC from young and elderly subjects were exposed to X-rays, or left unirradiated, and stimulated with PHA for 1, 3 and 5 h. Cytoplasmic and nuclear extracts were then prepared and tested by EMSA. Results show that the DNA-binding activity of nuclear, but not cytoplasmic, ku was increased by irradiation in PBMC from young (Fig. 5) but not from elderly (Fig. 6) subjects. In young subjects, the maximum radiation-induced DNA-binding activity of ku was attained after 1 h stimulation with PHA. The radiation-induced activation of ku in PBMC from young subjects results from the increased concentrations of ku 80 and DNA-PKcs in the cytoplasm of PBMC from young but not elderly subjects, leading to a higher concentration of phosphorylated ku 80 which readily migrates to the nucleus where, after dimerization with ku 70, it binds to DNA breaks (Frasca et al., 2001).

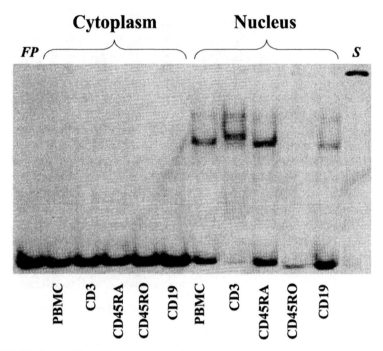

Fig. 4. DNA-binding activity of ku in nuclear and cytoplasmic extracts of PBMC, CD3+, CD45RA+, CD45RO+ and CD19+ cells from a young subject. Supershift was performed using the nuclear extract from PBMC.

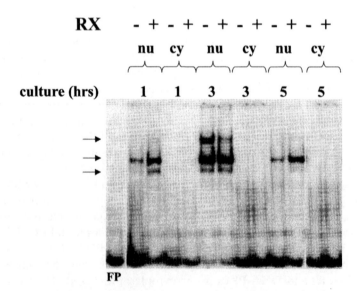

Fig. 5. Kinetic study of DNA-binding activity of ku in nuclear and cytoplasmic extracts of PBMC from a young subject. PBMC were irradiated and, immediately after irradiation, were PHA-activated for 1, 3 or 5 h. After these times, nuclear (nu) and cytoplasmic (cy) extracts were prepared.

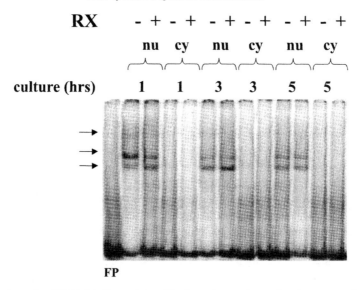

Fig. 6. Kinetic study of DNA-binding activity of ku in nuclear and cytoplasmic extracts of PBMC from an elderly subject. PBMC were irradiated and, immediately after irradiation, were PHA-activated for 1, 3 or 5 h. After these times, nuclear (nu) and cytoplasmic (cy) extracts were prepared.

4. Role of DNA-PK in signal transduction events in cells of the ageing immune system

Ionizing radiation activates not only signalling pathways in the nucleus as a result of DNA damage, but also signalling pathways initiated at the level of the plasma membrane. In irradiated cells, we have previously demonstrated that DNA-PK activity can be up-regulated in human PBMC exposed to X-rays in vitro by an IL-6-type cytokine, as DNA-binding of nuclear ku was increased by the cytokine treatment of cells from young and, to a negligible extent, from elderly subjects (Frasca et al., 2000). This cytokine effect was correlated with a higher amount of phosphorylated ku 80, rather than increased expression of ku 70 and ku 80. As to the mechanisms whereby ku and gp130 signalling are coupled in PBMC, we have recently performed experiments of co-immunoprecipitation (Frasca et al., 2002) in which ku was shown to be associated, in the cytoplasm of PBMC from young but not from elderly subjects, with Tyk-2, a kinase involved in signal transduction events after gp130 triggering by IL-6-type cytokines regardless of cell activation. After gp130 signalling, both Tyk-2 and ku 80 are phosphorylated, suggesting their activation by the IL-6-type cytokines.

Also in unirradiated lymphocytes, several lines of experimental evidence have indicated that the DNA-PK complex is involved in signal transduction events. In resting B lymphocytes (Morio et al., 1999), ku 80 is associated with the cytoplasmic tail of the CD40 molecule. After stimulation by triggering the CD40 receptor, ku 80 dissociates from CD40 and translocates into the nucleus where it activates the DNA-PK complex. In T lymphocytes (Adam et al., 2000), ku 80 has also been described to be involved in signal transduction events, as it has been found physically associated with Tyk2, suggesting that it may regulate the T cell response to cytokine signalling through this tyrosine kinase, and to p95vav, a kinase with multiple effects in T lymphocytes. The significance of ku 80 in signal transduction events is presently unclear.

Ku may be involved in Tyk-2-mediated activation by phosphorylation of other molecules operating in gp130 signal transduction events. As already demonstrated in the IFN-α-responsive human cell line U266 (Adam et al., 2000), ku functions by favouring the interaction between Tyk-2 and p95vav, a proto-oncogene product, acting as a signal transduction element in hematopoietic cells. Following this interaction, p95vav is phosphorylated by Tyk-2. Moreover, ku may act by protecting phosphorylated Tyk-2 from phosphatase attack, thus leading to prolonged activation of the mitogen-activated protein kinase (MAPK) cascade. In addition, ku associated with Tyk-2 has been reported to be phosphorylated and activated by this kinase, resulting in translocation to the nucleus, where phosphorylated ku exhibits DNA-binding activity. In elderly subjects, we were unable to demonstrate a physical association between ku 80 and Tyk-2. This may depend on the fact that ku 80 is significantly reduced in PBMC from ageing subjects, as previously shown in our laboratory (Frasca et al., 2001). It is unknown whether ageing also reduces Tyk-2. Nevertheless, PBMC from elderly subjects still display DNA-binding activity of ku, suggesting that Tyk-2 represents only one of several pathways for ku activation.

The dependence of DNA repair processes on signal transduction events has already been demonstrated in different cell types. In human prostate carcinoma cells, irradiation induces a time-dependent, MAPK-dependent, increase in DNA repair enzyme levels and DNA repair. Radiation-induced protein expression appears to require de novo transcription, suggesting a significant role for MAPK signalling in the early response to DNA damage caused by ionizing radiation (Yacoub et al., 2001). In irradiated human and murine cells the physical interaction between DNA-PKcs and Abl, Lyn, or protein kinase Cδ has been described, and this interaction results in the dissociation of DNA-PKcs from the heterodimer ku. The fact that Abl-mediated inactivation of DNA-PKcs occurs 4–8 h after irradiation, whereas the bulk of DNA repair takes place between 2 and 4 h after irradiation (Muller et al., 1999; Smith and Jackson, 1999), suggests that Abl could facilitate the release of the heterodimer ku from sites of dsb in the slow phase of dsb repair (Shangary et al., 2000). The phosphorylation-dependent inactivation of DNA-PKcs is a feature shared with the other members of the PI3-K family of enzymes (Douglas et al., 2001), such as the p110δ subunit of the class II PI3-K which down regulates its lipid kinase activity upon autophosphorylation.

5. Conclusions

Maintenance of DNA integrity is fundamental for normal immune functions, as supported by the lack of V(D)J recombination in lymphocytes of knock-out mice deficient in ku 70, ku 80 or DNA-PKcs proteins. The reduced DNA repair capacity in human lymphocytes with ageing could play a key role in the deterioration of immune reactivity and contribute to the development of age-associated immune dysfunctions which, in turn, may affect life span. The role of the DNA-PK complex in the transduction events following gp 130 activation by inflammatory cytokines is presently unclear but deserves investigation to unravel further mechanisms linking DNA repair, immune functions and ageing.

Acknowledgements

This work was supported in part by CNR Progetto Finalizzato Biotecnologie.

References

Adam, L., Bandyopadhyay, D., Kumarn, R., 2000. Interferon-alpha signaling promotes nucleus-to-cytoplasmic redistribution of p95Vav, and formation of a multisubunit complex involving Vav, Ku80, and Tyk2. Biochem. Biophys. Res. Commun. 267, 692–696.

Beamish, H., Lavin, M.F., 1994. Radiosensitivity in ataxia-telangiectasia: anomalies in radiation-induced cell cycle delay. Int. J. Radiat. Biol. 65, 175–184.

Boubnov, N.V., Weaver, D.T., 1995. Scid cells are deficient in Ku and replication protein A phosphorylation by the DNA-dependent protein kinase. Mol. Cell. Biol. 15, 5700–5706.

Boulton, S.J., Jackson, S.P., 1998. Components of the Ku-dependent non-homologous end-joining pathway are involved in telomeric length maintenance and telomeric silencing. EMBO J. 15, 1819–1828.

Candeias, S.M., Durum, S.K., Muegge, K., 1997. p53-dependent apoptosis and transcription of p21waf/cip1/sdi1 in SCID mice following gamma-irradiation. Biochimie 79, 607–612.

Cary, R.B., Peterson, S.R., Wang, J., Bear, D.G., Bradbury, E.M., Chen, D.J., 1997. DNA looping by ku and the DNA-dependent protein kinase. Proc. Natl. Acad. Sci. USA 94, 4267–4272.

Casciola-Rosen, L.A., Anhalt, G.J., Rosen, A., 1995. DNA-dependent protein kinase is one of a subset of autoantigens specifically cleaved early during apoptosis. J. Exp. Med. 182, 1625–1634.

Cimprich, K.A., Shin, T.B., Keith, C.T., Schreiber, S.L., 1996. cDNA cloning and gene mapping for a candidate human cell cycle checkpoint protein. Proc. Natl. Acad. Sci. USA 93, 2850–2855.

Critchlow, S.E., Bowater, R.P., Jackson, S.P., 1997. Mammalian DNA double-strand break repair protein XRCC4 interacts with DNA ligase IV. Curr. Biol. 7, 588–598.

Devries, D., 1989. HeLa nuclear protein recognizing DNA termini and translocating on DNA forming a regular DNA multimeric protein complex. J. Mol. Biol. 208, 65–78.

Difilippantonio, M.J., Zhu, J., Chen, H.T., Meffre, E., Nussenzweig, M.C., Max, E.E., Ried, T., Nussenzweig, A., 2000. DNA repair protein Ku80 suppresses chromosomal aberrations and malignant transformation. Nature 404, 510–514.

Doria, G., Frasca, D., 1997. Genes, immunity, and senescence: looking for a link. Immunol. Rev. 160, 159–170.

Douglas, P., Moorhead, G.B.G., Ye, R., Lees-Miller, S.P., 2001. Protein phosphatases regulate DNA-dependent protein kinase activity. J. Biol. Chem. 276, 18,992–18,998.

Enoch, T., Norbury, C., 1995. Cellular responses to DNA damage: cell-cycle checkpoints, apoptosis and the roles of p53 and ATM. Trends Biochem. Sci. 20, 426–430.

Featherstone, C., Jackson, S.P., 1999. Ku, a DNA repair protein with multiple cellular functions? Mutat. Res. 434, 3–15.

Franceschi, C., Monti, D., Sansoni, P., Cossarizza, A., 1995. The immunology of exceptional individuals: the lesson of centenarians. Immunol. Today 16, 12–16.

Frasca, D., Barattini, P., Tirindelli, D., Guidi, L., Bartoloni, C., Errani, A., Costanzo, M., Tricerri, A., Pierelli, L., Doria, G., 1999. Effect of age on DNA binding of the ku protein in irradiated human peripheral blood mononuclear cells (PBMC). Exp. Gerontol. 34, 645–658.

Frasca, D., Barattini, P., Tocchi, G., Guidi, F., Scarpaci, S., Guidi, L., Bartoloni, C., Errani, A., Costanzo, M., Doria, G., 2000. Modulation of X-ray-induced damage recognition and repair in ageing human peripheral blood mononuclear cells by an interleukin-6-type cytokine. Mech. Ageing Dev. 121, 5–19.

Frasca, D., Barattini, P., Tocchi, G., Guidi, L., Pierelli, L., Doria, G., 2001. Role of DNA-dependent protein kinase in recognition of radiation-induced DNA damage in human peripheral blood mononuclear cells. Int. Immunol. 13, 791–797.

Frasca, D., Scarpaci, S., Barattini, P., Bartoloni, C., Guidi, L., Costanzo, M., Doria, G., 2002. The DNA repair protein ku is involved in gp130-mediated signal transduction events in PBMC from young but not from elderly subjects. Exp. Gerontol. 37, 321–328.

Friedberg, E.C. 1995. DNA Repair and Mutagenesis. ASM Press, Washington.

Gottlieb, T.M., Jackson, S.P., 1993. The DNA-dependent protein kinase requirement for DNA ends and association with ku antigen. Cell 72, 131–142.

Goytisolo, F.A., Samper, E., Edmonson, S., Taccioli, G.E., Blasco, M.A., 2001. The absence of the DNA-dependent protein kinase catalytic subunit in mice results in anaphase bridges and in increased telomeric fusions with normal telomere length and G-strand overhang. Mol. Cell. Biol. 21, 3642–3651.

Grawunder, U., Wilm, M., Wu, X., Kulesza, P., Wilson, T.E., Mann, M., Lieber, M.R., 1997. Activity of DNA ligase IV stimulated by complex formation with XRCC4 protein in mammalian cells. Nature 388, 492–495.

Haffner, R., Oren, M., 1995. Biochemical properties and biological effects of p53. Curr. Opin. Genet. Dev. 5, 84–90.

Han, Z., Malik, N., Carter, T., Reeves, W.H., Wyche, J.H., Hendrickson, E.A., 1996. DNA-dependent protein kinase is a target for a CPP32-like apoptotic protease. J. Biol. Chem. 271, 25,035–25,040.

Hart, R.W., Setlow, R.B., 1974. Correlation between deoxyribonucleic acid, excision repair and life span in a number of mammalian species. Proc. Natl. Acad. Sci. USA 71, 2169–2173.

Hartley, K.O., Gell, D., Smith, G.C., Zhang, H., Divecha, N., Connelly, M.A., Admon, A., Lees-Miller, S. P., Anderson, C.W., Jackson, S.P., 1995. DNA-dependent protein kinase catalytic subunit: a relative of phosphatidylinositol 3-kinase and the ataxia telangiectasia gene product. Cell 82, 849–856.

Houldsworth, J., Lavin, M.F., 1980. Effect of ionizing radiation on DNA synthesis in ataxia-telangiectasia cells. Nucl. Acid Res. 8, 3709–3720.

Jackson, S.P., Jeggo, P.A., 1995. DNA double-strand break repair and V(D)J recombination: involvement of DNA-PK. Trends Biomed. Sci. 20, 412–415.

Jackson, S.P., 1996. The recognition of DNA damage. Curr. Opin. Genet. Dev. 6, 19–25.

Jeggo, P.A., Taccioli, G.E., Jackson, S.P., 1995. Menage a trois: double strand break repair, V(D)J recombination and DNA-PK. Bioessays 17, 949–957.

Jeggo, P.A., 1998. DNA breakage and repair. Adv. Genet. 38, 185–218.

Kanaar, R., Troelstra, C., Swagemakers, S.M., Essers, J., Smit, B., Franssen, J.H., Pastink, A., Bezzubova, O.Y., Buerstedde, J.M., Clever, B., Heyer, W.D., Hoeijmakers, J.H., 1996. Human and mouse homologs of the Saccharomices cerevisiae RAD54 DNA repair gene: evidence for functional conservation. Curr. Biol. 6, 828–838.

Kanaar, R., Hoeijmakers, J.H.J., 1997. Recombination and joining; different means to the same ends. Genes Funct. 1, 165–174.

Kastan, M.B., Zhan, Q., el-Deiry, W.S., Carrier, F., Jacks, T., Walsh, W.V., Plunkett, B.S., Vogelstein, B., Fornace, A.J., Jr., 1992. A mammalian cell cycle checkpoint pathway utilizing p53 and GADD45 is defective in ataxia-telangiectasia. Cell 71, 587–597.

Khanna, K.K., Beamish, H., Yan, J., Hobson, K., Williams, R., Dunn, I., Lavin, M.F., 1995. Nature of G1/S cell cycle checkpoint defect in ataxia-telangiectasia. Oncogene 11, 609–618.

Kupper, J.H., Muller, M., Jacobson, M.K., Tatsumi-Miyajima, J., Coyle, D.L., Jacobson, E.L., Burkle, A., 1995. Trans-dominant inhibition of poly(ADP-ribosyl)ation sensitizes cells against gamma irradiation and N-methyl-nitro-N-nitrosoguanidine but does not limit DNA replication of a polyoma virus replicon. Mol. Cell. Biol. 15, 3154–3163.

Kurimasa, A., Ouyang, H., Dong, L.J., Wang, S., Li, X., Cordon-Cardo, C., Chen, D.J., Li, G.C., 1999. Catalytic subunit of DNA-dependent protein kinase: impact on lymphocyte development and tumorigenesis. Proc. Natl. Acad. Sci. USA 96, 1403–1408.

Lambert, B., Ringborg, U., Skoog, L., 1979. Age-related decrease of UV-induced DNA repair synthesis in human peripheral leukocytes. Cancer Res. 39, 2792–2795.

Lansdorp, P.M., 2000. Repair of telomeric DNA prior to replicative senescence. Mech. Ageing Dev. 118, 23–34.

Leber, R., Wise, T.W., Mizuta, R., Meek, K., 1998. The XRCC4 gene product is a target for and interacts with the DNA-dependent protein kinase. J. Biol. Chem. 273, 1794–1801.

Lieber, M.R., Grawunder, U., Wu, X., Yaneva, M., 1997. Tying loose ends: roles of ku and DNA-dependent protein kinase in the repair of double-strand breaks. Curr. Opin. Genet. Dev. 7, 99–104.

Lindahl, T., Satoh, M.S., Poirier, G.G., Klungland, A., 1995. Post-translational modification of poly (ADP-ribose) polymerase induced by DNA strand breaks. Trends Biochem. Sci. 20, 405–411.

Manis, J.P., Gu, Y., Lansford, R., Sonoda, E., Ferrini, R., Davidson, L., Rajewsky, K., Alt, F.W., 1998. Ku70 is required for late B cell development and immunoglobulin heavy chain class switching. J. Exp. Med. 187, 2081–2089.

Morio, T., Hanissian, S.H., Bacharier, L.B., Teraoka, H., Nonoyama, S., Seki, M., Kondo, J., Nakano, H., Lee, S.K., Geha, R.S., Yata, J., 1999. Ku in the cytoplasm associates with CD40 in human B cells and translocates into the nucleus following incubation with IL-4 and anti-CD40 mAb. Immunity 11, 339–348.

Muller, C., Calsou, P., Frit, P., Salles, B., 1999. Regulation of the DNA-dependent protein kinase (DNA-PK) activity in eukaryotic cells. Biochimie (Paris) 81, 117–125.

Muris, D.F., Bezzubova, O., Buerstedde, J.M., Vreeken, K., Balajee, A.S., Osgood, C.J., Troelstra, C., Hoeijmakers, J.H., Ostermann, K., Schmidt, H., 1994. Cloning of human and mouse genes homologous to RAD52, a yeast gene involved in DNA repair and recombination. Mutat. Res. 315, 295–305.

Nette, E.G., Xi, Y.P., Sun, Y.K., Andrews, A.D., King, D.W., 1984. A correlation between aging and DNA repair in human epidermal cells. Mech. Dev. 24, 283–292.

Nussenzweig, A., Chen, G., Soares, V.D.C., Sanchez, M., Sokol, K., Nussenzweig, M.C., Li, G.C., 1996. Requirement for ku 80 in growth and immunoglobulin V(D)J recombination. Nature 382, 551–554.

Ogawa, T., Yu, X., Shinohara, A., Egelman, E.H., 1993. Similarity of the yeast Rad51 filament to the bacterial RecA filament. Science 259, 1896–1899.

Paillard, S., Strauss, F., 1991. Analysis of the mechanism of interaction of simian ku protein with DNA. Nucl. Acid Res. 19, 5619–5624.

Painter, R.B., Young, B.R., 1980. Radiosensitivity in ataxia-telangiectasia: a new explanation. Proc. Natl. Acad. Sci. USA 77, 7315–7317.

Puel, A., Mevel, J.C., Bouthillier, Y., Feingold, N., Fridman, W.H., Mouton, D., 1996. Toward genetic dissection of high and low antibody responsiveness in Biozzi mice. Proc. Natl. Acad. Sci. USA 93, 14,742–14,746.

Ramsden, D.A., Gellert, M., 1998. Ku protein stimulates DNA end joining by mammalian DNA ligases: a direct role for ku in repair of DNA dsb. EMBO J. 17, 609–614.

Roth, D.B., Lindahl, T., Gellert, M., 1995. How to make ends meet. Curr. Biol. 5, 496–499.

Rotman, G., Shiloh, Y., 1998. ATM: from gene to function. Hum. Mol. Genet. 7, 1555–1563.

Savitsky, K., Sfez, S., Tagle, D.A., Ziv, Y., Sartiel, A., Collins, F.S., Shiloh, Y., Rotman, G., 1995. The complete sequence of the coding region of the ATM gene reveals similarity to cell cycle regulators in different species. Hum. Mol. Genet. 4, 2025–2032.

Schreiber, V., Hunting, D., Trucco, C., Gowans, B., Grunwald, D., De Murcia, G., De Murcia, J.M., 1995. A dominant-negative mutant of human poly(ADP-ribose) polymerase affects cell recovery, apoptosis and sister chromatid exchange following DNA damage. Proc. Natl. Acad. Sci. USA 92, 4753–4757.

Shangary, S., Brown, K.D., Adamson, A.W., Edmonson, S., Ng, B., Pandita, T.K., Yalowich, J., Taccioli, G.E., Baskaran, R., 2000. Regulation of DNA-dependent protein kinase activity by ionizing radiation-activated abl kinase is an ATM-dependent process. J. Biol. Chem. 275, 30,163–30,168.

Smith, G.C., Jackson, S.P., 1999. The DNA-dependent protein kinase. Genes Dev. 13, 916–934.

Smith, G.S., Walford, R.L., 1977. Influence of the main histocompatibility complex on ageing in mice. Nature 270, 727–729.

Song, Q., Burrows, S.R., Smith, G., Lees-Miller, S.P., Kumar, S., Chan, D.W., Trapani, J.A., Alnemri, E., Litwack, G., Lu, H., Moss, D.J., Jackson, S., Lavin, M.F., 1996. Interleukin-1 beta-converting enzyme-like protease cleaves DNA-dependent protein kinase in cytotoxic T cell killing. J. Exp. Med. 184, 619–626.

Ting, N.S.Y., Kao, P.N., Chan, D.W., Lintott, L.G., Lees-Miller, S.P., 1998. DNA-dependent protein kinase interacts with antigen receptor response element binding proteins NF90 and NF45. J. Biol. Chem. 273, 2136–2145.

Troelstra, C., Jaspers, N.G.J., 1994. Ku starts at the end. Curr. Biol. 4, 1149–1151.

Tsukamoto, Y., Kato, J., Ikeda, H., 1997. Silencing factors participate in DNA repair and recombination in Saccharomyces cerevisiae. Nature 388, 900–903.

Walker, J.R., Corpina, R.A., Goldberg, J., 2001. Structure of the Ku heterodimer bound to DNA and its implications for double-strand break repair. Nature 412, 607–614.

Weaver, D.T., 1995. What to do at an end: DNA double-strand-break repair. Trends Genet. 11, 388–392.

Yacoub, A., Park, J.S., Qiao, L., Dent, P., Hagan, M.P., 2001. MAPK dependence of DNA damage repair: ionizing radiation and the induction of expression of the DNA repair genes XRCC1 and ERCC1 in DU145 human prostate carcinoma cells in a MEK1/2 dependent fashion. Int. J. Radiat. Biol. 77, 1067–1078.

Yaneva, M., Kowalewski, T., Lieber, M.R., 1997. Interaction of DNA-dependent protein kinase with DNA and with ku: biochemical and atomic-force microscopy studies. EMBO J. 16, 5098–5112.

Zhang, W.W., Yaneva, M., 1992. On the mechanisms of Ku protein binding to DNA. Biochem. Biophys. Res. Commun. 186, 574–579.

Linker histone H1o gene expression during ageing and after the effect of histone deacetylase inhibitors in human diploid fibroblasts and T lymphocytes

T.G. Sourlingas and K.E. Sekeri-Pataryas

National Centre for Scientific Research "DEMOKRITOS", Institute of Biology, Aghia Paraskevi, Athens, Greece
Correspondence address: Tel.: +30-1-650-3572; fax: +30-1-651-1767.
E-mail: ksek@mail.demokritos.gr (K.E.S.-P.)

Contents

Advances in Cell Aging and Gerontology, vol. 13, 231–242

1. Introduction

1.1. Why study histones during the ageing process?

The purpose of this review is to present accumulated recent investigative work from our lab related to changes in the gene expression of the H1 linker histone variant, H1o, as well as changes in the histone acetylation pattern of histone H4 during ageing and after treatment with histone deacetylase inhibitors. Why study histones in relation to ageing in the first place? Ageing is an intrinsic, genetically programmed process during which numerous biochemical pathways are altered, leading to G1 arrest of proliferating cells and inhibition of entry into the S phase (Pignolo et al., 1998). A cascade of genetically programmed events take place leading to this post-mitotic senescent state. Many molecular and genetic markers of senescence have been found, such as reduction of the activity of cyclin/CDK complexes, inability of Rb to be phoshorylated, inability of the E2F transcription factor to induce expression of its target genes, involvement of p53-antiproliferative effector pathways, overexpression of p21, as well as a number of other CDK inhibitors, etc. (Stein and Dulic, 1995; Wong and Riabowol, 1996; Cristofalo and Pignolo, 1996; Stein et al., 1999). This list of markers, which are involved in cell cycle related events, is continually increasing. Similar anti-proliferative cell cycle related events also occur during terminal differentiation (Rifkind et al., 1996; Stein, et al., 1999; Das et al., 2000; Martinez et al., 1999). Chromatin remodelling is a hallmark of both processes. More specifically, reorganization of the genome, i.e. changes in heterochromatin domains, reorganization of active and inactive chromatin domains and maintenance of repressive chromatin structures, occur during both processes (Macieira-Coelho, 1991; Howard, 1996; Chaly et al., 1996; Qumsiyeh, 1999) irrespective of the fact that the mode in which these remodelling events take place and the possible underlying factors which may perhaps be common to both senescence and differentiation, remain obscure.

Of the vast number of molecules that take part in the processes of chromatin remodelling, the histone proteins, are not only the main packaging element, but are also crucial determinants of virtually all aspects of genomic function. Changes in the histone variant constitution of chromatin have been implicated as taking place during chromatin remodelling which must inevitably occur for the progression of differentiation-associated and/or senescence-related events. Moreover, in more recent years, histone post-translational modifications have also been implicated in changes which take place during chromatin remodelling. Specifically, it has been shown that acetylation of the N-terminal tails of nucleosomal core histones is required to maintain the unfolded nucleosome structure associated with transcribing DNA (Walia et al., 1998). Evidence exists indicating that histone acetylation is not a consequence of transcription but a prerequisite and that it may be responsible not only for maintaining, but also for generating the open structure of poised and active genes (Crane-Robinson et al., 1997).

For many years, the histone constitution of chromatin has been extensively studied in many cell and tissue systems. In fact, during terminal differentiation, many changes in the histone variant constitution of chromatin have been observed and documented (Zweidler, 1984; Pina and Suau, 1987; Brown et al., 1988; Zlatanova and Doenecke, 1994; Bosch and Suau, 1995; Scaturro et al., 1995).

One especially important differentiation-associated histone gene, changes in the synthetic pattern of which have been well documented in numerous cell and tissue systems undergoing differentiation, is the linker histone variant, H1o. This specific linker histone variant was first detected in cells with low rates of proliferation (Panyim and Chalkley 1969). However, since then many investigators using different cell systems have shown that this linker histone variant accumulates during the process of terminal differentiation (Lea, 1987; Rousseau et al., 1991; Zlatanova and Doenecke, 1994; Scaturro et al., 1995).

Though senescence and differentiation have similarities with respect to gene expression and chromatin remodelling events, little attention has been given to changes that may occur in the differentiation-associated histone variant constitution of chromatin during ageing. In the framework of our studies of ageing-associated changes in the histone variant chromatin constitution, our most recent work has focused on the unique linker histone variant, H1o and the changes that may occur in its gene expression during the ageing process.

1.2. Histone deacetylase inhibitors: induced differentiation, ageing or apoptosis?

For the past two decades histone deacetylase inhibitors such as sodium butyrate and more recently trichostatin A (TSA), were used as inducers of terminal differentiation in numerous cell and/or tissue systems. For example, sodium butyrate induces biochemical and/or morphological differentiation in numerous tumour cell lines such as MEL (Leder and Leder, 1975), neuroblastoma (Prasad and Sinha, 1976), teratocarcinoma cells (Nishimune et al., 1983) as well as in *Xenopus laevis* (Khochbin and Wolffe, 1993) in millimolar concentrations. TSA induces differentiation in nanomolar concentrations also in numerous tumour cell lines such as MEL, F9, HeLa and other transformed cells (Yoshida et al., 1990; Hoshikawa et al., 1991, 1994) as well as in starfish (Ikegami, 1993) and *Xenopus* fertilized eggs (Almounzi et al., 1994). In these systems, along with differentiation, H1o gene expression was also induced. This is not surprising since it was known that the H1o protein accumulates during the normal process of terminal differentiation (Zlatanova and Doenecke, 1994). By the use of these histone deacetylase inhibitors and especially with the use of TSA, a correlation between the histone acetylation and deacetylation cycle and the execution of differentiation programs is surmised (Yoshida et al., 1995). In these systems, a striking correlation between the degree of histone acetylation and the induced expression of H1o was found (Girardot et al., 1994). In fact, full induction of H1o was observed with hyperacetylation of histones (Girardot et al., 1994).

These observations suggest that the acetylation of chromatin can, in some way, modulate H1o gene expression. Moreover, it seems highly possible that the histone H1o gene promoter is specifically responsive to chromatin acetylation because neither the H1 gene promoter nor the H4 gene promoter, although they share common regulatory elements with H1o, were found to be responsive to chromatin acetylation (Girardot et al., 1994). From these observations, one could consider that the responsiveness of the H1o gene promoter in cell or tissue systems physiologically committed to differentiation is mediated by histone acetylation (Girardot et al., 1994).

Aside from induced differentiation, work by Ogryzko et al. (1996) showed that histone deacetylase inhibitors also induce a senescent-like state or phenotype in human diploid fibroblasts (HDF). Thus, fibroblasts treated with histone deacetylase inhibitors were shown not only to have a senescent-like phenotype but were also found to resemble naturally senescent HDF in that they exit the cell cycle with a G1 DNA content and in that they exhibit pRb hypophosphorylation. From their results, Ogryzko et al. (1996) postulated that perhaps the anti-proliferative pathways activated by histone deacetylase inhibitors may be similar to those operative in naturally senescent HDF. Additionally they presented the view that the hastened entry of HDF into a senescent-like state may indicate that a major component of HDF senescence may be mechanistically related to differentiation.

Another surprising effect of histone deacetylase inhibitors in certain other cell types, such as for example rat thymocytes (Lee et al., 1996), neuronal cells (Salminen et al., 1998), lymphoblasts (Bernhard et al., 1999) and certain leukemic cell types (Santini et al., 1999) is the induction of DNA fragmentation leading to apoptosis. Moreover, in the investigation of Lee et al. (1996), butyrate-induced apoptosis was correlated with the hyperacetylation of histone H4. Thus it would seem to be that the specific effect of inhibitors of histone deacetylases is dependent on the cell type and that these seemingly diverse effects may have common underlying mechanisms at the level of histone acetylation and the chromatin remodelling events that they invoke.

Perhaps the analysis of specific histone variants such as H1o, in all these circumstances can shed further light on the matter. Indeed, as already mentioned above, analysis of induced H1o expression after treatment with histone deacetylase inhibitors in cell systems where these also induce differentiation, delineated the correlative relationship amongst H1o gene expression and histone acetylation (Yoshida et al., 1995). In fact, Girardot et al. (1994) observed full induction of H1o when histone H4 is tetra-acetylated by TSA.

Along these lines, another aspect of the more recent work from our lab to be presented in this review, is the relationship of H1o expression and histone acetylation induced by the histone deacetylase inhibitors, sodium butyrate and TSA, during apoptosis and ageing.

2. Experimental systems

2.1. Human diploid fibroblasts

Since 1961, when Hayflick and Moorhead (1961) showed that HDF have a limited in vitro lifespan, the fibroblast system has been widely used as an in vitro model system with which to study ageing and the senescent state. Fibroblasts undergo a specific number of cell divisions before exhausting their mitotic potential and enter into a viable, but stationary phase known as the post-mitotic state (Cristofalo and Pignolo, 1993) which is considered by most researchers to be a G1-like arrested state (Pignolo et al., 1998).

Here we review studies of the linker histone variant, H1o, during ageing as a function of increasing population doublings (PD) using the fibroblast in vitro model ageing cell system and more specifically using the human diploid lung embryonic cell system (FLOW 2002). This cell line undergoes more than 50–55 PD before entering a post-mitotic senescent state. The studies of both the H1o relative synthesis rates and the H1o mRNA levels were accomplished in mitotic proliferative fibroblasts of increasing in vitro age and in post-mitotic senescent cultures. In the case of post-mitotic senescent cultures, our studies used both naturally senescent fibroblast populations, i.e. populations that had reached senescence after serial subcultivations and fibroblast populations that had reached a senescent-like state in the continuous presence of the histone deacetylase inhibitor, sodium butyrate. In the latter case, as mentioned above, sodium butyrate was previously shown to induce a senescent-like phenotype in fibroblasts by Ogryzko et al. (1996). Under these conditions, cells of 27–28 cumulative PD (CPD) ceased to proliferate after 6 weekly subcultivations in the continuous presence of low concentrations of sodium butyrate (0.5 mM). Cell monolayers were kept in culture for an additional 2–3 weeks before being used as experimental material. During these last 3 weeks, these cultures were in a viable but non-proliferative stationary state (Tsapali et al., 2001). Because in the presence of sodium butyrate, fibroblast cell cultures age after having gone through a much smaller number of PD, the comparison of H1o expression in these two phenotypically similar, but perhaps in essence quite different, post-mitotic cell populations was carried out in order to determine differences which could eventually be related to chromatin remodelling events which take place as a function of CPD and which might also be informative for other cell types, i.e. T lymphocytes.

2.2. Peripheral blood lymphocytes

Another major part of our work presented in this review involves the use of peripheral blood lymphocytes. The effect of histone deacetylase inhibitors on peripheral blood lymphocytes and in relation to histones, had not been previously investigated. More specifically, in this system, we were interested in studying, the effects of sodium butyrate and TSA with respect to induced H1o expression, induced histone acetylation and apoptosis. The results from this study (see Section 6) opened the way to the study of the effects of histone deacetylase inhibitors on peripheral blood lymphocytes as a function of ex vivo age of the donor. Specifically we investigated whether the effects of histone deacetylase inhibitors, i.e. the degree of induced H1o gene expression and induced histone acetylation, can in some way be influenced by the increasing age of the donor.

3. Mitotically active HDF and the differentiation-associated H1 linker histone variant, H1o

3.1. H1o relative synthesis rates as a function of increasing in vitro age of the culture

Initially we analysed the H1o protein's relative synthesis rate as a function of increasing CPD of mitotically active HDF. Briefly, HDF cell cultures of different in vitro ages are synchronized by serum deprivation and after re-addition of serum cell populations in the Go, G1 and S phases were obtained. The entire synchronization procedure was monitored by the use of ^{3}H-thymidine (Tsapali et al., 2000a). For this cell line, we found that the S phase peak of DNA synthesis occurred 22 h after re-addition of serum. These populations (Go, G1 and S phase populations) were incubated with radioactive protein precursors and the isolated total histone fraction subjected to SDS-PAGE under conditions where the H1o variant is optimally resolved. The H1o relative synthesis rate is the percent incorporated radioactivity of the H1o protein band with respect to the radioactivity incorporated into the entire H1 fraction.

Analysis of the H1o relative synthesis rate as a function of the phases of the cell cycle and increasing CPD showed that there were cell cycle phase-specific changes, i.e. maximal synthesis rates were found during the Go phase while minimal synthesis rates were found during the S phase. However, no changes were found as a function of increasing age of the culture, i.e. in cell cultures of CPD 30, 40 and 50 (Tsapali et al., 2000a).

3.2. H1o mRNA levels as a function of increasing in vitro age of the culture

However, when the H1o mRNA levels were analysed, quite different results were obtained. Again, the analysis was accomplished as a function of increasing age of the culture (CPD 30, 40 and 50) and during the Go, G1 and S phases of the cycle. In CPD. 30, contrary to relative synthesis rates, maximal mRNA levels were found to occur during the S phase and not during the Go phase, whereas minimal H1o mRNA levels were found during the Go phase. For the G1 phase, intermediate amongst the S and Go phase mRNA levels were found. The exact same phase-related results were found in the case of CPD 40, except that the intensity of the bands for the Go and G1 phases were stronger as compared to the Go and G1 phases of CPD 30. However, again in CPD 40 the strongest signal was obtained in the S phase. Lastly, in cell cultures of CPD 50, no phase-related changes were observed. A strong signal, of more or less equal magnitude, was obtained for all three phases. In other words, at this very late cell culture age, where however, mitotic activity is still observed, there is a stabilization of H1o mRNA levels. Thus H1o mRNA levels change in a specific manner as a function of increasing age of the culture (Tsapali et al., 2000b).

4. Post-mitotic senescent HDF and the differentiation-associated H1 linker histone variant, H1o

As mentioned above, no differences in the relative synthesis rates of the H1o variant were found as a function of increasing age of the culture in mitotically active HDF. However, when the relative synthesis rates of H1o were analysed in post-mitotic senescent cell cultures, quite different results were obtained. In this latter case, an increase in the H1o relative synthesis rates were observed, both in the case of naturally aged, post-mitotic cell cultures, i.e. after serial subcultivations and in the case of sodium butyrate artificially induced post-mitotic senescent cultures (Tsapali et al., 2000a).

However, when the H1o mRNA levels of both these quite differently aged cultures were compared, only in the case of the naturally aged cultures were H1o mRNA levels increased as compared to mitotic cultures of CPD 33. In the case of the cultures aged in the presence of sodium butyrate, no increase the H1o mRNA levels were observed. These levels were on par with those of HDF in CPD 33 (Tsapali et al., 2001).

5. Conclusions regarding the relationship of the differentiation-associated linker histone variant, H1o, during the in vitro ageing of HDF

The relationship amongst H1o protein synthesis levels and H1o mRNA levels in mitotic and post-mitotic naturally senescent HDF populations parallels that which has been observed for this unique linker histone variant in differentiating cell and/or tissue systems. In differentiating systems, H1o mRNA accumulation is linked to DNA replication, whereas H1o protein accumulation was found to occur during the later non-proliferative stages of terminal differentiation. In other words, an uncoupling of H1o protein accumulation from that of its coding mRNA is observed in differentiating systems (Grunwald et al., 1991; Khochbin et al., 1991; Rousseau et al., 1992), as is observed in the in vitro model ageing cell system of HDF (Tsapali et al., 2000a,b, 2001). A hypothetical reason for this behaviour in both systems may be that H1o mRNA is accumulated in proliferating cells so as to be available to be used to direct H1o protein synthesis when needed by the cells when they eventually become arrested.

Moreover, an increase in H1o mRNA levels that was found in naturally senescent post-mitotic HDF, has also been observed during the last stages of terminal differentiation (Scaturro et al., 1995). Thus it would seem that the timing of expression is of importance not only in differentiating cell systems but in ageing cell systems as well.

The fact that sodium butyrate-induced post-mitotic senescent HDF cultures do not follow this same behaviour indicates that we must be careful in what we classify as being senescent. Though both types of senescent cultures appear phenotypically similar, underlying molecular mechanisms may in actuality be quite different. For the time being though, the underlying reasons for this discrepancy between H1o mRNA levels of naturally vs. artificially aged HDF cultures remain to be elucidated.

6. Peripheral blood lymphocytes and histone deacetylase inhibitors

As mentioned in Section 1, it is known that histone deacetylase inhibitors, such as sodium butyrate and TSA, induce histone acetylation and H1o accumulation in numerous differentiating cell/tissue systems. Peripheral blood lymphocytes do not normally express the H1o linker histone variant under any circumstances (Mannironi et al., 1987; Sourlingas et al., 1999). Thus they are an ideal model system with which to study the induction of H1o. As discussed above we showed that changes and/or increases in either H1o protein synthesis levels and/or H1o mRNA levels are associated with the process of ageing in the HDF in vitro model ageing cell system. Thus H1o induction has so far been associated with differentiation and from the very recent work reported in this review, with in vitro ageing. The question that now arises is whether in other cell systems in which the behaviour of this unique linker histone variant has not as yet been studied, such as peripheral blood lymphocytes, and under certain circumstances, H1o induction can be associated with other biological processes as well.

With the above in mind, recent work from our laboratory, showed that in lymphocytes, histone deacetylase inhibitors, namely, sodium butyrate and TSA, induced histone H4 acetylations and hyperacetylations, respectively, in a dose-dependent manner (Sourlingas et al., 2001). Concomitant to histone acetylation, the H1o protein and H1o mRNA were also induced (Sourlingas et al., 2001). Moreover, in this particular cell system, apoptosis was induced. This was demonstrated by pulsed field electrophoresis as shown by the presence of 50 kbp DNA fragmentation as well as by conventional DNA agarose electrophororesis as shown by the presence of internucleosomal DNA fragmentation (Sourlingas et al., 2001). Furthermore, apoptosis-induced by histone deacetylase inhibitors appears to be dose-dependent (Sourlingas et al., 2001).

7. Conclusions regarding the relationship of H1o induction, histone acetylation and apoptosis in peripheral blood lymphocytes after the use of histone deacetylase inhibitors

As mentioned previously, it was well known that histone deacetylase inhibitors, along with the induction of differentiation, also concomitantly induce H1o and histone acetylations in numerous cell or tissue systems (Leiter et al., 1984; Rousseau et al., 1991, 1992; Girardot et al., 1994; Seigneurin et al., 1995; Yoshida et al., 1995). However, the effects of these agents on lymphocytes had not been previously reported. In this particular cell system histone deacetylase inhibitors induce apoptosis. Moreover, from the work described above (Sourlingas et al., 2001), the induction of H1o (and also the acetylation of histones) has been correlated to the apoptotic phenomenon. Interestingly, Liu et al. (1999) have shown that histone H1 directly interacts with the DNA fragmentation factor and stimulates the nuclease activity of DFF40. However, at the moment, no information is available regarding the specificity and binding affinity of the nuclease to specific linker histone variants. Though no evidence as yet exists as to whether there is a direct relationship amongst the H1o variant and apoptosis, one may postulate the potential existence of such a relationship since chromatin remodelling events are necessary during the apoptotic phenomenon and the physicochemical properties of H1o (Roche et al., 1985), as well as this linker histone variant's close association with histone acetylation (recall that the H1o promoter is sensitive to histone acetylation and that this may probably be a central reason as to why histone deacetylase inhibitors can induce H1o expression) are indicative of a closer relationship with chromatin (Roche et al., 1985).

8. Can the effects of histone deacetylase inhibitors be influenced by an additional factor, i.e. that of age?

From what has been said so far, it would seem that the inhibition of histone deacetylases can play a key role not only during differentiation, but in numerous other biological processes, including apoptosis and ageing. Moreover, the close relationship of histone acetylation and H1o induction in all these processes have been overwhelmingly implicated from the work cited herein and elsewhere. The work described so far led us to the idea of ascertaining whether ageing may also influence the induced effects of histone deacetylase inhibitors, i.e. whether the degree of induced H1o expression and induced histone acetylation can be influenced as a function of increasing age of the donor. The investigation was initiated as a pilot study where peripheral blood lymphocytes, obtained from donors of different ages were placed in culture and mitogenically activated (PHA). Fourty-eight hours later TSA was added to the cultures. Cells were harvested 72 h after initiation of cultures, when PHA-activated lymphocytes are in the S phase. Control cell cultures were harvested simultaneously but without having been subjected to 24 h TSA treatment. The time of incubation with TSA, as well as the duration of treatment had been previously determined from our study with peripheral blood lymphocytes described above (Sourlingas et al., 2001). The results from this line of work proved to be somewhat unexpected. Because during the S phase the major portion of histone synthesis takes place, the effect of TSA on total histone synthesis was also ascertained. TSA blocks cell proliferation in the G1 phase (Ogryzko et al., 1996). Thus it is expected that it would inhibit to a certain extent, depending on the particular circumstances, total histone synthesis, since it was present during the G1-S phase (s). However, it was found that the degree of this total histone synthesis inhibition in the presence of TSA, increased with increasing age of the donor (Sourlingas et al., 2002).

When the degree of induced-by-TSA histone H4 acetylation was analysed, again it was found that with increasing age of the donor there was an increase in the histone H4 acetylation pattern. Finally, when the relative synthesis rate of H1o was ascertained, a discernible increase as a function of age was found (Sourlingas et al., 2002). This increase was more evident when comparing donors which had a major difference in age, i.e. young (25 years) vs. senior (60–65 years) or elderly (90–95 years).

9. Conclusions regarding the effects of histone deacetylase inhibitors in lymphocytes and the age of the donor

The findings of the above investigation (Sourlingas et al., 2002), i.e. that in peripheral blood lymphocytes, the degree of total histone synthesis inhibition by

TSA, the degree of induced histone H4 acetylation and induced H1o protein synthesis can be influenced by the age of the donor were both interesting and not anticipated. It would seem that with increasing age, there appears to be an enhanced sensitivity of these cells to histone deacetylase inhibitors. At present one can only speculate as to the possible underlying causes of this enhanced sensitivity with increasing age. It should however be mentioned that more than one type of histone deacetylase (HDAC) enzyme types exists (for example, HDAC1 (Taunton et al., 1996), HDAC 20 (Yang et al., 1996), HDAC3 (Dangond et al., 1998; Yang et al., 1997; Emiliani et al., 1998) etc.) and Dangond and Gullans (1998) have shown that although all histone deacetylases are regulated by histone deacetylase inhibitors, their response is differentially modulated in a time-dependent manner, suggesting differential sensitivity and different roles for the individual enzymes. Interestingly and in further support of the above findings are the recent results from the investigation of Wagner et al. (2001) where they set out to compare the activity and expression levels of HDACs in HDF undergoing senescence. They found a senescent-specific HDAC activity that appears only in senescent fibroblasts.

All these findings strongly implicate differential sensitivity and/or differential expression of different HDAC specifically associated with senescence and/or the possible existence of senescence-specific HDAC. Morever, the findings of the differential effect of histone deacetylase inbihitors in lymphocytes as a function of age, may in fact be especially important during immunosenescence in which chromatin remodelling may be an important event.

This differential responsiveness with increasing ex vivo age may in fact eventually turn out to be of potential benefit to the further elucidation of T cell decreased or altered response with increasing age (Dangond and Gullans, 1998). Moreover, elucidation of the differential responsiveness and potential specificities of these enzymes as a function of T cell age may possibly even be of benefit in therapeutic and/or antigeriatric regimens.

10. Concluding remarks

Differentiation, ageing and apoptosis are biological processes that, in order to proceed, need chromatin remodelling events to occur. It is well known that nucleosomal histones in general play a significant part during chromatin rearrangement. H1o has certain physicochemical properties that more closely associate this particular H1 linker histone variant with certain specific properties of chromatin (Roche et al., 1985; Gunjan et al., 1999) which may perhaps account for the necessity of H1o expression/and or increase in protein and/or mRNA levels during all three of these biological processes. Moreover, nothing further need be mentioned regarding the association of histone acetylation and its relationship to the more open and poised for transcription chromatin structures that this histone modification generates: the vast amount of data that has accumulated in recent years from this line of research are persuasive enough. Our interest stems from the intimate relationship of both H1o gene expression and histone acetylation. Future research goals should include a more in-depth analysis of the association of H1o gene expression in ageing cell systems as well as the differential sensitivity of histone deacetylases and the effect of histone deacetylase inhibitors on the apoptotic and ageing phenomena of lymphocytes as well as in other ageing cell systems.

References

Almounzi, G., Khochbin, S., Dimitrov, S., Wolffe, A.P., 1994. Dev. Biol. 165, 654–669.
Bernhard, D., Ausserlechner, M.J., Tonko, M., Loffler, M., Hartmann, B.L., Csordas, A., Kofler, R., 1999. FASEB J. 13, 1991–2001.
Bosch, A., Suau, P., 1995. Eur. J. Cell Biol. 68, 220–225.
Brown, D.T., Yang, Y.S., Sittman, D.B., 1988. Mol. Cell. Biol. 8, 4406–4415.
Chaly, N., Munro, S.B., Swallow, M.A., 1996. J. Cell Biochem. 62, 76–89.
Crane-Robinson, C., Hebbes, T.R., Clayton, A.L., Thorne, A.W., 1997. Methods 12, 48–56.
Cristofalo, V.P., Pignolo, R.J., 1993. Physiol. Rev. 73, 617–638.
Cristofalo, V.J., Pignolo, R.T., 1996. Exp. Gerontol. 31, 111–123.
Das, D., Pintucci, G., Stern, A., 2000. FEBS Lett. 472, 50–52.
Dangond, F., Gullans, S.R., 1998. Biochem. Biophys. Res. Commun. 247, 833–837.
Dangond, F., Hafler, D.A., Tong, J.K., Randall, J., Kojima, R., Utka, N., Gullans, S.R., 1998. Biochem. Biophys. Res. Commun. 242, 648–652.
Emiliani, S., Fischle, W., Van Lint, C., Al-Abed, Y., Verdin, E., 1998. Proc. Natl. Acad. Sci. USA 95, 2795–2800.
Girardot, V., Rabilloud, T., Yoshida, M., Beppu, T., Lawrence, J.J., Khochbin, S., 1994. Eur. J. Biochem. 224, 885–892.
Grunwald, D., Khochbin, S., Lawrence, J.J., 1991. Exp. Cell Res. 194, 174–179.
Gunjan, A., Alexander, B.T., Sittman, D.B., Brown, D.T., 1999. J. Biol. Chem. 274, 37,950–37,956.
Hayflick, L., Moorhead, P.S., 1961. Exp. Cell Res. 25, 585–621.
Hoshikawa, Y., Kijima, M., Yoshida, M., Beppu, T., 1991. Agric. Biol. Chem. 55, 1491–1495.
Hoshikawa, Y., Kwon, H.J., Yoshida, M., Horinouchi, S., Beppu, T., 1994. Exp. Cell Res. 214, 189–197.
Howard, B.H., 1996. Exp. Gerontol. 31, 281–293.
Ikegami, S., 1993. Roux's Arch. Dev. Biol. 202, 144–151.
Khochbin, S., Gorka, C., Lawrence, J.J., 1991. FEBS Lett. 283, 65–67.
Khochbin, S., Wolffe, A.P., 1993. Gene (Amst.) 128, 173–180.
Lea, M.A., 1987. Cancer Biochem. Biophys. 9, 199–209.
Leder, A., Leder, P., 1975. Cell 5, 319–322.
Lee, E., Furukubo, T., Miyabe, T., Yamauchi, A., Kariya, K., 1996. FEBS Lett. 395, 183–187.
Leiter, J.M.E., Helliger, W., Puschendorf, B., 1984. Exp. Cell Res. 155, 222–231.
Liu, X., Zou, H., Widlak, P., Garrard, W., Wang, X., 1999. J. Biol. Chem. 274, 13,836–13,840.
Macieira-Coelho, A., 1991. Mutat. Res. 256, 81–104.
Mannironi, C., Rossi, V., Biondi, A., Ubezio, P., Massara, G., Barbui, T., D'Incalci, M., 1987. Blood 70, 1203–1207.
Martinez, L.A., Chen, Y., Fischer, S.M., Conti, C.J., 1999. Oncogene 18, 397–406.
Nishimune, Y., Kume, A., Ogiso, Y., Matsushiro, A., 1983. Exp. Cell Res. 146, 439–444.
Ogryzko, V.V., Hirai, T.H., Russanova, V.R., Barbie, D.A., Howard, B.H., 1996. Mol. Cell. Biol. 16, 5210–5218.
Panyim, S., Chalkley, R.G., 1969. Biochem. Biopys. Res. Commun. 37, 1042–1049.
Pignolo, R.T., Martin, B.G., Horton, J.H., Kalbach, A.N., Cristofalo, V.J., 1998. Exp. Gerontol. 33, 67–80.
Pina, B., Suau, P., 1987. Dev. Biol. 123, 51–58.
Prasad, K., Sinha, P.K., 1976. In vitro 12, 125–132.
Qumsiyeh, M.B., 1999. Cell Mol. Life Sci. 55, 1129–1140.
Rifkind, R.A., Richon, V.M., Marks, P.A., 1996. Pharmacol. Ther. 69, 97–102.

Roche, J., Girardet, J.L., Gorka, C., Lawrence, J.J., 1985. Nucl. Acids Res. 13, 2843–2853.

Rousseau, D., Khochbin, S., Gorka, C., Lawrence, J.J., 1991. J. Mol. Biol. 217, 85–92.

Rousseau, D., Khochbin, S., Gorka, C., Lawrence, J.J., 1992. Eur. J. Biochem. 208, 775–779.

Salminen, A., Tapiola, T., Korhonen, P., Suuronen, T., 1998. Brain Res. Mol. Brain Res. 61, 203–206.

Santini, V., Gozzini, A., Scappini, B., Caporale, R., Zoccolante, A., Rigacci, L., Gelardi, E., Grossi, A., Alterini, R., Ferrini, P.R., 1999. Haematologica 84, 897–904.

Scaturro, M., Cestelli, A., Castiglia, D., Natasi, T., DiLiegro, I., 1995. Neurochem. Res. 20, 969–976.

Seigneurin, D., Grunwald, D., Lawrence, J.J., Khochbin, S., 1995. Int. J. Dev. Biol. 39, 597–603.

Sourlingas, T.G., Steger, M., Grubeck-Loebenstein, B., Tsapali, D.S., Sekeri-Pataryas, K.E., 1999. Exp. Gerontol. 34, 59–67.

Sourlingas, T.G., Tsapali, D.S., Kaldis, A.D., Sekeri-Pataryas, K.E., 2001. Eur. J. Cell Biol. 80, 726–732.

Sourlingas, T.G., Kypreou, K.P., Sekeri-Pataryas, K.E., 2002. Exp. Gerontol. 37, 341–348.

Stein, G.H., Dulic, V., 1995. BioEssays 17, 537–543.

Stein, G.H., Drullinger, L.F., Soulard, A., Dulic, V., 1999. Mol. Cell. Biol. 19, 2109–2117.

Taunton, J., Hassig, C.A., Schreiber, S.L., 1996. Science 272, 408–411.

Tsapali, D.S., Sekeri-Pataryas, K.E., Sourlingas, T.G., 2000a. Ann. N.Y. Acad. Sci. 908, 336–340.

Tsapali, D.S., Sekeri-Pataryas, K.E., Sourlingas, T.G., 2000b. Mech. Ageing Dev. 121, 101–112.

Tsapali, D.S., Sekeri-Pataryas, K.E., Sourlingas, T.G., 2001. Exp. Gerontol. 36, 1649–1661.

Wagner, M., Brosch, G., Zwerschke, W., Seto, E., Loidl, P., Jansen-Durr, P., 2001. FEBS Lett. 499, 101–106.

Walia, H., Chen, H.Y., Sun, J.M., Holth, L.T., Davie, J.R., 1998. J. Biol. Chem. 273, 14,516–14,522.

Wong, H., Riabowol, K., 1996. Exp. Gerontol. 31, 311–325.

Yang, W.M., Inouye, C., Zeng, Y.Y., Bearss, D., Seto, E., 1996. Proc. Natl. Acad. Sci. USA 93, 12,845–12,850.

Yang, W.M., Yao, Y.L., Sun, J.M., Davie, J.R., Seto, E., 1997. J. Biol. Chem. 272, 28,001–28,007.

Yoshida, M., Hoshikawa, Y., Koseki, K., Mori, K., Beppu, T., 1990. J. Antibiot. 43, 1101–1106.

Yoshida, M., Horinouchi, S., Beppu, T., 1995. BioEssays 17, 423–430.

Zlatanova, J., Doenecke, D., 1994. FASEB J. 8, 1260–1268.

Zweidler, A., 1984. Biochemistry 23, 4436–4443.

Zinc and the immune system of elderly

Klaus-Helge Ibs[a], Philip Gabriel[b] and Lothar Rink[a]

[a]*Institute of Immunology, University Hospital, Technical University of Aachen, Pauwelsstrasse 30, D-52074 Aachen, Germany*
Correspondence address: Tel.: +49-241-80-80208; fax: +49-241-80-82613.
E-mail: lrink@ukaachen.de (L.R.)
[b]*Institute of Immunology and Transfusion Medicine, University of Lübeck School of Medicine, Ratzeburger Allee 160, D-23538 Lübeck, Germany*

Contents

1. Introduction

Within the last two centuries zinc has been characterized as essential for all growing organisms. The first observations by Raulin (1869) showed that zinc is needed for the cultivation of *Aspergillus niger*. Based on these results further investigations discovered the importance of zinc for the mammalian organism as shown in a rat model by Todd et al. (1934) and later Prasad et al. (1963) even established that growth in the human organism was zinc-dependent. Symptoms like skin lesions, hypogonadism, growth and mental retardation, anaemia, hepatosplenomegaly and severe immune defects were also observed in zinc deficiency. Acrodermatitis enteropathica (a rare autosomal recessive inheritable disease) was discovered to be a zinc-specific malabsorption syndrome (Neldner and Hambidge, 1975), which made it obvious that these symptoms are related to zinc deficiency. Zinc supplementation reverses all these symptoms (Neldner and Hambidge, 1975) and prevents patients from succumbing to infections, demonstrating strict zinc dependency. Clinical investigations revealed that chronic diseases such as renal insufficiency are accompanied by decreased zinc levels. In the elderly, a reduced serum zinc level (Lee et al., 1993; Buokaiba et al., 1993) and immune defects have been described (Saltzman and Peterson, 1987; Sindermann et al., 1993; Pawelec et al., 1995).

2. Zinc physiology

The human body contains 2–3 g of zinc, mostly intracellular. Only a small amount occurs in the plasma and is predominantly bound to albumin, α2-macroglobulin and transferrin (Scott and Bradwell, 1983). The plasma concentration (12–16 μM) is influenced by different factors, e.g. age and gender.

There are more than 300 enzymes from all six enzyme classes that are known to carry zinc as a cofactor (Coleman, 1992; Vallee and Falchuk, 1993). In these enzymes zinc fulfils three functions: (1) it is used for structural stability; (2) it influences the catalytic activity by being a coactive factor; (3) it is needed as the central ion for catalytic activity. In some enzymes zinc may perform more than one of these functions (Table 1). This is possible because of the properties of zinc. It is almost non-toxic (Bertholf, 1988), highly flexible in its coordination and makes stable associations with macromolecules (Vallee and Aulid, 1990). Zinc plays an important role in cell proliferation, since there are zinc finger motifs present in a wide variety of transcription and replication factors. Consequently, in zinc deficiency, cell proliferation does not take place, so that highly-proliferating cells (e.g. cells of the immune system or the skin) are considered to be sensitive indicators for a lack of zinc. Furthermore, zinc plays a distinct role in apoptosis (Zalewski and Forbes, 1993) and signal transduction (Beyersmann and Haase, 2001).

3. Nutritional status of the elderly

There are different groups that are predisposed to zinc deficiency for a variety of reasons (Table 2). As mentioned above, we would like to focus on the elderly

Table 1
Function of zinc in some enzymes

Function	Enzyme	Enzyme class
Structural role	Sorbitol dehydrogenase	Oxidoreductase
	Protein kinase C	Hydrolase
Catalytic role	RNA polymerase	Transferase
	Carbonic anhydrase	Lyase
	Aldolase II (funghi)	Isomerase
	tRNA synthase	Ligase
Coactive role	Fructose-1,6-biphosphatase	Transferase
Catalytic + coactive role	Aminopeptidase	Transferase
	Alkaline phosphatase	Hydrolase
Catalytic + structural role	Alcohol dehydrogenase	Oxidoreductase

Table 2
Reasons for zinc deficiency

Reason	Group
Insufficient intake	Elderly
	Alcoholics
Decreased absorption	Vegetarians
Increased requirements	Children
	Adolescents
	Pregnant
	Nursing
Increased elimination	Alcoholics
	Diabetes patients
	Renal insufficient patients

population. Recent surveys confirmed the findings of Greger and Sciscoe (1977), Vir and Love (1979). Aged people are at risk of malnutrition. Their food intake is decreased, because their energy requirements are reduced, but the need for vitamins and minerals is constant or even increased (Chernoff, 1995).

Briefel et al. (2000) analyzed data from the Third National Health and Nutrition Examination Survey (1988–1994), which examined 29,103 participants to estimate their daily zinc intake. The study pointed out that only 51.1% of adults aged 51–70 years and only 42.5% of adults aged 71 years and older had an "adequate" zinc intake. This "adequate" intake is based on a total zinc intake at or above 77% of the 1989 RDA (recommended dietary allowance) age/sex-specific values. The data from this survey were also used by Lee et al. (2001) in order to assess food-insecurity among elderly U.S. citizens. They revealed that food-insecure elderly persons have poorer dietary intake, nutritional status and health status than food-secure elderly persons do.

Other surveys discovered marginally to moderately low content of zinc in the typical American diet (Ma and Betts, 2000). These low zinc intakes in the elderly might be associated with age, low income and less education. Furthermore, there seems to be a difference between free-living persons and non-free-living persons, since zinc levels of female and male nursing home residents were significantly lower than levels of non-nursing home residents (Worwag et al., 1999). At times of high nutritional requirements such as acute or chronic illnesses, composition of food is important and malnutrition prolongs hospital stays and possibly causes a higher rate of mortality (Seiler, 1999). This is not only a problem of hospitalized persons, but also of the elderly who partake of communal meals (Gilbride et al., 1998).

Information on the free-living elderly are contradictory. Some authors found that free-living elderly have an adequate mineral intake (Del Corso et al., 2000) while others restrict this only to free-living women (Kant and Schatzkin, 1999). Furthermore, there are differences between elderly members of distinct ethnic groups (Cid-Ruzafa et al., 1999).

These results show that inadequate zinc intake is prevalent among the aged population and that this problem is not restricted to a single group of elderly, but is a general phenomenon.

4. Immune status and the role of zinc

The composition of immune cells and the way the immune system works is different in the aged compared to healthy young people. The aged organism suffers from illnesses or chronic diseases that alter immune functions. Therefore, it is important to have guidelines that characterize healthy elderly people in order to reveal only age-dependent changes. Ligthart et al. developed the SENIEUR-Protocol in 1984, which defined immunogerontological criteria to study the immune system in the elderly (Ligthart et al., 1984, 1990). Although the differences between the status of the immune system in healthy young and SENIEUR-elderly are significant, these changes are mostly within the normal ranges of the appropriate parameters (Rink and Seyfarth, 1997).

As mentioned above, malnutrition is common in aged populations and it seems to be one of the main factors that can induce lower immune responses in this population, particularly in cell-mediated immunity (Lesourd et al., 1994; Lesourd, 1997; High, 1999). Other authors suggested that nutrition does not play a role in immunosenescence (Goodwin and Garry, 1988). However, besides zinc deficiency, lack of other micronutrients like selenium (Kiremidjian-Schumacher and Stotzky, 1987) or vitamins like folic acid (Nauss, 1986), vitamin B-6 (Gridley et al., 1988) and vitamin E (Meydani and Hayek, 1992) were shown to influence immune responses negatively. It was reported that aged subjects with low nutritional status showed similar but more marked changes in immune response compared to well-nourished elderly, particularly in terms of lower CD4 T cell counts and lower T cell functions associated with poorer nutritional status (Mazari and Lesourd, 1998). Thus, these findings indicate a cumulative effect of aging and malnutrition in aged persons.

The deviation between healthy elderly and healthy young in some leukocyte cell-populations as well as the effects of zinc deficiency are shown in Table 3.

4.1. Innate immunity

The number of monocytes, basophils and eosinophils in SENIEUR-elderly remains unchanged compared with the healthy young population (Fagiolo et al., 1993; Born et al., 1995; Franceschi et al., 1995; Cakman et al., 1997), but neutrophils increase significantly (Cakman et al., 1997). The immune system is a highly proliferating organ, and consequently it is markedly influenced by zinc availability, even in the first stages of an immune response. In vivo, natural killer cell activity, macrophage and neutrophil phagocytosis and generation of the oxidative burst are all impaired by decreased zinc levels (Allen et al., 1983; Keen and Gershwin, 1990). In neutrophils, recruitment as well as chemotaxis are affected. Because zinc is required by pathogens for proliferation, decreasing zinc in the plasma is an acute phase response of the human organism. Furthermore, zinc is chelated by the S-100

Ca^{2+} binding protein calprotectin which is released on degradation of neutrophils. Consequently, reproduction of bacteria and *Candida albicans* is inhibited by zinc chelation (Sohnle et al., 1991; Murthy et al., 1994; Clohessy and Golden, 1995).

Natural killer (NK) cells play a role in immunity against infections and tumors. The absolute and relative NK cell count is increased in aging (Sansoni et al., 1993; Xu et al., 1993; Born et al., 1995; Cakman et al., 1996). However, only the number of NK cells increases, but not the cytotoxic activity (Franceschi et al., 1995). The number and activity of NK cells is dependent on serum zinc level, even in the oldest old (Ravaglia et al., 2000). Zinc is needed by NK cells for the recognition of major histocompatibility cell complex (MHC) class I molecules by p58 killer cell inhibitory receptors on NK cells (Rajagopalan et al., 1995) in order to inhibit the killing activity. Only the inhibitory signal is zinc-dependent and zinc deficiency might therefore evoke non-specific killing. However, zinc deficiency decreases NK cell activity, and this was suggested to be a consequence of decreased interleukin (IL)-2 production in zinc-deficient individuals (because IL-2 augments NK activity) (Prasad, 2000b).

4.2. Specific immunity

4.2.1. T cells

The number of lymphocytes is decreased significantly in the SENIEUR-elderly (Sansoni et al., 1993; Born et al., 1995; Franceschi et al., 1995; Cakman et al., 1997). The absolute T cell count (CD3-positive) decreases significantly in aging (Sansoni et al., 1993; Xu et al., 1993; Born et al., 1995), whereas changes in the percentage of CD3 T cells are not significant in all studies (Sansoni et al., 1993; Xu et al., 1993; Born et al., 1995; Cakman et al., 1996). The percentage of T-helper cells (CD4-positive) is unchanged in SENIEUR-elderly, but not the absolute number (Sansoni et al., 1993; Xu et al., 1993; Born et al., 1995; Cakman et al., 1996). On the other hand, CD8 T cells (cytotoxic T cells) are decreased in absolute cell count as well as in percentage (Sansoni et al., 1993; Sindermann et al., 1993; Xu et al., 1993; Born et al., 1995; Cakman et al., 1996). This results in an increased CD4/CD8 ratio (Sindermann et al., 1993). Within the T cell population, memory T cells (CD45RO-positive), which have already had contact with antigens, are generally increased with age, and naive T cells (CD45RA+) decreased (Xu et al., 1993; Franceschi et al., 1995; Cakman et al., 1996).

Zinc influences not only NK cell mediated killing as mentioned above but could also modulate cytolytic T cell activity (Mingari et al., 1998). It was found that the percentage of $CD8^+CD73^+$ lymphocytes was decreased during zinc depletion (Prasad, 2000b). These cells were predominantly precursors of cytotoxic T lymphocytes and CD73 is known to be a molecule needed on these cytotoxic T cells for antigen recognition, proliferative process, and for the generation of cytolytic processes (Beck et al., 1997b). Further, zinc is involved in the development of T cells: zinc deficiency is responsible for thymic atrophy, but thymic changes can be reversed by zinc supplementation (Mocchegiani et al., 1995). The thymus produces a hormone called thymulin which is released by thymic epithelial cells

Table 3
Comparison of the influence of aging and zinc deficiency on some leukocyte cell-populations

Subpopulation	Elderly	References	Zinc deficiency	References
Lymphocytes	Cells →	1,2,3,4	Cells →	12
Monocytes	Cells ↕ Phagocytosis *	2,3,4,5	Cells * Phagocytosis →	13
Neutrophils	Cells ↑ Oxidative burst *	4,6	Cells → Oxidative burst →	12 13
Eosinophils	Cells ↕ Oxidative burst *	3,4,6	Cells → Oxidative burst →	12 13
Basophils	Cells ↕ Oxidative burst *	3,4,6	Cells → Oxidative burst →	12 13
T cells	CD3 (%) in parts not significant ↘ CD3 (●) →	1,2,6,7 1,2,6,7	T cells (total) → CD3 *	12
	CD4 (%) ↕ CD4 (●) in parts not significant ↘	1,2,7 1,2,7	CD4 (%) * CD4 (●) *	
	CD8 (● and %) % in parts not significant →	1,6,7,8	CD8 CD73 (%) →	14,15
	CD4/CD8-ratio ↑	8	CD4/CD8-ratio →	14
	CD45 RA (%) → CD45 RA (●) →	3,6,7 7	CD45 RA (%) * CD45 RA (●) *	
	CD45 RO (%) in parts not significant decrease ↗ CD45 RO (●) ↑	3,6,7 7	CD45 RO (%) * CD45 RO (●) *	
	CD4 CD45 RA/CD4 CD45 RO-ratio →	3,6,7	CD4 CD45 RA/CD4 CD45 RO-ratio →	14,15
	Imbalance of Th1 and Th2 (Th1/Th2) →	6	Imbalance of Th1 and Th2 (Th1/Th2) →	14,15,16
	IL-2 and IFN-γ production →	6,9,10	IL-2 and IFN-γ production →	12,15,16

	Parameter	Change	Ref.	Parameter	Change	Ref.
B cells	CD19 (● and %)	→	1,2,6	B cells CD19 (● and %)	→	17
	CD45 Ig^-		*	CD45 Ig^-	→	17
	CD45 IgM^+		*	CD45 IgM^+	→	17
	CD45 IgM^+IgD^+		*	CD45 IgM^+IgD^+	→	17
	IgM, IgA IgG levels (not IgG4)	↑	3,11, X	Antibody production	↓	18
NK cells	CD16 (● and %) in parts not significant	↑	1,2,7	CD16 (%)	→	19
	CD56 (%)	↑	1,6	CD56(%)	→	19
	CD56 (●)	⟷	1	CD56(●)	*	
	CD57 (● and %)	↑	1	CD57	*	
	Lytic activity	↓	1	Lytic activity	→	16,19,20

References: 1, Sansoni et al. (1993); 2, Born et al. (1995); 3, Franceschi et al. (1995); 4, Cakman et al. (1997); 5, Fagiolo et al. (1993); 6, Cakman et al. (1996); 7, Xu et al. (1993); 8, Sindermann et al. (1993); 9, Gillis et al. (1981); 10, Paganelli et al. (1994); 11, Paganelli et al. (1992); 12, Prasad (2000a); 13, Keen and Gershwin (1990); 14, Beck et al. (1997a); 15, Prasad (1998); 16, Prasad (2000b); 17, Fraker et al. (1995); 18, Fraker (1983); 19, Ravaglia et al. (2000); 20, Allen et al. (1983). %, percentage; ●, absolute cell count; ↑, significantly increased; ↗, increased; ⟷, unchanged; ↘, decreased; ↓, significantly decreased; *, no data available; X, unpublished results.

(Dardenne et al., 1982; Hadden, 1992) and for which zinc is an essential cofactor. Thymulin regulates the differentiation of immature T cells in the thymus and the function of mature T cells in the periphery. Moreover, thymulin modulates cytokine release by peripheral blood mononuclear cells (PBMC), induces proliferation of CD8 T cells in combination with IL-2 (Coto et al., 1992; Safie-Garabedian et al., 1993) and ensures the expression of the high-affinity receptor for IL-2 on mature T cells (Tanaka et al., 1989). All this is consistent with the observation that zinc deficiency results in decreased T cell proliferation after mitogen stimulation (Dowd et al., 1986; Crea et al., 1990).

Furthermore, T helper (Th) cells are influenced by zinc. Zinc deficiency leads to decreased production of the Th1 cell cytokines interferon (IFN)-γ and IL-2, whereas the Th2 products IL-4, IL-6 and IL-10 remain unchanged (Prasad, 1998, 2000a,b). These results suggest that zinc deficiency mainly affects Th1 cytokines and that the cell-mediated immune dysfunction in human zinc deficiency may be due to an imbalance between Th1 and Th2 cells (Prasad, 2000b).

4.2.2. B cells

The elderly show significant changes in humoral immunity. Investigations revealed an increase of the serum level of the immunoglobulins IgM and IgA as well as the IgG subclasses IgG1, 2, 3, but not IgG4 (Paganelli et al., 1992; Franceschi et al., 1995, unpublished results), whereas a decrease of circulating B cells was observed (Paganelli et al., 1992). Flow cytometric analysis showed that they decrease in absolute as well as in relative number during life (Sansoni et al., 1993; Born et al., 1995; Cakman et al., 1996). Single B cell clones of irrelevant specificity and high Ig production rate might be responsible for this phenomenon, like those observed in benign gammopathy in the elderly.

Although B lymphocyte development is influenced negatively during zinc deficiency (Fraker et al., 1993, 1995; Fraker and Telford, 1997), B cells are less dependent on zinc for proliferation than T cells (Zazonico et al., 1981; Flynn, 1984). B-Lymphocytes and their precursors are reduced in absolute number during zinc deficiency. Predominantly pre-B and immature B cells are affected by zinc deficiency, whereas mature B lymphocytes are only slightly influenced. However, low zinc levels have no influence on the cell cycle status of precursor B cells and only modest influence on cycling pro-B cells. The cycling status of cells of myeloid series is, however, changed significantly (King and Fraker, 2000). Thus, in zinc deficiency there are less naive B cells which could react on neoantigens. Taking into account that the response to most antigens is T cell-dependent and that the number of T cells is also reduced in zinc deficiency, it is very probable that in deficiency and aging the body is unable to respond properly with antibody production in response to neoantigens. Moreover, B lymphocyte antibody production was inhibited during a lack of zinc (Fraker et al., 1978; Fraker, 1983; DePasquale-Jardieu and Fraker, 1984). Studies show that antibody production, as a response to T cell-dependent antigens, is more sensitive to zinc deficiency than antibody production during the response to T cell-independent antigens (Fraker et al., 1978; Moulder and Steward, 1989). Interestingly, immunological memory is also influenced by zinc, because zinc-deficient mice show reduced antibody recall responses to antigens to which they had been immunized (Fraker, 1983; Fraker et al., 1986, 1987; DePasquale-Jardieu and Fraker, 1984). This effect was observed in T-cell independent as well as T-cell dependent systems.

4.3. Regulation of the immune system

Cytokines are important factors for regulating the immune system. Mitogen stimulation of lymphocytes from elderly persons results in a reduced response in comparison to stimulation of lymphocytes from young donors (Gillis et al., 1981). This observation cannot be attributed only to the reduced number of CD3 cells in the SENIEUR-elderly, as mentioned above, but an altered response of the leukocytes must also be taken into account. The balance between cellular (Th1) and humoral (Th2) immune responses is disturbed in the elderly. This can be observed in the production of IL-2, a Th1 cytokine and the most important T cell growth factor, after stimulation. Lymphocytes from healthy elderly persons produced significantly less IL-2 than lymphocytes from young people (Gillis et al., 1981; Huang et al., 1992; Sindermann et al., 1993; Born et al., 1995). Not only IL-2 but also its counterpart, the soluble IL-2 receptor (sIL-2R, a marker of T cell activation), is released in smaller amounts by leukocytes from the elderly (Sindermann et al., 1993; Cakman et al., 1996). Besides the total amount, the sIL-2R release kinetics also changed: in the elderly the maximum sIL-2R release was reached earlier and the level decreased already, while the amount of sIL-2R in lymphocyte cultures from young subjects was still rising. However, the expression of membrane-bound IL-2R is the same in the elderly and the young (Cakman et al., 1996) or even increased in the elderly (Born et al., 1995).

Further evidence for a dysregulation between Th1 and Th2 subsets is that IL-4 and IL-10, both Th-2 cytokines, are produced in greater amounts in the elderly (Paganelli et al., 1994, 1996; Cakman et al., 1996), and release kinetics of IL-10 after stimulation differ in the elderly and the young. IL-10 and IL-4 also induce immunoglobulins and are involved in immunoglobulin class switching. Higher levels of IL-10 and IL-4 result in predominantly IgM production, by inducing B1 cells. IgM binds numerous different ligands with relatively low affinity (Weksler, 2000). This could be an explanation for higher levels of immunoglobulins in the elderly.

Another cytokine produced by Th1 cells is IFN-γ, which is negatively influenced by high levels of Th2 cytokines. Thus, in the healthy elderly the level of IFN-γ is decreased because of the dominance of Th2 cells (Paganelli et al., 1994, 1996; Born et al., 1995; Cakman et al., 1996) and its reduced secretion by Th1 cells. On the other hand, Th1 cells are not the only source of IFN-γ because NK cells (Trichneri, 1989) and cytotoxic T memory cells (Pawelec et al., 1995) can also produce it. The latter are decreased in the elderly. Although NK cells are increased in aged subjects, their cytotoxic function is reduced (Franceschi et al., 1995), so that the production of IFN-γ might be disturbed also. Decreased IFN-γ production may result in a higher incidence of neoplasia and infections in aged persons (Saltzman and Peterson, 1987), since this cytokine is the major factor for inducing killing activity in T and NK cells as well as an inducer of NO-production and respiratory burst in phagocytes (Ginaldi et al., 1999).

The decreased level of IFN-γ in the elderly was established using whole blood samples and determining their cytokine concentration by ELISA after stimulation (Cakman et al., 1996). On the other hand, findings of other groups were contradictory. Studies showed an increase in IFN-γ-positive cells in the elderly by using flow cytometry (Brandes et al., 2000; Sandmand et al., 2002). Furthermore, using spleen cells from mice and determining IFN-γ concentration by ELISA, as well as IFN-γ-positive cells by flow cytometry, older mice showed an increase in IFN-γ in both cases after stimulation (Wakikawa et al., 1999). These contradictory findings might be due to a pre-activation in vivo. Similar effects were found for in vitro IL-2 production by T cells of aged individuals. Huang et al. (1992) suggested dividing the elderly into two groups, one with low in vitro IL-2 production and the other group with normal or slightly decreased IL-2 production. The first group showed a higher level of sIL-2R in the serum compared to the second. Sindermann (1992) observed a lower production of IL-2 and a higher production of sIL-2R in aged persons with higher amounts of CRP in the serum. Thus, these results support the hypothesis that pre-activation in vivo could lead to contradictory findings during in vitro investigations.

The monocyte cell count is normal in the SENIEUR-elderly compared to young subjects. It is astonishing, however, that monokine release is increased significantly. These cytokines are IL-1, IL-6, IL-8 and tumor necrosis factor-α (TNF-α) (Fagiolo et al., 1993; Born et al., 1995; Franceschi et al., 1995; Cakman et al., 1996, 1997). All researchers used stimulation by lipopolysaccharide (LPS) to activate monocytes in these studies; consequently, monocytes in the elderly are able to produce higher amounts of proinflammatory cytokines or the cells of elderly respond to LPS in a hypersensitive way. A recent study showed that when whole blood samples from healthy elderly were stimulated with LPS, IL-1 and IL-6 secretion were significantly elevated, but stimulation of their PBMCs showed lower amounts of produced cytokines compared to the PBMCs of healthy young people (Gabriel et al., 2002). This phenomenon might be explained by an interaction of LPS with a serum factor or transcellular metabolism of other mediators of immunity in the blood of the elderly. IFN-α is responsible for antiviral and anti-tumoral activity. IFN-α production is decreased in the elderly (Sindermann et al., 1993; Katschinski et al., 1994; Cakman et al., 1997). This is consistent with the fact that aged persons are more prone to viral infections and neoplasia as mentioned before.

Zinc is also important for immune regulation: it stimulates PBMC to release IL-1, IL-6, TNF-α, soluble IL-2 receptor and IFN-γ (Salas and Kirchner, 1987; Scuderi, 1990; Driessen et al., 1994). IL-1, IL-6 and TNF-α are directly induced in monocytes by zinc in the absence of lymphocytes (Driessen et al., 1994). It was shown that TNF-α release after PBMC stimulation by zinc was due to the induction of mRNA transcription and is not the result of the enhanced translation of already-expressed mRNA (Wellinghausen et al., 1996a). T cells are not stimulated directly, since at least the induction of IFN-γ is dependent on the presence of monocytes (Rühl and Kirchner, 1978; Salas and Kirchner, 1987; Driessen et al., 1994; Wellinghausen et al., 1997). The monokines IL-1 and IL-6 and cell–cell contact between monocytes and

T cells are necessary for T cells to release IFN-γ and sIL-2R (Driessen et al., 1994; Wellinghausen et al., 1997). Zinc is not capable of inducing cytokine production in isolated and monocyte-depleted T-cells (Hadden, 1995; Wellinghausen et al., 1997), B cells (Crea et al., 1990), NK cells (Crea et al., 1990) or neutrophils (unpublished results). The activation of T cells and monocytes is dependent on the amount of free zinc ions and protein composition in the culture medium. Transferrin and insulin specifically enhance zinc-induced monocyte stimulation by means of a non-receptor-dependent mechanism (Phillips and Azari, 1974; Crea et al., 1990; Driessen et al., 1995; Wellinghausen et al., 1996b). However, high levels of serum proteins in the culture medium inhibit monocyte activation because the proteins bind free zinc and the free zinc concentration is low. In serum-free culture medium, zinc concentrations > 100 μM stimulate monocytes but actually prevent T cell activation due to T cells having a lower intracellular zinc concentration and being more susceptible to increasing zinc levels than monocytes (Bulgarini et al., 1989; Goode et al., 1989; Wellinghausen et al., 1997). This effect is attributed to the inhibition of the IL-1 type I receptor-associated kinase by zinc, since T cell activation is indirectly dependent on IL-1 which is secreted by monocytes (Driessen et al., 1994; Wellinghausen et al., 1996b, 1997). In conclusion, T cell activation by zinc takes place when zinc concentrations are high enough for monokine induction but not exceeding the crucial concentrations for T cell suppression.

5. Zinc therapy

Malnutrition and inadequate intake of micronutrients are prevalent among the elderly as mentioned above. Although even SENIEUR-elderly show marginal zinc deficiency compared to healthy young persons, they are within the normal range of plasma zinc, but at the lower limit.

Studies revealed that supplementation and optimal intake of essential micronutrients restored an impaired immune response in the elderly (Chandra, 1992; Girodon et al., 1997). The number of T cells and NK cells were elevated, production of IL-2 and sIL-2R increased, and lymphocyte responses to PHA stimulation, as well as NK cell activity, improved significantly compared to the placebo group. Incidence of infections was reduced and the use of antibiotics was lowered in that group which received zinc together with selenium (Girodon et al., 1997). Vitamins had no effects. A similarly designed second, larger, trial provided similar results, but in this case the reduction of infection rates was not significant (Girodon et al., 1999). Other studies showed that CD4 + T cells and cytotoxic T lymphocytes were increased significantly after zinc supplementation and cell-mediated immune response was improved (Fortes et al., 1998).

However, the optimal therapeutic dosage remains unclear. Thus, in order to reverse zinc deficiency, the pharmacological zinc dose should be adapted to the actual requirements to avoid negative effects on the immune system. Therefore, zinc plasma level should not exceed 30 μM. On the other hand, zinc is non-toxic, even in dosages exceeding recommended dietary intake (Fosmire, 1990). Zinc concentrations at three to four times the physiological level did not decrease T cell proliferation in vitro nor show immunosuppressive effects in vivo, but they did inhibit alloreactivity of T cells in the mixed lymphocyte culture (MLC) (aCampo et al., 2001). Thus, zinc might be used for the treatment of T cell-mediated reactions such as graft rejection in transplantation medicine. However, high dose zinc supplementation, representing seven to eight times the physiological zinc level, blocked IFN-α induction in elderly persons (Cakman et al., 1997).

There is evidence that specific antibody titres after immunization are decreased in the elderly (Gergen et al., 1995; Steger et al., 1996), at least for tetanus-specific antibodies. As many as 60% of the elderly group were unprotected against tetanus in contrast to 8% of the young control group (Steger et al., 1996). This was consistent with findings that zinc deficient people respond rather poorly to vaccination (Fraker et al., 1986; Cakman et al., 1996). Furthermore, trials showed a correlation between humoral responses and zinc supplementation when zinc was provided as an adjuvant in vaccination. Low-dose supplementation with zinc showed an improvement in the humoral response after vaccination in the elderly (Girodon et al., 1999), whereas supplementation with high dosages did not improve the antibody response (Provinciali et al., 1998). Kreft et al. (2000) showed that non-responders for diphtheria vaccination had a significantly lower serum zinc level than responders in aged haemodialysis patients.

Furthermore, other nutrients are also of benefit for the immune system of the elderly: selenium supplementation was found to restore immune functions (Girodon et al., 1997, 1999) and vitamin E was shown to improve delayed-type hypersensitivity responses and to augment primary immunization responses to hepatitis B (Meydani et al., 1990, 1997). Other vitamins, such as vitamin A, showed no beneficial effect (Murphy et al., 1992; Fortes et al., 1998).

6. Conclusions

Various groups of people are predisposed to zinc deficiency, particularly the aged population. Even in the very healthy elderly, zinc levels are decreased compared with healthy young individuals. The immune system changes during our lifespan, such that alterations in immune cell subsets and dysfunctions appear. An ongoing zinc deficiency can amplify these effects, followed by higher incidences of infections and mortality. Supplementation with zinc can restore impaired immune functions and zinc levels can be elevated to values comparable to those of young persons. Because there is no agreed standard therapeutic dosage yet, zinc administration must be adjusted according actual individual requirements and should not exceed two times the physiological zinc level. Since most immune defects are the same in zinc deficiency and aging, zinc application might be a simple and cost-effective way to improve immune functions in the elderly.

References

Allen, J.I., Perri, R.T., McClain, C.J., Kay, N.E., 1983. Alterations in human natural killer cell activity and monocyte cytotoxicity induced by zinc deficiency. J. Lab. Clin. Med. 102, 577–589.

Beck, F.W., Prasad, A.S., Kaplan, J., Fitzgerald, J.T., Brewer, G.J., 1997a. Changes in cytokine production and T cell subpopulations in experimentally induced zinc-deficient humans. Am. J. Physiol. 272(6 Pt 1), E1002–E1007.

Beck, F.W., Kaplan, J., Fine, N., Handschu, W., Prasad, A.S., 1997b. Decreased expression of CD73 (ecto-5'-nucleotidase) in CD8+ subset is associated with zinc deficiency in human patients. J. Lab. Clin. Med. 130, 147–156.

Bertholf, R.L., 1988. Zinc. In: Seiler, H.G., Sigel, H. (Eds.), Handbook on Toxicity of Inorganic Compounds. Dekker, New York, pp. 788–800.

Beyersmann, D., Haase, H., 2001. Functions of zinc in signalling, proliferation and differentiation of mammalian cells. Biometals 14, 331–341.

Born, J., Uthgenannt, D., Dodt, C., Nünninghoff, D., Ringvolt, E., Wagner, T., Fehm, H.L., 1995. Cytokine production and lymphocyte subpopulation in aged humans. An assessment during nocturnal sleep. Mech. Ageing Dev. 84, 113–126.

Brandes, E., Merino, J., Vazquez, B., Inoges, B., Moreno, C., Subira, M.L., Sanchez-Ibarrola, A., 2000. The increase of IFN-gamma production through aging correlates with the expanded CD8(+high) CD28(−) CD57(+) subpopulation. Clin. Immunol. 96(3), 230–235.

Briefel, R.R., Bialostosky, K., Kennedy-Stephenson, J., McDowell, M.A., Ervin, R.B., Wright, J.D., 2000. Zinc intake of the U.S. population: findings from the Third National Health and Nutrition Examination Survey, 1988–1994. J. Nutr. 130, 1367–1373.

Bulgarini, D., Habetswallner, D., Boccoli, G., Montesoro, E., Camagna, A., Mastroberardino, G., Rosania, C., Testa, U., Peschle, C., 1989. Zinc modulates the mitogenic activation of human periperal blood lymphocytes. Ann. Inst. Super. Sanita 25, 463–470.

Buokaiba, N., Flament, C., Acher, F., Chappuis, P., Monier, D., 1993. A physiological amount of zinc supplementation: effects on nutritional, lipid, and thymic status an elderly population. Am. J. Clin. Nutr. 57, 566–572.

aCampo, C., Wellinghausen, N., Faber, C., Fischer, A., Rink, L., 2001. Zinc inhibits the mixed lymphocyte culture. Biol. Trace Elem. Res. 79, 15–22.

Cakman, I., Rohwer, J., Schütz, R.M., Kirchner, H., Rink, L., 1996. Dysregulation between Th1 and Th2 T cell subpopulations in the elderly. Mech. Ageing Dev. 87, 197–209.

Cakman, I., Kirchner, H., Rink, L., 1997. Reconstitution of interferon-γ production by zinc-supplementation of leukocyte cultures of elderly individuals. J. Interferon Cytokine Res. 13, 15–20.

Chandra, R.K., 1992. Effect of vitamin and trace-element supplementation on immune. Responses and infection in the elderly. Lancet 340, 1124–1137.

Chernoff, R., 1995. Effects of age on nutrient requirements. Clin. Geriatr. Med. 11, 641–651.

Cid-Ruzafa, J., Caulfield, L.E., Barron, Y., West, S.K., 1999. Nutrient intakes and adequacy among an older population of the eastern shore of Maryland: the Salisbury Eye Evaluation. J. Diet. Assoc. 99, 564–571.

Clohessy, P.A., Golden, B.E., 1995. Calprotectin-mediated zinc chelation as a biostatic mechanism in host defence. Scand. J. Immunol. 42, 551–556.

Coleman, J.E., 1992. Zinc proteins: enzymes, storage proteins, transcriptions factors and replication proteins. Annu. Rev. Biochem. 16, 897–946.

Coto, J.A., Hadden, E.M., Sauro, M., Zorn, N., Hadden, J.W., 1992. Interleukin 1 regulates secretion of zinc-thymulin by human thymic epithelial cells and its action on T-lymphocyte proliferation and nuclear protein kinase C. Proc. Natl. Acad. Sci. USA 89, 7752–7756.

Crea, A., Guérin, V., Ortega, F., Hartemann, P., 1990. Zinc et système immunitaire. Ann. Med. Interne 141, 447–451.

Dardenne, M., Pléau, J.M., Nabarra, B., Lefrancier, P., Derrien, M., Choay, J., Bach, J.F., 1982. Contribution of zinc and other metals to the biological activity of the serum thymic factors. Proc. Natl. Acad. Sci. USA 79, 5370–5373.

Del Corso, L., Pastine, F., Protti, M.A., Romanelli, A.M., Moruzzo, D., Ruocco, L., Pentimore, F., 2000. Blood zinc, copper and magnesium in aging. A study in healthy home-living elderly. Panminerva Med. 42, 273–277.

DePasquale-Jardieu, P., Fraker, P.J., 1984. Interference in the development of a secondary immune response in mice by zinc deprivation: persistence of effects. J. Nutr. 114, 1762–1769.

Dowd, P.S., Kelleher, J., Guillou, P.J., 1986. T-lymphocyte subsets and interleukin-2 production in zinc-deficient rats. Br. J. Nutr. 55, 59–69.

Driessen, C., Hirv, K., Rink, L., Kirchner, H., 1994. Induction of cytokines by zinc ions in human peripheral blood mononuclear cells and separated and monocytes. Lymphokine Cytokine Res. 13, 15–20.

Driessen, C., Hirv, K., Wellinghausen, N., Kirchner, H., Rink, L., 1995. Influence of serum on zinc, toxic shock syndrome toxin-1, and lipopolysaccharide-induced production of IFN-γ and IL-1β by human mononuclear cells. J. Leukoc. Biol. 57, 904–908.

Fagiolo, U., Cossarizza, A., Scala, A., Fanales-Belasio, E., Ortolani, C., Cozzi, E., Monti, D., Franceschi, C., Paganelli, R., 1993. Increased cytokine production in mononuclear cells of healthy elderly people. Eur. J. Immunol. 23, 2375–2378.

Flynn, A., 1984. Control of in vitro lymphocyte proliferation by copper, magnesium and zinc deficiency. J. Nutr. 114, 2034–2042.

Fosmire, G.J., 1990. Zinc toxicity. Am. J. Clin. Nutr. 15, 225–227.

Fortes, C., Forastiere, F., Agabiti, N., Fano, V., Pacifici, R., Virgili, F., Piras, G., Guidi, L., Bartoloni, C., Tricerri, A., Zuccaro, P., Ebrahim, S., Percucci, C.A., 1998. The effect of zinc and vitamin A supplementation on immune response in an older population. J. Am. Geriatr. Soc. 46, 19–26.

Fraker, P.J., DePasquale-Jardieu, R., Zwickl, C.M., Luecke, R.W., 1978. Regeneration of T-cell helper function in zinc-deficient adult mice. Proc. Natl. Acad. Sci. USA 75, 5660–5664.

Fraker, P.J., 1983. Zinc deficiency: a common immunodeficiency state. Surv. Immunol. Res. 2, 155–163.

Fraker, P.J., Gershwin, M.E., Good, R.A., Prasad, A.S., 1986. Interrelationships between zinc and immune function. Fed. Proc. 45, 1474–1479.

Fraker, P.J., Jardieu, P., Cook, J., 1987. Zinc deficiency and immune function. Arch. Dermatol. Res. 123, 1699–1701.

Fraker, P.J., King, L.E., Garvy, B.A., Medina, C.A., 1993. The immunopathology of zinc deficiency in humans and rodents: a possible role for programmed cell death. In: Klurfeld, D.M. (Ed.), Human Nutrition: A Comprehensive Treatise. Plenum Press, New York, pp. 267–283.

Fraker, P.J., Osatiashtiani, E., Wagner, M.A., King, L.E., 1995. Possible roles for glucocorticoids and apoptosis in the suppression of lymphopoiesis during zinc deficiency: a review. J. Am. Coll. Nutr. 14, 11–17.

Fraker, P.J., Telford, W.G., 1997. A reappraisal of the role of zinc in life and death decisions of cells. Proc. Soc. Exp. Biol. Med. 215, 229–236.

Franceschi, C., Monti, D., Sansoni, P., Cossarizza, A., 1995. The immunology of exceptional individuals: the lesson of the centenarians. Immunol. Today 16, 12–16.

Gabriel, P., Cakman, I., Rink, L., 2002. Overproduction of monokines by leukocytes after stimulation with lipopolysaccharide in the elderly. Exp. Gerontol. 37, 235–247.

Gergen, P.J., McQuillan, G.M., Kiely, M., Ezzati-Rize, T.M., Sutter, R.W., Viralla, G., 1995. A population-based serologic survey of immunity to tetanus in the United States. N. Engl. J. Med. 32, 761–766.

Gilbride, J.A., Amella, E.J., Breines, E.B., Mariano, C., Mezey, M., 1998. Nutrition and health status assessment of community residing elderly in New York: a pilot study. J. Am. Diet. Assoc. 98, 554–558.

Gillis, S., Kozak, R., Durante, M., Weksler, M.E., 1981. Immunological studies of aging. Decreased production of and response to T cell growth factor by lymphocytes from aged humans. J. Clin. Invest. 67, 937–942.

Ginaldi, L., De Martinis, M., D'Ostilio, A., Marini, L., Loeto, M.F., Quaglino, D., 1999. The immune system in the elderly: III. Innate immunity. Immunol. Res. 20, 117–126.

Girodon, F., Lombard, M., Galan, P., 1997. Effect of micronutrient supplementation on infection in institutionalized elderly subjects: a controlled trial. Ann. Nutr. Metab. 41(2), 98–107.

Girodon, F., Galan, P., Monget, A.L., Boutron-Ruault, M.C., Brunet-Lecomte, P., Preziosi, P., Arnaud, J., Manuguerra, J.C., Herchberg, S., 1999. Impact of trace elements and vitamin supplementation on immunity and infections in institutionalized elderly patients: a randomized controlled trial. MIN. VIT. AOX. Geriatric network. Arch. Intern. Med. 159, 748–754.

Goode, H.F., Kelleher, J., Walker, B.E., 1989. Zinc concentrations in pure populations of peripheral blood neutrophils, lymphocytes and monocytes. Ann. Clin. Biochem. 26, 89–95.

Goodwin, J.S., Garry, P.J., 1988. Lack of correlation between indices of nutritional status and immunologic function in elderly humans. J. Gerontol. 43, M46–M49.

Greger, J.L., Sciscoe, B.S., 1977. Zinc nutriture of aged participants in an urban Title VII feeding program. J. Am. Diet. Assoc. 70, 37–41.

Gridley, D.S., Shultz, T.D., Stickney, D.R., Slater, J.M., 1988. In vivo and in vitro stimulation of cell-mediated immunity by vitamin B6. Nutr. Res. 8, 210–217.

Hadden, J.W., 1992. Thymic endocrinology. Int. J. Immunopharmacol. 14, 345–352.

Hadden, J.W., 1995. The treatment of zinc is an immunotherapy. Int. J. Immunopharmacol. 17, 697–701.

High, H.P., 1999. Micronutrient supplementation and immune function in the elderly. Clin. Infect. Dis. 28, 717–722.

Huang, Y.P., Pechere, J.C., Michel, M., Cauthey, L., Loreto, M., Curran, J.A., Michel, J.P., 1992. In vivo T cell activation, in vitro defective IL-2 secretion, and response to influenza vaccination in elderly woman. J. Immunol. 148, 715–722.

Kant, A.K., Schatzkin, A., 1999. Relation of age and self-reported chronic medical condition status with dietary nutrient intake in the U.S. population. J. Am. Coll. Nutr. 18, 69–76.

Katschinski, D.M., Neustock, P., Klüter, H., Kirchner, H., 1994. Influence of various factors on interferon-α production of human leukocytes. J. Interferon Res. 14, 105–110.

Keen, C.L., Gershwin, M.E., 1990. Zinc deficiency and immune function. Annu. Rev. Nutr. 19, 415–431.

King, L.E., Fraker, P.J., 2000. Variations in the cell cycle status of lymphopoietic and myelopoietic cells created by zinc deficiency. J. Infect. Dis. 182(Suppl. 1), S16–S22.

Kiremidjian-Schumacher, L., Stotzky, G., 1987. Selenium and immune responses. Environ. Res. 42, 277–303.

Kreft, B., Fischer, A., Krüger, S., Sack, K., Kirchner, H., Rink, L., 2000. The impaired immune response to diphtheria vaccination in elderly chronic hemodialysis patients is related to zinc deficiency. Biogerontology 1, 61–66.

Lee, D.Y., Prasad, A.S., Haysrick-Adair, C., Brewer, E., Johnson, P.E., 1993. Homeostasis of zinc in marginal human zinc deficiency; role of absorption and endogenous excretion of zinc. J. Lab. Clin. Med. 122, 549–556.

Lee, J.S., Frongillo, E.A. Jr., 2001. Nutritional and health consequences are associated with food insecurity among U.S. elderly persons. J. Nutr. 131, 1503–1509.

Lesourd, B.M., Laisney, C., Salvatore, R., Meaume, S., Moulias, R., 1994. Decreased maturation of T-cell populations in the healthy elderly: influence of nutritional factors on the appearance of double negative CD4−CD8−CD2+ cells. Arch. Gerontol. Geriatr. Suppl. 4, 139–154.

Lesourd, B.M., 1997. Nutrition and the immunity in the elderly: modification of immune responses with nutritional treatments. Am. J. Clin. Nutr. 66, 478S–484S.

Ligthart, G.J., Corberand, J.X., Fournier, C., Glanaud, P., Hijmans, W., Kennes, B., Müller-Hermelink, H.K., Steinmann, G.G., 1984. Admission criteria for immunogerontological studies in man: the SENIEUR-protocol. Mech. Ageing Dev. 28, 47–55.

Ligthart, G.J., Corberand, J.X., Geertzen, H.G., Meinders, A.E., Knook, D.L., Hijmanns, W., 1990. Necessity of the assessment of the health status in human immunogerontological studies: evaluation of the SENIEUR-protocol. Mech. Ageing Dev. 55, 89–105.

Ma, J., Betts, N.M., 2000. Zinc and copper intakes and their major food sources for older adults in the 1994–96 continuing survey of food intakes by individuals (CSFII). J. Nutr. 130, 2838–2843.

Mazari, L., Lesourd, B.M., 1998. Nutritional influences on immune response in healthy aged persons. Mech. Ageing Dev. 104(1), 25–40.

Meydani, S.N., Barklund, M.P., Liu, S., Meydani, M., Miller, R.A., Cannon, J.G., Morrow, F.D., Rocklin, R., Blumberg, J.B., 1990. Vitamin E supplement enhances cell-mediated immunity in healthy elderly subjects. Am. J. Clin. Nutr. 52(3), 557–563.

Meydani, S.N., Hayek, M., 1992. Vitamin E and the immune response. In: Chandra, R.K. (Ed.), Nutrition and Immunology. pp. 105–128. ARTS Biomedical Publishers, St. John's, Canada.

Meydani, S.N., Meydani, M., Blumberg, J.B., Leka, L.S., Siber, G., Loszewski, R., Thompson, C., Pedrosa, M.C., Diamond, R.D., Stollar, B.D., 1997. Vitamin E supplementation and in vivo immune response in healthy elderly subjects. A randomized controlled trial. J. Am. Med. Assoc. 277(17), 1380–1386.

Mingari, M.C., Moretta, A., Moretta, L., 1998. Regulation of KIR expression in human T cells: a safety mechanism that may impair protective T-cell responses. Immunol. Today 19, 153–157.

Mocchegiani, E., Santarelli, L., Muzzioli, M., Fabris, N., 1995. Reversibility of the thymic involution and of age-related peripheral immune dysfunction by zinc supplementation in old mice. Int. J. Immunopharmacol. 17, 703–718.

Moulder, K., Steward, M.W., 1989. Experimental zinc deficiency: effects on cellular responses and the affinity of humoral antibody. Clin. Exp. Immunol. 77, 269–274.

Murphy, S., West, K.P. Jr., Greenough, W.B. III, Chernot, E., Katz, J., Clement, L., 1992. Impact of vitamin A supplementation on the incidence of infection in elderly nursing-home residents: a randomized controlled trail. Age Ageing 21(6), 435–439.

Murthy, A.R.K., Lehrer, R.I., Harwig, S.S.L., Miyasaki, K.T., 1994. In vitro candidastatic properties of the human neutrophil calprotectin complex. J. Immunol. 151, 6291–6301.

Nauss, K.M., 1986. Folican and immunity. Nutritional Diseases: Research and Direction in Comparative Pathology. Alan R Liss, New York, pp. 173–196.

Neldner, K.H., Hambidge, K.M., 1975. Zinc therapy in acrodermatitis enteropatica. N. Engl. J. Med. 292, 879–882.

Paganelli, R., Quinti, I., Fagiolo, U., Cossarizza, A., Ortolani, C., Guerra, E., Sansoni, P., Pucillo, L.P., Cozzi, E., Bertollo, L., Monti, D., Franceschi, C., 1992. Changes in circulating B cells and immunoglobulin class and subclasses in a healthy aged population. Clin. Exp. Immunol. 90, 351–354.

Paganelli, R., Scala, E., Quinti, I., Ansotegui, I.J., 1994. Humoral immunity in aging. Aging Milano 6, 143–150.

Paganelli, R., Scala, E., Rosso, R., Cossarizza, A., Bertollo, L., Barberi, D., Fabrizi, A., Lusi, E.A., Fagiolo, U., Franceschi, C., 1996. A shift to Th0 cytokine production by CD4+ cells in human longevity: studies on two healthy centenarians. Eur. J. Immunol. 26, 2030–2034.

Pawelec, G., Adibzadeh, M., Pohla, H., Schaudt, K., 1995. Immunosenescence: ageing of the immune system. Immunol. Today 16, 420–422.

Phillips, J.L., Azari, P., 1974. Zinc transferring: enhancement of nucleic acid synthesis in phytohemag-glutinin-stimulated human lymphocytes. Cell Immunol. 10, 31–37.

Prasad, A.S., Miaie, A. Jr., Farid, Z., Schulert, A., Sandsteadt, H.H., 1963. Zinc metabolism in patients with the syndrome of iron deficiency, hypogonadism and dwarfism. J. Lab. Clin. Med. 83, 537–549.

Prasad, A.S., 1998. Zinc and immunity. Mol. Cell. Biochem. 188, 63–69.

Prasad, A.S., 2000a. Effects of zinc deficiency on immune functions. J. Trace Elem. Exp. Med. 13, 1–20.

Prasad, A.S., 2000b. Effects of zinc deficiency on Th1 and Th2 cytokine shifts. Infect. Dis. 182(Suppl. 1), S62–S68.

Provinciali, M., Montenovo, A., DiStefano, G., Colombo, M., Daghetta, L., Cairati, M., Veroni, C., Cassino, R., Della Torre, F., Fabris, N., 1998. Effect of zinc or zinc plus arginine supplementation on antibody titre and lymphocyte subsets after influenza vaccination in elderly subjects: a randomized controlled trail. Age Ageing 27, 715–722.

Rajagopalan, S., Winter, C.C., Wagtmann, N., Long, E.O., 1995. The Ig-related killer cell inhibitory receptor binds zinc and requires zinc for recognition of HLA-C on target T cells. J. Immunol. 155, 4143–4146.

Raulin, J., 1869. Etudes Chimique sur la vegetation (chemical studies on plantes). Annales des Sciences Nautrelles Botanique et Biologie Vegetale 11, 293–299.

Ravaglia, G., Forti, P., Maioli, F., Bastagali, L., Facchini, A., Erminia, M., Savarino, L., Sassi, S., Cucinotta, D., Lenaz, G., 2000. Effect of micronutrient status on natural killer cell immune function in healthy free-living subjects aged ≥90 y. Am. J. Clin. Nutr. 71, 590–598.

Rink, L., Seyfarth, M., 1997. Features of immunolgical investigations in the elderly. Z. Gerontol. Geriatr. 30, 220–225.

Rühl, H., Kirchner, H., 1978. Monocyte-dependent stimulation of human T-cells by zinc. Clin. Exp. Immunol. 32, 484–488.

Safie-Garabedian, B., Ahmed, K., Khamashta, M.A., Taub, N.A., Hughes, G.R.V., 1993. Thymulin modulates cytokine release by peripheral blood mononuclear cells: a comparison between healthy volunteers and patients with systemic lupus erythermatodes. Int. Arch. Allergy Immunol. 101, 126–131.

Salas, M., Kirchner, H., 1987. Induction of interferon-γ in human leukocyte cultures stimulated by Zn^{2+}. Clin. Immunol. Immunopathol. 45, 139–142.

Saltzman, R.L., Peterson, P.K., 1987. Immunodeficiency of the elderly. Rev. Infect. Dis. 9, 1127–1139.

Sandmand, M., Bruunsgaard, H., Kemp, K., Andersen-Ranberg, K., Pedersen, A.N., Skinhoj, P., Pedersen, B.K., 2002. Is ageing associated with a shift in the balance between Type 1 and Type 2 cytokines in humans. Clin. Exp. Immunol. 127(1), 104–114.

Sansoni, P., Cossarizza, A., Brianti, V., Fagnoni, F., Snelli, G., Monti, D., Marcato, A., Passeri, G., Ortolani, C., Forti, E., Fagiolo, U., Passeri, M., Franceschi, C., 1993. Lymphocyte subsets and natural killer cell activity in healthy old people and centenarians. Blood 82, 2767–2773.

Scott, B.J., Bradwell, A.R., 1983. Identification of the serum binding proteins for iron, zinc, cadmium, nickel and calcium. Clin. Chem. 29, 629–633.

Scuderi, P., 1990. Differential effects of copper and zinc on human peripheral blood monocyte cytokine secretion. Cell. Immunol. 126, 391–405.

Seiler, W.O., 1999. Nutritional status of ill elderly patients. Z. Gerontol. Geriatr. 32(Suppl. 1), I7–I11.

Sindermann, J., Kruse, A., Frercks, H.J., Schütz, R.M., Kirchner, H., 1993. Investigations of the lymphokine system in elderly individuals. Mech. Ageing Dev. 70, 149–159.

Sindermann, J., 1992. Dissertation, Medizinsche Fakultät, Medizinische Universität zu Lübeck.

Sohnle, P.G., Collins-Lech, C., Wiessner, J.H., 1991. The zinc-reversible antimicrobial activity of neutrophil lysates and abscess fluid supernatants. J. Infect. Dis. 164, 137–142.

Steger, M.M., Maczek, C., Berger, P., Grubeck-Loebenstein, B., 1996. Vaccination against tetanus in the elderly: do recommended vaccination strategies give sufficient protection? Lancet 348, 762.

Tanaka, Y., Shiozawa, S., Morimoto, I., Fujita, T., 1989. Zinc inhibits pokeweed mitogen-induced development of immunoglobulin-secreting cells through augmentation of both CD4 and CD8 cells. Int. J. Immunopharmacol. 11, 673–679.

Todd, W.K., Elvelym, A., Hart, E.B., 1934. Zinc in the nutrition of the rat. Am. J. Physiol. 107, 146–156.

Trichneri, G., 1989. Biology of natural killer cells. Adv. Immunol. 47, 187–376.

Vallee, B.L., Aulid, D.S., 1990. Zinc coordination, function and structure of zinc enzymes and other proteins. Biochemistry 29, 5647–5659.

Vallee, B.L., Falchuk, K.H., 1993. The biochemical basis of zinc physiology.. Physiol. Rev. 73, 79–118.

Vir, S.C., Love, A.H.G., 1979. Zinc and copper status of the elderly. Am. J. Clin. Nutr. 32, 1472–1476.

Wakikawa, A., Utsuyama, M., Wakabayashi, A., Kitagawa, M., Hirokawa, K., 1999. Age-related alteration of cytokine production profile by T cell subsets in mice: a flow cytometric study. Exp. Gerontol. 34(2), 231–242.

Weksler, M.E., 2000. Changes in the B cell repertoire with age. Vaccine 18, 1624–1628.

Wellinghausen, N., Driessen, C., Rink, L., 1996a. Stimulation of human peripheral blood mononuclear cells by zinc and related cations. Cytokine 18, 767–771.

Wellinghausen, N., Fischer, A., Kirchner, H., Rink, L., 1996b. Interaction of zinc ions with human peripheral blood mononuclear cells. Cell. Immunol. 171, 255–261.

Wellinghausen, N., Martin, M., Rink, L., 1997. Zinc inhibits IL-1 dependent T cell stimulation. Eur. J. Immunol. 27, 2529–2535.

Worwag, M., Classen, H.G., Schumacher, E., 1999. Prevalence of magnesium and zinc deficiencies in the nursing home residents in Germany. Magnes. Res. 12, 181–189.

Xu, X., Beckman, I., Ahern, M., Bradley, J., 1993. A comprehensive analysis of peripheral blood lymphocytes in healthy aged humans by flow cytometry. Immunol. Cell Biol. 71, 549–557.

Zalewski, P.D., Forbes, I.J., 1993. Intracellular zinc and the regulation of apoptosis. In: Laviri, M., Watters, D. (Eds.), Programmed Cell Death: The Cellular and Molecular Biology of Apoptosis. Harword Academic Press, Melbourne, pp. 73–85.

Zazonico, R., Fernandes, G., Good, R.A., 1981. The differential sensitivity of T cell and B cell mitogenesis to in vitro Zn deficiency. Cell. Immunol. 60, 203–211.

Advances in
Cell Aging and
Gerontology

Altered zinc binding by metallothioneins in immune-neuroendocrine senescence: a vicious circle between metallothioneins and chaperones?

Eugenio Mocchegiani, Robertina Giacconi, Mario Muzzioli
and Catia Cipriano

*Immunology Ctr. (Section Nutrition, Immunity and Ageing) Research Department "N. Masera", Italian
National Research Centres on Ageing (INRCA), Via Birarelli 8, 60121, Ancona, Italy
Correspondence address: Tel.: +39-071-8004216; fax: +39-071-206791.
E-mail: e.mocchegiani@inrca.it (E.M.)*

Contents

Abbreviations

AAS: atomic absorption spectrophotometry; AIDS: acquired immunodeficiency syndrome; Ca-A/K: AMPA/kainate channels; CNS: central nervous system;

Advances in Cell Aging and Gerontology, vol. 13, 261–281

CRH: corticosterone releasing hormone; ER: endoplasmic reticulum; Erp: endoplasmic reticulum protein; GIF: growth inhibitory factor; GM-CSF: granulocytes-macrophage colony stimulating factor; Grps: glucose regulated proteins; GSH: glutathione peroxidase; GSSG: glutathione reductase; Hsp: heat shock protein; Kd: constant of dissociation; INOS: inducible nitric-oxide synthase; IFN: interferon; IL: interleukin; MHC: major histocompatibility complex; MMPs: matrix metallo-proteases; MTF-1: metallothionein transcription factor-1; MTs: metallothioneins; NLS: nuclear localization signal; NO: nitric oxide; PARP: poly (ADP-ribose) poly-merase; Phx: hepatectomy; PKC: protein kinase-C; PTU: propyl-thiouracil; SOD: superoxide dismutase; TEC: thymic epithelial cell; Tg: thyroglobulin; TIMPs: tissue inhibitors metalloproteases; T_4: thyroxine; T_3: triiodothyronine; TNF: tumour necrosis factor; TRH: thyroid releasing hormone; VSSC: voltage-sensitive Ca^{++} channels; ZIP: zipper interacting proteins; Zn-MTs: zinc-bound metallothioneins.

1. Introduction

Metallothioneins (MTs) are low-molecular-weight metal-binding proteins with 61–68 amino acids. Among these, there are 20 are cysteine residues in conserved positions (Kagi and Schaffer, 1988). MTs play pivotal roles in metal-related cell homeostasis in cell growth and development because of their high affinity for metals, in particular zinc and copper (Vallee and Falchuck, 1993), and Cd and Hg by helping to prevent toxicity of the latter (Viarengo et al., 2000). The 20 cysteines are present in a reduced form and bind seven zinc atoms through mercaptide bonds forming metal thiolate clusters (Maret and Vallee, 1998). The zinc/cysteine interactions form two clusters, of two differing types, either as bridging or as terminal cysteine thiolate. Three bridging and six terminal cysteine thiolates provide an identical co-ordination environment for each of the three zinc atoms in a β-domain cluster. Two different zinc sites are present in α-domain clusters; two of them have one terminal ligand and three bridging ligands, respectively, while the other two have two terminal and two bridging ligands (Kagi and Schaffer, 1988; Maret and Vallee, 1998). Four different isoforms of MTs exist, characterised by amino acid chain length. The genes encoding each isoform are clustered on mouse chromosome 8 and human chromosome 16 and share a similar intron/exon structure (Palmiter et al., 1993) also with complex polymorphisms (West et al., 1990). Isoforms I and II are expressed in virtually all tissues (Palmiter et al., 1992), including brain (Hidalgo et al., 1997). Isoform III, called GIF, is restricted to the central nervous system (CNS) and isoform IV in tissues containing stratified squamous epithelia (Palmiter et al., 1992). MTs bind zinc with a high binding affinity (kd = 1.4×10^{-13} M) (Kagi and Schaffer, 1988) and also distribute cellular zinc because zinc undergoes rapid inter- and intracluster exchange (Otvos et al., 1993). In general, all isoforms of MTs are antioxidant agents because zinc–sulphur clusters are sensitive to changes in cellular redox state and oxidizing sites, inducing transfer of zinc from its binding sites in MTs to those of lower affinity in other proteins (Maret and Vallee, 1998), such as occurs in superoxide dismutase (SOD) activation

(Suzuki and Kuroda, 1995). Thereby the redox properties of MTs are crucial for their protective role against the cytotoxic effect of reactive oxygen species, ionizing radiation, electrophilic anti-cancer drugs and mutagens, as well as heavy metal toxicity (Kagi, 1993). Consistent with this role, MTs gene expression (MT I, II isoforms) induced by means of transcription factor MTF-1 (Palmiter, 1998), is present in organisms as diverse as molluscs and humans (Kagi, 1993). By contrast, MT III and IV seem to be expressed only in mammals (Kagi, 1993). Because of this particular antioxidant task, MTs are also transferred by means of chaperones from the cytosol into the nucleus to protect DNA from oxidative and chemical damage (Chubatsu and Meneghini, 1993) as well as against DNA fragmentation (apoptosis) induced by free radical toxicity or various protoncogenes (c-myc, ras and BAX) or transcription factors (NF-κB) (Apostolova et al., 1999). It appears, in these cases, that MTs are functioning as zinc shuttles between the cytoplasm and zinc-dependent structures in the nucleus in response to changes in nuclear zinc demand (Roesijadi, 2000).

This role is pivotal in young adult-age, but it can be questioned during ageing. Indeed this protection occurs in young adults during transient stress-like conditions (Kagi, 1993). By contrast, the stress-like condition is constant in ageing, as documented by high levels of glucocorticoids during the whole circadian cycle in old mice (Mocchegiani et al., 1998b, 2002a,b). This phenomenon allows continuous sequestering of zinc by MTs with subsequent limited zinc release, as shown by a greater number of zinc ions bound to MTs in the liver from old mice in comparison with young (Mocchegiani et al., 2002a). A persistent low zinc ion bioavailability is present in ageing (Mocchegiani et al., 1998a). Therefore, the role of MTs may switch from a protective role to a detrimental one. Indeed, despite increased levels of MTs in the liver (Apostolova and Cherian, 2000), thymus (Mocchegiani et al., 1998a) and testis (Suzuki et al., 1998) from old animals, cellular oxidative damage increases in ageing (free radical theory of ageing) (Ashok and Ali, 1994). Moreover, increased MTs are also observed in cancer (Ebadi and Swanson, 1988), infections (Grider et al., 1989) and in type 2 diabetes mellitus (Minami et al., 1999b), which are age-related diseases exhibiting low zinc ion bioavailability and immune-endocrine damage (Fabris and Mocchegiani, 1995). Since ageing also displays low zinc ion bioavailability (Mocchegiani et al., 1998a), MTs are not donors of zinc in ageing, but rather sequesters of zinc with a task in conserving residual zinc ions: a characteristic of MTs in zinc deficiency as observed in pups with a zinc-deprived diet (Vruwink et al., 1988). This phenomenon may induce low bioavailability of zinc ions for immune and endocrine responses, because zinc is required for the maintenance of the immune-neuroendocrine network (Mills, 1989). Therefore, zinc-bound MTs (Zn-MTs) may switch from having protective roles as zinc donors in young adults to a detrimental one in ageing as zinc sequesters, resulting in low zinc bioavailability and associated neuroendocrine diseases and autoimmune phenomena (Mocchegiani et al., 2000c).

Such a change may be related to the trafficking of MTs in moving zinc within the cytosol and from the cytosol to the nucleus involving transporter intracellular proteins (ZnT1–T4) (Palmiter and Findley, 1995) and nuclear localisation signals (NLS)

(Deshaies et al., 1988; Ogra and Suzuki, 2000). Their activity seems decreased in zinc deficiency and stress (Roesijadi, 2000) that, in turn, are characteristic events during ageing (Mocchegiani et al., 1998a). Therefore, the presence of high Zn-MTs levels in ageing may be related to a defect in cellular trafficking and folding of MTs, preventing release of zinc by MT with subsequent derangement in neuroendocrine and immune responses.

Here, we review the trafficking of zinc and MTs within the cytosol and nucleus as well as the possible mechanisms involved in this trafficking and in MT folding, in particular chaperone activity (Ogra and Suzuki, 2000). Because some chaperones (Hsp33, Hsp70, Hsp90) are also zinc-dependent, having "zinc finger" motifs (Jakob et al., 2000), a vicious circle between Zn-MTs and chaperones may develop in ageing. The different roles of Zn-MTs in young and old age and the efficacy of zinc supplementation are discussed in relation to this Zn-MTs/chaperone vicious circle in ageing.

2. Trafficking of zinc, of Zn-MTs within the cytosol and nucleus and role of chaperones

2.1. Trafficking of zinc

Regulation and homeostasis of zinc are considered to be a basic function of MT, underlying fundamental interactions between MT and zinc-dependent macromolecules. By means of various mechanisms, zinc ions can enter from the extracellular matrix into the cytosol and then bind to MT. First of all, observations in brain and yeast have shown $Zn++$ translocation through $Ca++$ permeable AMPA/kainate (Ca-A/K) channels (Carriedo et al., 1998). Alternatively, $Zn++$ can enter cells through voltage-sensitive $Ca++$ channels (VSCC) (Yin et al., 1998), or $Na+-Ca++$ exchangers (which can transport $Zn++$ bidirectionally across the plasma membrane) (Colvin et al., 2000) or ZIP family (Zrt1-3p) or Nramp2 protein (natural resistance-associated macrophage protein 2) (Roesijadi, 2000) (Fig. 1). Within the cytosol, $Zn++$ is captured by Nramp1 transporter protein produced in phagosomes (Supek et al., 1997) and transferred to MTs (Roesijadi, 2000) (Fig. 1). The apoform of MT is a potent zinc acceptor due to the exclusive co-ordination of zinc with cysteine sulphur ligands (Kagi and Schaffer, 1988), which are highly reactive and also determine the release of zinc from MT. In this way, MT becomes a zinc donor (Roesijadi, et al., 1998). The release of zinc may occur by means of the nitric-oxide pathway (Pearce et al., 2000; Zangger et al., 2001) or glutathione reductase (GSSG) (Fig. 1) (Jacob et al., 1998). Subsequently, some zinc ions are transferred by ZnT2 and ZnT3 transporter proteins into mitochondria, vesicles or the endosomal/lysososmal compartment in order to activate zinc-dependent enzymes, including superoxide dismutase (SOD) or glutathione peroxidase (GSH) (McMahon and Cousins, 1998). The efflux of zinc ions from the cell into the extracellular matrix and, subsequently, again becoming bioavailable for activating extracellular matrix proteins (tissue inhibitors of metalloproteases TIMPs 1–4, α-2 macroglobulin and growth factors in order to regulate matrix metalloproteases

Extracellular Matrix

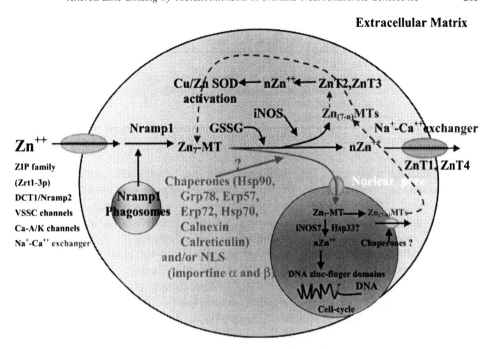

Fig. 1. Trafficking of zinc and Zn-MTs within the cytosol and from the cytosol to the nucleus in normal condition. Chaperones activity may be involved in the trafficking of MTs and in the release of zinc by MT within the nucleus because zinc is required for DNA zinc-finger domains including chaperones themselves, DNA protection by apotosis and correct phases of cell cycle. For explanations see the text. Question marks are under investigation.

MMPs activity) (Vallee and Falchuck, 1993) occurs by means of other zinc transporter proteins (ZnT1, ZnT4) (Palmiter and Findley, 1995; McMahon and Cousins, 1998 (Fig. 1).

2.2. Trafficking of Zn-MTs within the cytosol and nucleus

MTs are pivotal in cellular zinc homeostasis (Palmiter, 1998). For this process, Zn-MTs trafficking is very important because Zn-MTs are also transferred into and stored in the nucleus for specific functions (Table 1) (Ogra and Suzuki, 2000; Cherian and Apostolova, 2000). This is particularly important, considering the requirement for zinc for the transcription in DNA zinc-finger motifs (Dreosti, 2001). In addition, "in vitro" and "in vivo" studies have also shown that MTs protect DNA from oxidative and chemical damage and prevent apoptosis (Cai et al., 1995). An important consequence of MT nuclear localization is related to the proper functioning of the cell cycle. Indeed, the nuclear localization of MT occurs with increased Zn levels in the nucleus during progression of cells into G1- and S-phases of the cell cycle, and MT is retained in the nucleus if the cell cycle is blocked at S-phase (Apostolova and Cherian, 2000). These results suggest that nuclear localization

Table 1
Functions of Nuclear Zn-MT as zinc donor

Protection of DNA from apoptosis
Protect the nucleus from oxidative and chemical damage
Donate zinc to transcription factors and to DNA zinc finger domains
Promote normal phases of cell cycle
Control of DNA-repair

of MT is closely linked to the requirement for zinc during cell proliferation and regeneration, as illustrated by the slow repair of skin lesions in MT-null mice (Hanada et al., 1998). The high requirement for zinc in cells under certain conditions (for example during stress) could act a signal for MT nuclear localization, along with effects of growth factors and pro-inflammatory cytokines (IL-1, IL-6 and TNF-α) (Cherian and Apostolova, 2000). Another special task of the nuclear localization of MT is related to inducing DNA-repair by means of activating the nuclear enzyme poly (ADP-ribose) polymerase-1 (PARP-1) (Mocchegiani et al., 2000c). Indeed, PARP-1 is zinc-dependent and its role in base excision DNA-repair is strictly related to the oxidative status. In a state of high oxidative stress, as occurs in ageing, PARP-1 induces cell-death, via caspase-3, rather than DNA-repair, and the important point here is that this is due to low zinc ion bioavailability (Mocchegiani et al., 2000c). The mechanism of nuclear transport of MTs and their retention in the nucleus are not yet well defined and remains poorly understood. Retention may occur via polymerization of MTs, as suggested in some oxidative stress conditions (heat shock or H_2O_2) (Suzuki et al., 1983). Recent data in Balb/c3T3 cells have shown that nuclear translocation and retention of MTs occur by means of nuclear localization signals (NLS) (importine α, β), via GTP-ase activity (Nagano et al., 2000). However, other mechanisms have also been suggested. In this context, chaperones may have a crucial role. Indeed, they are involved in the correct folding of proteins in order to confer protein activity. Moreover, they transport proteins into the nucleus through nuclear pores with subsequent retention in the nucleus (Cherian and Apostolova, 2000). Increased activity of some chaperones (Hsp90, Grp78, Erp57, Erp72, Hsp70, calnexin, calreticulin) is associated with increased MTs within the cytosol (Cigliano et al., 1996; Abe et al., 2000; Murata et al., 1999; Theocharis et al., 2000). It has been suggested that chaperones of unknown origin may be involved in the correct folding of MT and also its release of zinc in the nucleus (Cherian and Apostolova, 2000). In this latter context, Hsp33 may be involved because it has a Kd for zinc of less than $(10^{-17} M)$ whereas MTs bind more strongly $(10^{-13} M)$ (Jakob et al., 2000). Subsequently, Hsp33 forms a dimer realising zinc under stress condition (H_2O_2) (Graumann et al., 2001). Alternatively, iNOS may also be involved in zinc release by the β-domain of MT within the nucleus, as occurs in mitochondria of endothelial cells (Pearce et al., 2000; Zangger et al., 2001). In this complex picture of zinc release by MTs (in the cytosol and nucleus), it is evident that chaperones play a pivotal role in MT trafficking and hence are critical for correct biological functions of the cells. Such an assumption acquires more relevance during ageing in endocrine and immune responses, which are known to be strongly influenced by zinc ion bioavailability (Mocchegiani et al., 1998a). Indeed, as mentioned above, the stress-like condition in ageing is constant, not transient. Other than high levels of glucocorticoids (Mocchegiani et al., 2002a,b), pro-inflammatory cytokines, such as IL-6, are found at constantly high levels in ageing (Franceschi et al., 1999). IL-6 is a potent activator of the hypothalamic–pituitary–adrenal-axis and plays a pathogenic role in conditions related to chronic stress and physiological ageing (Path et al., 2000). IL-6 and glucocorticoids affect MTmRNA induction (Grider et al., 1989). Thus, increased levels of MT-bound zinc and MTmRNA gene expression are expected in ageing, but result in gradual lowering of zinc ion bioavailability (Mocchegiani et al., 2002a,b,c). In this context, the activity of chaperones may be relevant because some of them are zinc-dependent (Jakob et al., 2000) and involved in the trafficking and correct folding of proteins, including MTs (Apostolova and Cherian, 2000). An age-related decline in chaperone-mediated autophagy has been recently reported (Cuervo and Dice, 2000). Moreover, chaperone Gpr78 activity is reduced in the liver of old rats (Heydari et al., 1995; Dhahbi et al., 1997). These findings are very suggestive of a possible defect in chaperone activity in preventing release of zinc by MTs in ageing. Increased levels of MTs may occur during zinc deficiency as a compensatory phenomenon in order to capture residual zinc ions (Vruwink et al., 1988) because the consistent and persistent loss of zinc ions is detrimental to the functioning of many homeostatic mechanisms in the body (Mills, 1989; Prasad, 1993). It has been shown that overexpression of MT characterises a mouse fibroblast cell line that was selected for its highly unusual ability to survive under zinc-deficient conditions (Suhy et al., 1999). In this case, the ability to overexpress MT appears to have been a fitness trait that was favoured during selection by low zinc conditions. Thus, the problem in ageing is presumably more closely related to the complex mechanisms of zinc release by MT, rather than increments of Zn-MTs themselves.

3. Zn-metallothioneins, immune responses, ageing and the role of chaperones

3.1. Zn-MTs, immune system, ageing

MTs protect cells from stresses by inducing macrophage secretion of IL-1, IL-6, IFN-α, TNF-α for a prompt immune response. In turn, these cytokines are involved in new synthesis of MTs in the liver (Cousins and Leinart, 1988; Cui et al., 1998) and IL-1 induces alterations in zinc status and in liver MT concentrations (Bui et al., 1994). These findings suggest the existence of an interplay between zinc, MTs and the immune system. On the other hand, IL-1 induces MTmRNA gene expression in thymic epithelial cells (TEC) of young humans via protein kinase C, which is zinc-dependent (Coto et al., 1992) and participates in the process of metal-induced MT gene expression (Yu et al., 1997). Moreover, MTs are donors of zinc for the reactivation of thymulin, a zinc-dependent thymic hormone, in TEC (Savino et al., 1984). Experiments in MT-null mice or in mice exposed to endotoxins or radiation support the existence of links between MTs and the immune system. GM-CSF and

GM-CSF receptor decrements are present in MT-null mice, inducing glial cell-death (Penkowa et al., 1999). MTmRNA, pro-inflammatory cytokines and chemokines are increased in the lungs of mice exposed to endotoxins (LPS) (Johnston et al., 1998). Moreover, thymocyte apoptosis is augmented in irradiated MT-null mice (Deng et al., 1999; Kondo et al., 1997). Indeed nuclear MTs inhibit apoptosis induced by free radical toxicity in the G1/S phase of the cell cycle by mechanisms involving increments of c-fos, p53 and BAX and decrements of c-myc or bcl-2 transcripts (Apostolova et al., 1999). A particular role in apoptosis prevention can also be exerted by transcription factor NF-κB that is under the control of MTs (Adbel-Magged and Agrawal, 1999) and TNF-α (Beg and Baltimore, 1996), which is also involved in MTmRNA expression (Cousins, 1998). These links are relevant in inflammation and ageing characterised by high MT and pro-inflammatory cytokine levels, impaired immune responses, low zinc ion bioavailability and increased immune cell-death (apoptosis) (Mocchegiani et al., 1998a). MTs protect against stress (Kagi, 1993) during zinc deficiency and zinc toxicity by acting as zinc reservoir or sequestering excess of zinc, respectively (Kelly et al., 1996). However, recent findings have shown high Zn-MTs protein levels and zinc content in the liver and atrophic thymus of old mice (Mocchegiani et al., 1998a). It has been also reported that MTmRNA gene expression is augmented in lymphocytes of old people and Down's syndrome subjects as compared to young adults. Surprisingly, however, the level of MTmRNA is low in lymphocytes of centenarians and resembles that of young adults (Mocchegiani et al., 2002b). Low MTmRNA levels are also observed in the livers of extremely old mice (30 months of age), thus mimicking human centenarians (Mocchegiani et al., 2002b). These findings are clear evidence that MTs may have different roles in immunosenescence: from protective to dangerous with a task of potential biomarkers of ageing (Mocchegiani et al., 1998a, 2000c, 2001). Such an assumption is supported by the fact that the oldest old individuals (centenarians) maintain good natural killer (NK) cell cytotoxicity, a young ratio of naive/memory T cells and normal activation of T cells bearing TCR γ/δ (Franceschi et al., 1995; Mariani et al., 1999; Borrego et al., 1999; Ibs and Rink, 2001; Colonna-Romano et al., 2002). Moreover, the number of NKT cells bearing TCR γ/δ increases in nonagenarians in comparison with younger elderly (Mocchegiani et al., 2002c). "In vitro" activated T cell clones from near-centenarians display IFN-γ production quite similar to young individuals (Pawelec et al., 2000). These immune functions (NK cell cytotoxicity, IFN-γ production and T cell proliferation) are related to zinc ion bioavailability (Wellinghausen et al., 1997), confirming the relevance of the interrelationship between zinc ion bioavailability and MTs homeostasis required to achieve healthy longevity and successful ageing (Mocchegiani et al., 2002b). The existence of a significant positive or inverse correlation in old mice between increased MTmRNA and enhanced corticosterone, and augmented IL-6 or decreased liver NK cell activity, respectively, during the whole circadian cycle (Mocchegiani et al., 2002a,b) is in line with this interpretation. The same significant positive or inverse correlation between high MTs and enhanced IL-6 or decreased NKT cell numbers bearing TCR γ/δ, respectively, is also observed in healthy elderly as well as in aged patients with infections

(Mocchegiani et al., 2002c). Additionally, increments of IL-6 and MTs may be also involved in the lack of restoration of immune dysfunction in old patients affected by myasthenia gravis after thymectomy (Mocchegiani et al., 2000a), as well as in old patients with rheumatoid arthritis after treatment with anti-inflammatory drugs (Winters et al., 1997). However, the role played by the IL-6 receptor (gp130) must also be taken into account. Despite increased levels of IL-6 commonly found in centenarians and thought to be due to constant high-level inflammation (Franceschi et al., 1999), gp130 activity is decreased in centenarians (Giuliani et al., 2001) as well as in very old mice (Mocchegiani et al., 2002c). Such a decrease in centenarians is associated with normal plasma levels of zinc and satisfactory anti-oxidant and immune-neuroendocrine activities, as compared to old people (Ravaglia et al., 1999; Mecocci et al., 2000). These intriguing findings suggest on the one hand the presence of inactive IL-6 in centenarians and, on the other pin-point the special role played by Zn-MTs over the entire life span, with different roles in the young and old (Mocchegiani et al., 2001). The increments of MTmRNA in lymphocytes from old subjects (Yurkow and Makhijani, 1998; Mocchegiani et al., 2002b) and old people affected by infections (Mocchegiani et al., 2002c), coupled with decreased immune responses and low zinc ion bioavailability (Fabris and Mocchegiani, 1995; Mocchegiani et al., 2000b) further support this possible reversal of the effect of MTs in normal ageing with implications for immune damage. Indeed, data from Kelly et al. (1996) showing a protective role of MT were obtained in young adults during nutritional zinc deprivation over a short period, which may mimic a transient stress-like condition. By contrast, zinc deficiency, impaired immune responses and stress-like conditions are constant in ageing (Mocchegiani et al., 1998a). Therefore, MTs may constantly retrieve zinc from plasma and tissues and fail to release it normally for immune efficiency. Such an assumption is strongly supported by the following findings: (i) the necessity for "in vitro" zinc addition to old plasma samples for thymulin reactivation (Mocchegiani et al., 1998a); (ii) the presence of constant high liver MT protein levels and impaired immune responses in old hepatectomized (phx) mice during liver regeneration, as compared to young phx mice (Mocchegiani et al., 1997); (iii) the preferential binding of zinc over copper by MTs in ageing (Hamer, 1986; Mocchegiani et al., 2002b); (iv) the requirement for exogenous zinc for T cell growth in old individuals and for TEC restoration in the old thymus (Mocchegiani et al., 1998a); (v) the prevention of apoptosis of old thymocytes by physiological zinc levels (Provinciali et al., 1998); (vi) the necessity of "in vitro" zinc in recovering mitogen responsiveness and NK activity by old splenocytes (Mocchegiani et al., 1995) as well as cytokine production (IFN-γ and IL-2) by old lymphocytes (Wellinghausen et al., 1997). Consistent with these find-ings, MTs cannot be zinc donors in ageing, but rather sequesters of intracellular zinc. This phenomenon has been proposed for thymic involution in ageing (Mocchegiani et al., 1998a). On the other hand, increased MT levels induce down-regulation of many other biological functions related to zinc, such as metabo-lism, gene expression and signal transduction (Kagi, 1993). The zinc-binding affinity (kd) of MTs is higher than that of other zinc-dependent molecules (thymulin, SOD and some cytokines) (Kagi and Schaffer, 1988; Mocchegiani et al., 1998a) and atomic absorption spectrophotometry (AAS) detects bound and unbound zinc (Mocchegiani et al., 1998a). These findings account for the presence of both high zinc-bound MTs and enhanced zinc content in the thymus and liver from old mice (Mocchegiani et al., 1998a), but also suggest that high levels of zinc-bound MTs lead to low free zinc ion bioavailability, which decreases the efficiency of the endo-crine-immune network in ageing. The larger amount of zinc ions, rather than copper, bound by MTs in the liver from old mice (HPLC analysis) (Mocchegiani et al., 2002a) supports this interpretation. Therefore, different roles of MTs (from protective to dangerous) may be further supported because of zinc deficiency, not copper, in ageing (Jacob et al., 1985).

3.2. Role of chaperones

MTs capture zinc during zinc deficiency in order to conserve residual zinc ions (Vruwink et al., 1988). Thus, such a potentially dangerous role may be more relevant to the lack of release of zinc by MT due to possible altered activity of molecular chaperones. Indeed, chaperones (Hsp70, Hsp90) are also involved in MHC class I antigen processing (Binder et al., 2001). In turn, abnormal increments of MTs during inflammation decrease the level of MHC class I and CD8 molecules detectable on the surface of lymphocytes resulting in immunosuppression of cell-mediated immunity (Youn and Lynes, 1999). Since the MHC complex requires zinc in its structure (Petersson et al., 2001), the abnormal increments of Zn-MTs are unable to release zinc for normal MHC function in inflammation and stress. Taking into account that chaperone activity: (i) is altered during stress (Feder and Hofmann, 1999; Moseley, 2000) and in ageing (Soti and Csermely, 2000) affecting the immune system; (ii) requires zinc for "zinc finger" motifs (Jakob et al., 2000); and (iii) is involved in the "danger signal" to alert the immune system to the death of a cell under stress (Danger Theory of Immunity) (Gallucci and Matzinger, 2001), the limited release of zinc by MT may create a vicious circle between zinc-bound MTs and chaperones. In other words, reduced zinc-dependent chaperone activity may provoke an incorrect folding of MTs resulting in even more limited zinc release. Such a limited zinc release may in turn be insufficient for the chaperone activity itself. A vicious circle occurs in the release of copper by MTs in copper-diet deficient mice. Copper deficiency impairs the activity of Hsp70 with a subsequent reduction in release of free copper ions by MT for the activity of Cu-SOD1 (Rae et al., 1999; Suzuki et al., 2002). The same mechanism might also occur for Zn-MTs and chaper-ones in ageing. Moreover, altered activity of molecular chaperones has been found in Down's syndrome patients (Yoo et al., 2001) and it is implicated in inflammation, infections, cancer and neurodegenerative diseases (for review see Macario, 1995; Soti and Csermely, 2000), which are pathologies related to the ageing process. In addition, chaperone activity is also involved in thymocyte maturation (Wiest et al., 1995), which requires zinc and decreases with ageing (Mocchegiani et al., 1998a). Therefore, the reduced zinc-dependent chaperone activity may be also responsible for the limited zinc release by MT with subsequent damage to immune responses during ageing.

4. Zn-metallothioneins, the endocrine system, ageing and the role of chaperones

4.1. Zn-MTs, the endocrine system, and ageing (peripheral glands)

Little is known about the influence of MTs on the endocrine system. It is well established, however, that stress hormones, i.e. corticosterone, as well as releasing factors from the pituitary gland, such as adrenocorticotropin hormone (ACTH), induce synthesis of MTs (Sato et al., 1996). In order to better provide resistance to stress (toxic agents, reactive oxygen species, ionizing radiation), MT induction by stress hormones may synergistically occur together with IL-6 gene expression and production (Sato et al., 1996). The resistance mediated by MTs, via stress hormones, occurs by means of the transfer of zinc from MT to other molecules or antioxidant enzymes through redox/oxidation mechanisms (Maret and Vallee, 1998). Conversely, a downregulation of MT with antisense expression vectors attenuates the ability of the glucocorticoid receptor (a CCCC zinc finger transcription factor) to respond to dexamethasone and initiate transcription under zinc-deficient conditions (DeMoor and Koropatnick, 2000). Zn-MTs activate the expression of the glucocorticoid receptor gene, indicating a reciprocal influence of Zn-MTs and glucocorticoids (Jacob et al., 1999).

This paracrine influence is the basis of MTs induction by stress hormones in metal toxicity or during stress (DeMoor and Koropatnick, 2000). Such a mechanism may also be active during shock: a condition of cardiovascular and/or respiratory collapse in which ACTH acts upon the adrenals to produce steroids, that induce MTs synthesis causing cell-death by apoptosis and necrosis, via the MTs-NO pathway (Sato et al., 1996; Simpkins, 2000). This occurs because NO could combine with O^{2-} forming the highly destructive ^-OONO (Simpkins, 2000). Such an effect is also evident in prolonged hemorrhagic (aortic aneurysm) and cardiogenic shock (in coronary artery diseases, such as atherosclerosis), in which increased MT induction by steroids does not protect against apoptosis induced by Fas (Xu, G. et al., 1999). Low zinc ion bioavailability is a usual event in cardiovascular diseases (Henning et al., 1996), which are also characteristic pathologies of the elderly (McLaughlin, 2001). At the same time, stress hormones and Zn-MTs are constantly high during ageing (Mocchegiani et al., 2000d). These findings, taken together, while on the one hand confirming that MT induction is under the control of stress hormones, on the other hand are consistent with the interpretation discussed above in the case of the immune system; that increased Zn-MTs are harmful in prolonged stress. Moreover, also taking into account an increased apoptosis of thymocytes and Peyer's patches during shock (Xu, Y.X. et al., 1997), the link among zinc ion bioavailability, stress hormones and Zn-MTs is crucial for endocrine-immune response.

Other hormones are suggested to affect MT induction as well. Indirect evidence in experimental animals exposed to cadmium show that MTs protect various endocrine glands from oxidative damage. The testes (Suzuki et al., 1998; Cai et al., 2000) and the islet cells of the pancreas (Tomita, 2000) are protected from cadmium-induced tumorigenesis by the release of zinc from MTs via a mechanism involving p53 gene expression (Xu et al., 1999). However, increments of Zn-MTs and zinc supplementation in young alloxanized animals (mimicking type 1 juvenile diabetes in humans) are not protective against alloxan (Minami et al., 1999b). No protection via increased MTs is seen in type 2 diabetes (Minami et al., 1999b), although here a specific role of augmented MTs in diabetes remains unclear. Renal failure and coronary artery disease seem to be involved in increased MT in diabetes (Minami et al., 1999b). In any case, low zinc ion bioavailability has been reported in diabetes (Type 1 and 2) (DiSilvestro, 2000). Exogenous zinc is capable of restoring insulin secretion (Cunningham, 1998). These findings suggest that low zinc and increased Zn-MTs, may be implicated in altered insulin secretion. This might occur because of limited zinc release by MTs, as occurs in alloxan diabetic mice (Minami et al., 1999a).

With regard to the reproductive system, there is evidence of intimate associations between MTs and testis (Suzuki et al., 1998), seminal plasma (Suzuki et al., 1994) and ovary (Levadoux-Martin et al., 2001) function. Orchidectomy decreases MTmRNA expression in the epididymis and prostate (Cry et al., 2001). Testosterone injections restore MTmRNA to control value (Cry et al., 2001). In this context, it is noteworthy that, other than MT-1 and MT-2, orchidectomy also causes decreased MT-3mRNA, which is restored by testosterone injection (Moffatt and Seguin, 1998). This fact is very intriguing because MT-3 is present, other than in brain (Ebadi et al., 1995), also in peripheral organs and tissues (liver, testis, prostate, epididymis, tongue, ovary, uterus, stomach, heart, seminal vesicles) (Moffatt and Seguin, 1998). Therefore, experiments performed in mice with disrupted liver MT-1 and MT-2 genes are not completely MT knockout mice, suggesting further studies before allocating a definitive protective role of liver MT against stress using this model of mice.

Despite this consideration, MTs are increased in the testis of old rats in order to preserve spermatozoa by nuclear chromatin decondensation (Suzuki et al., 1995). Concomitantly, testes of old individuals show low zinc content (Oldereid et al., 1993) and decreased testosterone (Rolf and Nieschlag, 2001), which is restored by exogenous zinc (Prasad et al., 1997). This fact is clear evidence that MTs are not donors of zinc in the old testis, but rather sequesters of zinc ions with subsequent failure in testosterone activity and production.

Young propyl-thiouracil (PTU)-treated mice (experimental model of hypothyroidism) also exhibit increased liver MT protein concentrations, low zinc ion bioavailability and depressed immune functions (Mocchegiani et al., 2002a). Such a failure in zinc and immunity and augmented MTmRNA expression also occurs in clinical hypothyroidism (Down's syndrome) (Mocchegiani et al., 2002b). Therefore, Zn-MTs are not protective as zinc donors in some endocrine diseases, even at young-adult age. On the other hand, the stress condition is constant both in hypothyroidism (Thoei et al., 1997) and diabetes (Kramer et al., 2000), which are frequent age-related pathologies (Fabris and Mocchegiani, 1995). Therefore, the proposal that MT is not protective in immunity during constant stress like-conditions may also be extended to the endocrine system. This consideration is important in ageing, which exhibits high MTmRNA, augmented corticosterone and hypothyroidism during the whole circadian cycle (Mocchegiani et al., 2000d;

Mocchegiani et al., 1998a). Since zinc ion bioavailability is pivotal for the efficiency of neuroendocrine network (Fabris et al., 1997), high Zn-MTs may also lead to low zinc ion bioavailability for endocrine functions during ageing. Transgenic mice overexpressing MTs showing low zinc bioavailability, depressed immune efficiency and altered thyroid hormones turnover (Mocchegiani et al., 2002a) and thus support such a hypothesis.

4.2. Zn-MTs, the endocrine system, and ageing (brain glands)

In this entire context, it is necessary also to take into account the complex network of hormone-releasing factors and some hormones secreted by endocrine glands in the brain, in particular melatonin secreted by the pineal gland in the suprachiasmatic nucleus. Such a consideration is of great relevance for two reasons. First, because MTs (isoform III) also bind zinc and are implicated in zinc release in various brain areas to enhance antioxidant enzyme activity in order to protect cerebral cells from oxidative damage (Mocchegiani et al., 2001; Ebadi et al., 1995). Therefore, the expression of MT-3 is extremely important in terms of maintaining the steady-state level of zinc and controlling redox potential. Second, pineal gland contains MT isoforms (Ebadi, 1991) and zinc (Demmel et al., 1982). Melatonin secretion is also under the control of zinc turnover (Mocchegiani et al., 1998b). Since melatonin secretion is reduced in ageing (Mocchegiani et al., 1998b), MTs in the pineal gland may not be zinc donors for melatonin secretion in old age. Although further studies are required to elucidate this point, pineal gland is atrophic in old mice (Boya and Calvo, 1984) as the thymus, and atrophic thymus contains high Zn-MTs (Mocchegiani et al., 1998a). On the other hand, the signal transduction (protein kinase C phosphorilation), via zinc, is involved in melatonin production by pinealocytes (Ou and Ebadi, 1992), and PKC activity is zinc dependent (Coto et al., 1992) and reduced in ageing (Wang et al., 2000). In addition, increased brain Zn-MTs levels have been suggested to be involved in some neurodegenerative diseases, including Alzheimer's (Zambenedetti et al., 1998), which, in turn, exhibits neuroendocrine alterations (melatonin and cortisol) (Ferrari et al., 2000) and decreased hypothalamic peptides (CRH and TRH) production (Sadow and Rubin, 1992). Indeed, such a production is also controlled by zinc turnover (Pekary et al., 1991), which in turn is reduced in the hypothalamus of old mice with concomitant increments of MTs isoforms (Mocchegiani et al., 2001). Therefore, increments of Zn-MTs may be deleterious for the neuroendocrine network during ageing. However, the few data in the literature regarding ageing of the brain due to the possible presence of compensatory phenomena among various isoforms of MTs (I–II and III) remain contradictory (Hidalgo et al., 1997). Therefore, further studies are necessary in old brain.

4.3. Role of chaperones

As discussed in Section 3 for the immune system, the major problem of increased MTs during ageing and, in general, in constant stress-like conditions, is the limited release of zinc for neuroendocrine functions, also under the control of chaperone activity (Dannies, 2000). Molecular chaperones of the Hsp70 or glucose-regulated protein (Grps) families produced in the endoplasmic reticulum (ER) are involved in the storage of secretory proteins (prohormones) within the ER of endocrine cells (spermatocytes, beta islets of the pancreas, thymocytes and thyroid cells) (Jain et al., 2000; Arvan et al., 1997; Wiest et al., 1995; Yoshinaga et al., 1999; Kobayashi et al., 2000). A prolonged retention of chaperones (Grps family) and prohormones (thyroglobulin) (Tg) within the ER may cause reduced synthesis of active thyroid hormones (T_3 and T_4), and hypothyroidism (Medeiros-Neto et al., 1996; Kim et al., 1996). In this context, no data are reported on the role played by zinc. However, taking into account that some chaperones are zinc-dependent (Jakob et al., 2000) and hypothyroidism is a constant stress-like condition associated with low zinc ion bioavailability and increased Zn-MTs (Mocchegiani et al., 2002a), the release of zinc by MT might not occur in hypothyroidism due to altered chaperone activity and massive storage of Tg in ER. Such an interpretation is suggestive for a deleterious role of Zn-MTs because of altered chaperone activity, also regarding endocrine functions under constant stress-like conditions, including ageing.

5. Effect of zinc supply

5.1. Zinc supply and Zn-MTs in ageing, stress and inflammation

"In vitro" and "in vivo" physiological zinc levels during ageing restores the impaired neuroendocrine-immune response (Wellinghausen et al., 1997) and increases longevity in old mice and resistance to infections in elderly humans (Mocchegiani et al., 2000b). However, taking into account that nutritional zinc affects MTmRNA expression (Cousins, 1998), a relevant question arises: may nutritional zinc affect MTmRNA expression and MT protein levels with a subsequent deleterious role of Zn-MTs in immune and endocrine functions in ageing?

Data obtained in young adults show the beneficial effect of zinc supplementation against toxicity by metals or by ionizing radiation due to increased MTmRNA (Chung et al., 1996). Nutritional zinc increases liver MTmRNA also in old and PTU mice with, however, no further significant increments of already high liver MT protein concentrations (Fig. 2) (Mocchegiani et al., 2002a). Plasma zinc, immune responses and neuroendocrine functions are also normalized in old and in PTU zinc-treated mice (Fabris et al., 1997).

The same phenomenon occurs in partial hepatectomy (phx) during liver regeneration: a model of acute and constant stress-like condition (Mocchegiani et al., 1997). Young and old phx mice display high liver Zn-MT levels and low zinc ion biovailability at 48 h after hepatectomy. Old phx mice exhibit no modifications at day 15 post-hepatecotmy, as compared to a complete remodelling in young phx mice at the same time of observation (15th day) (Mocchegiani et al., 1997) (Fig. 2). These findings obtained in two different experimental models (zinc supplementation and phx) are clear evidence that Zn-MTs are completely saturated by pre-existing zinc ions in constant stress-like conditions. The induction of

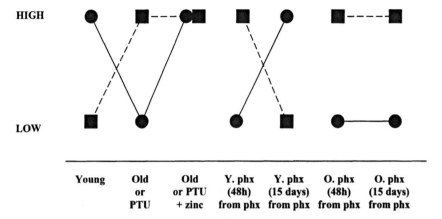

Fig. 2. Kinetics of zinc ions bioavailability (black circle) and Zn-MTs protein levels (black square) before and after zinc supply in old and PTU mice as well as in young (Y) and old (O) hepatectomized (phx) mice at 48 h and 15th day from partial hepatectomy. Adapted by Rae et al. (1999), Vallee and Falchuck (1993).

MTmRNA by nutritional zinc may act as a reservoir for prompt MT protein production for defence against continuous oxidative stress because of the faster degradation in lysosomes of Zn-MT, compared to other metal-bound MTs (Klaassen et al., 1994). Rapid Zn-MT degradation also occurs after zinc supplementation (Klaassen et al., 1994). The degradation of Zn-MTs in acid lysosomal compartments favours complete release of zinc from MT (Klaassen et al., 1994). Thus, zinc supplementation in constant stress-like conditions has two pivotal roles. First, it induces major zinc ion bioavailability by faster degradation of Zn-MTs; second, it prevents the continuous depletion of intracellular zinc by MTs. Therefore, a plateau in MT protein production is maintained, with subsequent retention of the role of MTs as zinc donors (Mocchegiani et al., 2000c). Larger amounts of free zinc ions are thus available for maintaining endocrine and immune efficiency in ageing and hypothyroidism. Indeed, old and PTU zinc-treated mice show increased survival with significant reductions in the number of infectious episodes (Mocchegiani et al., 1998a, 2002a). Zinc supplementation is also beneficial for reducing infectious episodes in old humans, in Down's syndrome subjects, and in AIDS patients (Mocchegiani et al., 2000b). Inflammation and stress conditions are usual events in these pathologies (Mocchegiani et al., 2000b). MTs are increased in inflammation to protect against stress (Borghesi and Lynes, 1996). Therefore, the zinc-facilitated resistance to infections may be due to better homeostasis between zinc and Zn-MTs (Mocchegiani et al., 2000c). Additionally, relapse of viral (herpes simplex II, cytomegalovirus) or bacterial (bronchitis, bronchopneumonia, mycobacterium tuberculosis) infections are reduced by 50% in old humans, in Down's syndrome and AIDS patients after one month of zinc treatment (Mocchegiani et al., 2000b). Appropriate zinc supplementation (physiological dose and short periods of treatment) is beneficial for preventing competition between zinc and other metals (copper, calcium and iron) (Prasad, 1993), as also demonstrated in other infections by fungi and protozoa (Shankar and Prasad, 1998).

5.2. Role of chaperones

The intricate mechanisms involved in this beneficial role of zinc supply may be related to the trafficking of Zn-MTs within the cytosol and from the cytosol to the nucleus, as well as to the correct folding of MTs, involving chaperone-like activity (Fig. 1). Some chaperones (e.g. Hsp33, Hsp70, Hsp90) are zinc-dependent (Jakob et al., 2000) and their induction and activity are reduced in old fruit flies (Fleming et al., 1988), in old rats (Heydari et al., 1995) and in elderly humans (Derham and Harding, 1997; Rao et al., 1999). Reduced Hsp70 activity has been also found in senescent T lymphocytes (Effros et al., 1994) and in old fibroblasts from humans (Bonelli et al., 1999). These findings lead to the interpretation that altered chaperone activity may provoke an incorrect folding of MTs with subsequent limited zinc release, which in turn impairs the chaperone activity itself, thereby creating a vicious circle between Zn-MTs and chaperones within the cells. An appropriate zinc supply might restore chaperone like-activity with consequent correct folding and trafficking of MTs and zinc release. Thereafter, MTs may again release zinc with restoration of the neuroendocrine-immune network, as observed in old and PTU zinc-treated mice (Mocchegiani et al., 2002a). In healthy centenarians, the satisfactory nutritional-endocrine-immune profile (Ravaglia et al., 1999), moderate stress levels (Ravaglia et al., 1999), low MTmRNA expression and increased PARP activity in base excision DNA-repair in lymphocytes (Mocchegiani et al., 2002b,c) suggest that an adequate balance between free zinc ions and Zn-MTs is crucial for maintaining the immune-neuroendocrine network necessary for successful ageing. Although no data exist on chaperone-like activity in centenarians, chaperones may be involved in maintaining this balance. We are continuing to investigate this possibility in our laboratory.

6. Concluding remarks and future directions

The high levels of Zn-MTs found under constant stress-like conditions, such as during ageing or in hypothyroidism, are mainly due to a continuous sequestering of intracellular zinc. Since zinc deficiency is a common event in ageing and hypothyroidism, the continuous sequestering of zinc is accomplished by MTs in order to conserve zinc ions, because zinc is one of the most relevant trace elements for many of the body's homeostatic mechanisms, including the immune-neuroendocrine network. This task is also accomplished by MTs under transient stress-like conditions; such as occur in young adults. However, during transient stress, MTs release zinc via NO nitrosylation (Pearce et al., 2000) to certain antioxidant enzymes (SOD and GSH). This zinc release may not occur under constant stress-like conditions. This would cause continuous low zinc ion bioavailability for endocrine and immune responses. Therefore, Zn-MTs may switch from having a protective role during transient stress conditions at a young age to having a deleterious one under the constant stress of ageing and hypothyroidism, because of limited zinc release by MT. Physiological zinc supplementation corrects the defect and this is associated with normalization of the neuroendocrine-immune network. The cause of this dichotomy of MT between young and old age, other than due to alterations in iNOS activity (Mocchegiani et al., 2000c), may be more upstream, in particular related to a defect in the trafficking of MT within the cytosol and from the cytosol to the nucleus, via chaperone activity. The task of chaperones is to fold proteins in order to make them biologically active, as well as to transfer proteins from the cytosol to the nucleus. Therefore, chaperones may be involved in the trafficking of MTs within the cytosol and nucleus and in the correct folding of MTs. The activity of some chaperones is zinc-dependent and decreases in ageing. The limited zinc release by MTs in ageing might affect chaperone activity resulting in incorrect folding of MTs, and hence, also lead to a limited zinc release affecting chaperone activity itself. As such, a vicious circle may arise in ageing and, in general, under constant stress-like conditions (hypothyroidism, diabetes) between zinc-bound MTs and chaperones with negative influence on the efficiency of the immune-neuroendocrine network. Because zinc supply restores endocrine and immune functions with no substantial modifications of already high MT protein levels in old and PTU mice, it is tempting to assign chaperones a crucial role in correct MT functioning. Therefore, chaperones may be involved on the one hand in the trafficking of MTs, on the other hand in the correct folding of MTs with subsequent release of zinc which in turn affects the chaperones themselves. This paracrine feedback loop is pivotal during ageing representing, as such, an interesting future field of research in order to better understand the role of Zn-MTs-chaperone interactions in immune-neuroendocrine senescence.

Acknowledgements

This paper was supported by INRCA, Italian Health Ministry (RF No. 99/107 to E. Mocchegiani) and ECC (ImAginE project, No. QLK6-CT-1999-02031, Co-ordinator Professor G. Pawelec).

References

Abe, T., Yamamoto, O., Gotoh, S., Yan, Y., Todaka, N., Higashi, K., 2000. Arch. Biochem. Biophys. 382, 81–88.

Adbel-Magged, A., Agrawal, K.C., 1999. Cancer Res. 58, 2335–2338.

Apostolova, M.D., Cherian, M.G., 2000. J. Cell. Physiol. 183, 135–140.

Apostolova, M.D., Ivanova, I.A., Cherian, M.G., 1999. Toxicol. Appl. Pharmacol. 159, 175–184.

Arvan, P., Kim, P.S., Kuliawat, R., Prabakaran, D., Muresan, Z., Yoo, S., Abu-Hossain, S., 1997. Thyroid 7, 89–105.

Ashok, B.T., Ali, R., 1994. Exp. Gerontol. 34, 293–303.

Beg, A.A., Baltimore, D., 1996. Science 274, 782–784.

Binder, R.J., Blachere, N.E., Srivastava, P.K., 2001. J. Biol. Chem. 276, 17,163–17,171.

Bonelli, M.A., Alfieri, R.R., Petronini, P.G., Brigotti, M., Canpanini, C., Borghetti, A.F., 1999. Exp. Cell. Res. 252, 20–32.

Borghesi, L.A., Lynes, M.A., 1996. Cell Stress Chaperones 1, 99–108.

Borrego, F., Alonso, M.C., Galiani, M.D., Carracedo, J., Ramirez, R., Ostos, B., Pena, J., Solana, R., 1999. Exp. Gerontol. 34, 253–265.

Boya, J., Calvo, J., 1984. J. Pineal Res. 1, 83–89.

Bui, L.M., Dressendorfer, R.H., Keen, C.L., Summary, J.J., Dubick, M.A., 1994. Proc. Soc. Exp. Biol. Med. 206, 438–444.

Cai, L., Deng, X., Jiang, J., Chen, S., Zhong, R., Cherian, M.G., Chakrabarti, S., 2000. Urol. Res. 28, 97–103.

Cai, L., Koropatnick, J., Cherian, M.G., 1995. Chem. Biol. Interact. 96, 143–155.

Carriedo, S.G., Yin, H.Z., Sensi, S.L., Weiss, J.H., 1998. J. Neurosci. 18, 7727–7738.

Cherian, M.G., Apostolova, N.D., 2000. Cell. Mol. Biol. 46, 347–356.

Chubatsu, L.S., Meneghini, R., 1993. Biochem. J. 291, 193–198.

Chung, J., Nartey, N.O., Cherian, M.G., 1996. Arch. Environ. Health 41, 319–323.

Cigliano, S., Remondelli, P., Minichiello, L., Mellone, M.C., Martire, G., Bonatti, S., Leone, A., 1996. Exp. Cell Res. 228, 173–180.

Colonna-Romano, G., Potestio, M., Aqiuno, A., Candore, G, Lio, D., Caruso, G., 2002. Exp. Gerontol. 37, 205–211.

Colvin, R.A., Davis, N., Nipper, R.W., Carter, P.A., 2000. J. Nutr. 130, 1484S–1487S.

Coto, J.A., Hadden, E.M., Sauro, M., Zorn, M., Hadden, J.W., 1992. Proc. Natl. Acad. Sci. 89, 7752–7756.

Cousins, R.J., 1998. Proc. Nutr. Soc. 57, 307–311.

Cousins, R.J., Leinart, A.S., 1988. FASEB J. 2, 2844–2890.

Cry, D.G., Dufresne, J., Pillet, S., Alfieri, T.J., Hermo, J., 2001. J. Androl. 22, 124–135.

Cuervo, A.M., Dice, J.F., 2000. J. Biol. Chem. 275, 31,505–31,513.

Cui, L., Takagi, Y., Wasa, M., Iiboshi, Y., Inoue, M., Khan, J., Sando, K., Nezu, R., Okada, A., 1998. J. Nutr. 128, 1092–1098.

Cunningham, J.J., 1998. J. Am. Coll. Nutr. 17, 7–10.

Dannies, P.S., 2000. Vitam. Horm. 58, 1–26.

Demmel, U., Hock, A., Kasperek, K., Feinendegen, L.E., 1982. Sci. Total Environ. 24, 135–146.

DeMoor, J.M., Koropatnick, D.J., 2000. Cell. Mol. Biol. 46, 367–381.

Deng, D.X., Cai, L., Chakrabarti, S., Cherian, M.G., 1999. Toxicology 134, 39–49.

Derham, B.K., Harding, J.J., 1997. Biochem. J. 328, 763–768.

Deshaies, R.J., Koch, B.D., Werner-Washburne, M., Craig, E.A., Schekman, R., 1988. Nature 332, 800–805.

Dhahbi, J.M., Mote, P.L., Tillman, J.B., Walford, R.L., Spindler, S.R., 1997. J. Nutr. 127, 1758–1764.

DiSilvestro, R.A., 2000. J. Nutr. 130, 1509–1511.

Dreosti, I.E., 2001. Mutat. Res. 475, 161–167.

Ebadi, E., Swanson, S., 1988. In: Ebadi, M. (Ed.), Nutrition, Growth and Cancer. Alan. R. Liss, New York, pp. 161–175.

Ebadi, M., 1991. Methods Enzymol. 205, 363–387.

Ebadi, M., Iversen, P.L., Hao, R., Cerutis, D.R., Rojas, P., Happe, H.K., Murrin, L.C., Pfeiffer, R.F., 1995. Neurochem. Int. 27, 1–22.

Effros, R.B., Zhu, X., Walford, R.L., 1994. J. Gerontol. 49, B65–B70.

Fabris, N., Mocchegiani, E., 1995. Ageing Clin. Exp. Res. 7, 77–93.

Fabris, N., Mocchegiani, E., Provinciali, M., 1997. Exp. Gerontol. 32, 415–429.

Feder, M.E., Hofmann, G.E., 1999. Annu. Rev. Physiol. 61, 243–282.

Ferrari, E., Fioravanti, M., Magri, F., Solerte, S.B., 2000. Ann. N.Y. Acad. Sci. USA 917, 582–596.

Fleming, J.E., Walton, J.K., Dubitsky, R, Bensch, K.G., 1988. Proc. Natl. Acad. Sci. USA 85, 4099–4103.

Franceschi, C., Monti, D., Sansoni, C., Cossarizza, A., 1995. Immunol. Today 16, 12–16.

Franceschi, C., Valensin, S., Fahnoni, F., Barbi, C., Bobafè, M., 1999. Exp. Gerontol. 34, 911–921.

Gallucci, S., Matzinger, P., 2001. Curr. Opin. Immunol. 13, 114–119.

Giuliani, N., Sansoni, P., Girasole, G., Vescovini, R., Passeri, M., Pedrazzoni, M., 2001. Exp. Gerontol. 36, 547–557.

Graumann, J., Lilie, H., Tang, X., Tucker, K.A., Hoffmann, J.H., Vijayalakshmi, J., Saper, M., Bardwell, J.C., Jakob, U., 2001. Structure (Camb) 9, 377–387.

Grider, A., Kao, K.J., Klein, P.A., Cousins, R.J., 1989. J. Lab. Clin. Med. 113, 221–228.

Hamer, D.H., 1986. Ann. Rev. Biochem. 55, 913–951.

Hanada, K., Sawamura, D., Hashimoto, I., Kida, K., Naganuma, A., 1998. J. Invest. Dermatol. 110, 259–262.
Henning, B., Toboreck, M., McClain, C.J., 1996. Nutrition 12, 711–717.
Heydari, A.R., Conrad, C.C., Richardson, A., 1995. J. Nutr. 125, 410–418.
Hidalgo, J., Castellano, B., Campbell, I.L., 1997. Curr. Topics Neurochem. 1, 1–26.
Ibo, K.H., Rink, L., 2001. Gerontol. Geriatr. 34, 480–485.
Jacob, C., Maret, W., Vallee, B.L., 1998. Proc. Natl. Acad. Sci. 95, 3489–3494.
Jacob, R.A., Russel, R.N., Sandstead, H.H., 1985. In: Watson, R.R. (Ed.), Handbook of Nutrition in the Aged. CRC Press, Boca Raton CA, pp. 77–88.
Jacob, S.T., Ghoshal, K., Sheridan, J.F., 1999. Gene Exp. 7, 301–310.
Jain, R.K., Joyce, P.B., Gorr, S.U., 2000. J. Biol. Chem. 275, 27,032–27,036.
Jakob, U., Eser, M., Bardwell, J.C., 2000. J. Biol. Chem. 275, 38,302–38,310.
Johnston, C.J., Finkelstein, J.N., Gelein, R., Oberdoster, G., 1998. Toxicol. Sci. 46, 300–307.
Kagi, R.H.J., 1993. In: Suzuki, K.T., Imura, N., Kimura, M. (Eds.), Metallothioneins III. pp. 29–56. Birkhauser Verlag, Basel.
Kagi, R.H.J., Schaffer, A., 1988. Biochemistry 27, 8509–8515.
Kelly, E.J., Quaife, C.F., Froelik, G.J., Palmiter, R.D., 1996. J. Nutr. 126, 1782–1790.
Kim, P.S., Kwon, O.Y., Arvan, P., 1996. J. Cell Biol. 133, 517–527.
Klaassen, C.D., Choudhuri, S., McKim, J.M., Lehman-Mckeeman, L.D., Kershaw, W.C., 1994. Environ. Health Perspect. 102, 141–146.
Kobayashi, T., Ogawa, S., Yura, T., Yanagi, H., 2000. Biochem. Biophys. Res. Commun. 267, 831–837.
Kondo, Y., Rusnak, J.M., Hoyt, D.G., Settineri, C.E., Pitt, B.R., Lazo, J.S., 1997. Mol. Pharmacol. 52, 195–201.
Kramer, J.R., Ledolter, J., Manos, G.N., Bayless, M.L., 2000. Ann. Behav. Med. 22, 17–28.
Levadoux-Martin, M., Hesketh, J.E., Beattie, J.H., Wallace, H.M., 2001. Biochem. J. 355, 473–479.
Macario, A.J., 1995. Int. J. Clin. Lab. Res. 25, 59–70.
Maret, W., Vallee, B.L., 1998. Proc. Natl. Acad. Sci. 95, 3478–3482.
Mariani, E., Ravaglia, G., Forti, P., Meneghetti, A., Tarozzi, A., Maioli, F., Boschi, F., Pratelli, L., Pizzoferrato, A., Piras, F., Facchini, A., 1999. Clin. Exp. Immunol. 116, 19–27.
McLaughlin, M.A., 2001. Geriatrics 56, 45–49.
McMahon, R.J., Cousins, R.J., 1998. Proc. Natl. Acad. Sci. 95, 4841–4846.
Mecocci, P., Polidori, M.C., Troiano, L., Cherubini, A., Cecchetti, R., Pini, G., Straatman, M., Monti, D., Stahl, W., Sies, H., Franceschi, C., Senin, U., 2000. Free Radical Biol. Med. 28, 1243–1248.
Medeiros-Neto, G., Kim, P.S., Yoo, S.E., Vono, J., Tragovnik, H.M., Carmago, R., Hossain, S.A., Arvan, P., 1996. J. Clin. Invest. 98, 2838–2844.
Mills, C.F. (Ed.), 1989. Zinc in Human Biology. Springer-Verlag, London. 1–630.
Minami, T., Shimizu, M., Tanaka, H., Okazaki, Y., Cherian, M.G., 1999a. Toxicology 132, 33–41.
Minami, T., Tohno, S., Tohno, Y., Otaki, N., Kimura, M., Cherian, M.G., 1999b. In: Klaassen, C.D. (Ed.), Metallothionein IV. Birkhauser, Basel, pp. 535–539.
Mocchegiani, E., Giacconi, R., Cipriano, C., Gasparini, N., Orlando, F., Stecconi, R., Muzzioli, M., Isani, G., Carpenè, E., 2002a. Mech. Ageing Dev. 123, 675–694.
Mocchegiani, E., Giacconi, R., Cipriano, C., Muzzioli, M., Fattoretti, P., Bertoni-Freddari, C., Isani, C., Zambenedetti, P., Zatta, P., 2001. Brain Res. Bull. 55, 147–153.
Mocchegiani, E., Giacconi, R., Cipriano, C., Muzzioli, M., Gasparini, N., Morresi, F., Stecconi, R., Suzuki, H., Cavalieri, E., Mariani, E., 2002b. Exp. Gerontol. 37, 349–357.
Mocchegiani, E., Giacconi, R., Muzzioli, M., Gaetti, R., Cipriano, C., 2002c. 3rd Int. Conference on Basic Biology and Clinical Impact of Immunosenecence. (ImAginE), Palermo, April 10–14, (Abstract) p. 19.
Mocchegiani, E., Giacconi, R., Muzzioli, M., Gasparini, N., Provinciali, M., Spazzafumo, L., Licastro, F., 2000a. Mech. Ageing Dev. 117, 79–91.
Mocchegiani, E., Giacconi, R., Muzzioli, M., Cipriano, C., 2000b. Mech. Ageing Dev. 121, 21–36.
Mocchegiani, E., Muzzioli, M., Giacconi, R., 2000c. Trends Pharmacol. Sci. 21, 205–208.
Mocchegiani, E., Muzzioli, M., Giacconi, R., 2000d. Biogerontology. 1, 133–143.
Mocchegiani, E., Muzzioli, M., Cipriano, C., Giacconi, R., 1998a. Mech. Ageing Dev. 106, 183–204.
Mocchegiani, E., Santarelli, L., Muzzioli, M., Fabris, N., 1995. Int. J. Immunopharmacol. 17, 703–718.

Mocchegiani, E., Santarelli, L., Tibaldi, A., Muzzioli, M., Bulian, D., Cipriano, C., Olivieri, F., Fabris, N., 1998b. J. Neuroimmunol. 86, 111–122.

Mocchegiani, E., Verbanac, D., Santarelli, L., Tibaldi, A., Muzzioli, M., Radosevic-Stasic, B., Milin, C., 1997. Life Sci. 61, 1125–1145.

Moffatt, P., Seguin, C., 1998. DNA Cell Biol. 17, 501–510.

Moseley, P., 2000. Immunopharmacology 48, 299–302.

Murata, M., Gong, P., Suzuki, K., Koizumi, S., 1999. J. Cell Physiol. 180, 105–113.

Nagano, T., Itoh, N., Ebisutani, C., Takatani, T., Miyoshi, T., Nakanishi, T., Tanaka, K., 2000. J. Cell Physiol. 185, 440–446.

Ogra, Y., Suzuki, K.T., 2000. Cell. Mol. Biol. 46, 357–365.

Oldereid, N.B., Thomassen, Y., Attramadal, A., Olaisen, B., Purvis, K., 1993. J. Reprod. Fertil. 99, 421–425.

Otvos, J.D., Liu, X., Li, H., Shen, G., Basti, M., 1993. In: Suzuki, K.T., Imura, N., Kimura, M. (Eds.), Metallothionein III. Birkhauser Verlag, Basel, pp. 57–74.

Ou, C.Z., Ebadi, M., 1992. J. Pineal Res. 12, 17–26.

Palmiter, R.D., 1998. Proc. Natl. Acad. Sci. 95, 8428–8430.

Palmiter, R.D., Findley, S.D., 1995. EMBO J. 14, 639–649.

Palmiter, R.D., Findley, S.D., Whitemore, T.E., Durnam, D.M., 1992. Proc. Natl. Acad. Sci. 89, 6333–6337.

Palmiter, R.D., Sandgren, E.P., Koeller, D.M., Findley, S.D., Brinster, R.L., 1993. In: Susuki, K.D., Imura, N., Kimura, M. (Eds.), Metallothionein III. Birkhauser Verlag, Basel, pp. 399–406.

Path, G., Scherbaum, W.A., Bornstein, S.R., 2000. Eur. J. Clin. Invest. 30(Suppl. 3), 91–95.

Pawelec, G., Mariani, E., Bradley, B., Solana, R., 2000. Biogerontology 1, 247–254.

Pearce, L., Wasserloss, K., St. Croix, C.M., Gandley, R., Levitan, E.S., Pitt, B.R., 2000. J. Nutr. 130, 1467S–1470S.

Pekary, A.E., Lukaski, H.C., Mena, I., Hershman, J.M., 1991. Peptides 12, 1025–1032.

Penkowa, M., Giralt, M., Moss, T., Thomsen, P., Hernandez, J., Hidalgo, J., 1999. Exp. Neurol. 156, 149–164.

Petersson, K., Hakansson, M., Nilsson, H., Forsberg, G., Svensson, L.A., Liljas, A., Walse, B., 2001. EMBO J. 20, 3306–3312.

Prasad, A.S., 1993. Biochemistry of Zinc, Plenum Press: New York. 1–315

Prasad, A.S., Mantzoros, C.S., Beck, F.W., Hess, J.W., Brewer, G.J., 1997. Nutrition 12, 344–348.

Provinciali, M., Di Stefano, G., Stronati, S., 1998. Cytometry 32, 1–8.

Rae, T.D., Schmidt, P.J., Pufahl, R.A., Culotta, V.C., O'Halloran, T.V., 1999. Science 284, 805–808.

Rao, D.V., Watson, K., Jones, G.L., 1999. Mech. Ageing Dev. 107, 105–118.

Ravaglia, G., Forti, P., Maioli, F., Nesi, B., Pratelli, L., Savarino, L., Cucinotta, D., Cavalli, G., 1999. J. Clin. Endocrinol. Metab. 85, 2260–2265.

Roesijadi, G., 2000. Cell. Mol. Biol. 46, 393–406.

Roesijadi, G., Bogumil, R., Vasak, M., Kagi, J.H.R., 1998. J. Biol. Chem. 273, 17,425–17,432.

Rolf, C., Nieschlag, E., 2001. Exp. Clin. Endocrinol. Diabetes 109, 68–74.

Sadow, T.F., Rubin, R.T., 1992. Psyconeuroendocrinology 17, 293–314.

Sato, M., Yamaki, J., Hamaya, M., Hojo, H., 1996. Int. J. Immunopharmacol. 18, 167–172.

Savino, W., Huang, P.C., Corrigan, A., Berrih, S., Dardenne, M., 1984. J. Histochem. Cytochem. 32, 942–946.

Shankar, A.H., Prasad, A.S., 1998. Am. J. Clin. Nutr. 68, 447S–463S.

Simpkins, C.O., 2000. Cell. Mol. Biol. 46, 465–488.

Soti, C., Csermely, P., 2000. Biogerontology 1, 225–233.

Suhy, D.A., Simon, K.D., Linzer, D.I., O'Halloran, T.V., 1999. J. Biol. Chem. 274, 9183–9192.

Supek, F., Supekova, L., Nelson, H., Nelson, N., 1997. J. Exp. Biol. 200, 321–330.

Suzuki, J.S., Kodana, N., Molotov, A., Aoki, E., Tohyama, C., 1998. Biochem. J. 334, 695–701.

Suzuki, K.T., Kuroda, T., 1995. Res. Commun. Mol. Pathol. Pharmacol. 87, 287–296.

Suzuki, K.T., Ohnuki, R., Yaguchi, K., 1983. Toxicol. Lett. 16, 77–84.

Suzuki, K.T., Someya, A., Komada, Y., Ogra, Y., 2002. J. Inorg. Biochem. 88, 173–182.

Suzuki, T., Nakajima, K., Yamamoto, A., Yamanaka, H., 1995. Andrologia 37, 161–164.

Suzuki, T., Suzuki, K., Nakajiama, K., Otaki, N., Yamanaka, H., 1994. Int. J. Urol. 1, 345–348.

Theocharis, S.E., Kanelli, H., Margeli, A.P., Spiliopoulou, C.A., Koutselinis, A.S., 2000. Clin. Chem. Lab. Med. 38, 1137–1140.

Thoei, A., Akai, M., Tomabechi, T., Mamada, M., Taya, K., 1997. J. Endocrinol. 152, 147–154.

Tomita, T., 2000. Med. Pathol. 13, 389–395.

Vallee, B.L., Falchuck, K.H., 1993. Physiol. Rev. 73, 79–118.

Viarengo, A., Burlando, B., Ceratto, N., Pandolfi, I., 2000. Cell. Mol. Biol. 46, 407–418.

Vruwink, K.G., Hurley, L.S., Gershwin, M.E., Keen, C.L., 1988. Proc. Soc. Exp. Biol. Med. 188, 30–40.

Wang, H.Y., Bashorc, T.R., Tran, Z.V., Friedman, E., 2000. J. Gerontol. Biol. Sci. Med. Sci. 55, B545–B551.

Wellinghausen, N., Kirkner, H., Rink, L., 1997. Immunol. Today 18, 519–521.

West, A.K., Stallings, R., Hildebrand, C.E., Chiu, R., Karin, M., Richards, R.I., 1990. Genomics 8, 513–518.

Wiest, D.L., Bhandoola, A., Punt, J., Kreibich, G., McKean, D., Singer, A., 1997. Proc. Natl. Acad. Sci. 94, 1884–1889.

Wiest, D.L., Burgess, W.H., McKean, D., Kearse, K.P., Singer, A., 1995. EMBO J. 14, 3425–3453.

Winters, C., Jasani, B., Marchant, S., Morgan, A.J., 1997. Histochem. J. 29, 301–307.

Xu, G., Zhou, G., Jin, T., Zhou, T., Hammarstrom, S., Bergh, A., Nordberg, G., 1999. Biometals 12, 131–139.

Xu, Y.X., Ayala, A., Monfils, B., Cioffi, W.G., Chaudry, I.H., 1997. J. Surg. Res. 70, 55–60.

Yin, H.Z., Ha, D.H., Carriedo, S.V., an Weiss, J.H., 1998. Brain Res. 781, 45–55.

Yoo, B.C., Vlkolinsky, R., Engidawork, E., Cairns, N., Fountoulakis, M., Lubec, G., 2001. Electrophoresis 22, 1233–1241.

Yoshinaga, K., Tanii, I., Toshimori, K., 1999. Arch. Histol. Cytol. 62, 283–293.

Youn, J., Lynes, M.A., 1999. Toxicol. Sci. 52, 199–208.

Yu, C.W., Chen, J.H., Lin, L.Y., 1997. FEBS Lett. 420, 69–73.

Yurkow, E.J., Makhijani, P.R., 1998. J. Toxicol. Environ. Health 54, 445–457.

Zambenedetti, P., Giordano, R., Zatta, P., 1998. J. Chem. Neuroanat. 15, 21–26.

Zangger, K., Oz, G., Haslinger, E., Kunert, O., Armitage, I.M., 2001. FASEB J. 15, 1303–1305.

Advances in
Cell Aging and
Gerontology

T cell exhaustion and aging: is replicative senescence relevant?

Rita B. Effros

Department of Pathology and Laboratory Medicine, UCLA School of Medicine, and The UCLA Molecular Biology Institute, 10833 Le Conte Avenue, Los Angeles, CA 90095-1732, USA
Correspondence address: Tel.: + 1-310-825-0748; fax: + 1-310-206-5178. E-mail: reffros@mednet.ucla.edu

Contents

1. Overview

The > 65 age group is the fastest growing component of both the U.S. and world population. Two major clinical problems in the elderly are intimately linked to the aging immune system, namely, the increased morbidity and mortality due to infection, and the dramatic increase in cancer incidence. The component of the immune system that is most directly responsible for effective control over virally infected cells and tumor cells is the CD8 T cell subset. Our own research has documented a progressive increase with age in humans in a particular type of memory CD8 T cell that lacks expression of the CD28 costimulatory receptor and has other characteristics suggestive of replicative senescence.

Replicative senescence describes the intrinsic limit in proliferative potential that is a property of normal human somatic cells. Cells that reach replicative senescence in culture are in a state of irreversible growth arrest and show striking changes in cell function, such as apoptosis resistance and altered cytokine profiles. We propose that the changes associated with CD8 T cell senescence and the persistence of these putatively senescent CD8 T cells in vivo exerts a dramatic effect on immune function during aging. Although formal demonstration that replicative senescence is actually occurring during human aging may be impossible, the more similarities that can be identified between the stages along the senescence trajectory in cell culture and phenotypes present among virus-specific CD8 T cells in vivo, the stronger the evidence in support of our hypothesis.

2. The replicative senescence model

The term replicative senescence describes the irreversible state of growth arrest experienced by all mitotically-competent cells of human origin following a fairly predictable number of cell divisions in culture. The phenomenon was first identified by Hayflick in human fetal fibroblasts 40 years ago (Hayflick, 1992), and since that time, the characteristics of replicative senescence have been explored in a variety of cell types (Smith and Pereira-Smith, 1996). It is becoming increasingly clear that the striking changes in cell phenotype and function associated with replicative senescence may be just as important as the irreversible cell cycle arrest (Campisi, 2001). Indeed, senescent fibroblasts enhance the growth of tumor cells both in cell culture and in vivo, underscoring the potentially delirious outcome of senescent cells within the aging organism (Krtolica et al., 2001). Ironically, the replicative senescence model has only recently been applied to cells of the immune system, despite the fact that the ability to undergo rapid clonal expansion is absolutely essential to their function. Moreover, the well-documented decline in T cell immune function during aging suggests that T cells might be an ideal system to further explore the potential role of replicative senescence during in vivo aging. This review will summarize the results of research on T cell replicative senescence in cell culture and will demonstrate that cells from elderly people have undergone changes in vivo that are similar to those observed in the cell culture model. We propose that the occurrence of T cell replicative senescence in certain CD8 T cells in vivo is the outcome of past immunological history, and depending on the particular environmental pathogens encountered, can function to increase morbidity and/or mortality in the elderly.

3. Lymphocyte function is dependent on proliferation

The intricate genetic mechanism by which T cell antigen receptors are generated allows a limited number of T cell receptors (TCR) to create an immune system with an enormous range of specificities. The random series of gene segment juxta-positions and DNA recombination events that occurs during lymphocyte develop-ment results in a TCR that is unique to each T cell. The diversity so-generated is further amplified by the pairing of two different chains, each encoded by distinct sets of gene segments, to form a functional antigen receptor. By these mechanisms, a small amount of genetic material is utilized to generate at least 10^8 different specificities. Each lymphocyte bears many copies of its antigen receptor, and once generated, the receptor specificity of a lymphocyte does not change. The outcome of this elegant genetic process is that the number of cells that can recognize and respond to any single antigen is extremely small. Thus, to generate a sufficient quantity of specific effector cells to fight an infection, an activated lymphocyte must proliferate extensively before its progeny differentiate into effector cells. For this reason, a limitation on the process of cell division could potentially have devastating consequences on immune function. To determine if extensive rounds of cell division result in replicative senescence, we adapted the extensively-studied fibroblast model to human T cells, as will be described below.

4. The T cell model

We have developed a long-term T cell culture system in which human T cells are followed from the point of primary stimulation to the end-point of replicative senescence. The culture system involves the use of antigen-presenting cells (APC) rather than antibodies to CD3 and CD28, because of the difference in avidity of the two systems and the greater similarity of APC to physiological stimulation in vivo. An additional aspect of the culture system is that we use alloantigen as opposed to viral antigens, in order to elicit a larger proportion of responding cells (Reiser et al., 2000), and also to ensure that the cells undergoing expansion are more likely to be naïve cells.

The basic protocol involves the use of a telomerase-negative allogeneic lymphoblastoid cell line ("NOR 20"), which expresses both B7-1 and B7-2 (Valenzuela and Effros, 2002). Cultures are initiated by combining PBMC cells with irradiated Nor 20 (1:1) in Yssel Medium, and 20 U/ml recombinant IL-2. The cell concentration is determined every few days and cultures are routinely split back to a concentration of 5×10^4 whenever their concentration exceeds 1×10^6. Cultures are repeatedly restimulated with the same APC each time they reach quiescence, which occurs approximately 3–4 weeks after the previous stimulation. This procedure is repeated until the T cells cease undergoing any proliferation in response to at least two rounds of restimulation (Perillo et al., 1989). At this point, the culture is considered to have reached senescence. Senescent cultures are metabolically active, and although the cells are unable to enter cell cycle in response to antigen or increasing doses of IL-2, they nevertheless respond to restimulation by upregulation of the IL-2R α chain (CD25) in an antigen-specific manner (Perillo et al., 1993), further confirming published reports on a variety of cell types indicating that senescence does not constitute a generalized loss of function (Perillo et al., 1989, 1993).

The above cell culture system unexpectedly provided a possible clue to the origin of the CD28– T cells observed in vivo. During the course of our T cell studies, we made the surprising discovery that when T cells reach the irreversible state of cell cycle arrest associated with replicative senescence, they no longer express CD28 costimulatory molecule (Effros et al., 1994a). The observation that CD28 expression is totally absent in T cells that reach replicative senescence in cell culture is particularly significant, in light of the continued undiminished expression of all other surface markers tested, including those reflecting cell lineage, adhesion pathways, activation, and antigen receptor (Effros, 1997). The loss of CD28 gene expression constitutes a fundamental change in the biology of senescent T cells, since CD28 mediates the essential costimulatory signal required for T cell activation and a variety of other functions, including adhesion, IL-2 gene transcription and enhancement of telomerase induction.

We and others have shown that replicative senescence in T cells is quite distinct from "anergy", since the inability to enter cell cycle cannot be reversed by re-exposure to antigen, treatment with phorbol esters and calcium ionophores, or stimulation with a combination of monoclonal antibodies (mAb) specific for CD3 and CD28, or IL-2 (Beverly et al., 1992; Effros, 1998). Importantly, replicative senescence is also distinct from cell death caused either by apoptosis or necrosis. On the contrary, human CD8+ T cells that reach replicative senescence in culture are actually resistant to apoptosis induction by a variety of stimuli that do induce robust apoptosis in early passage cultures from the same donor (Spaulding et al., 1999). Interestingly, T cells with the same phenotype (i.e. non-proliferative, CD8+ CD28−) isolated from human peripheral blood are also more resistant to apoptosis than the CD28+ T cells from the same donor (Posnett et al., 1999). Finally, T cell replicative senescence in culture is associated with increased production of the pro-inflammatory cytokines TNFα and IL-6 (unpublished observations, Spaulding and Effros), consistent with observations on elevated levels of these same cytokines in the serum of aged mice (Spaulding et al., 1997).

5. Telomere/telomerase dynamics

Telomeres are repetitive DNA sequences at the ends of eukaryotic chromosomes that are critical for genomic stability, protection of chromosome ends from exonucleotytic degradation and prevention of aberrant end-to-end fusion (Greider and Blackburn, 1996). During DNA replication, somatic cells lose telomeric DNA at a rate of ~ 40–100 bp per population doubling. Telomerase is a specialized reverse transcriptase that can extend telomeric repeats at the ends of linear chromosomes. In the absence of telomerase, telomeres shorten with each cell division (Harley et al., 1990). Telomerase activity has been detected in more than 90% of human tumors, and is believed to enable such cells to divide indefinitely (Shay and Wright, 1996). Conversely, telomerase is absent from most normal human somatic cells, leading to the hypothesis that telomere shortening is the ultimate cause of replicative senescence. The recent landmark study by Bodnar et al. (1998) demonstrating that gene transfer of the catalytic component of telomerase (hTERT) into normal fibroblasts and pigmented retinal epithelial cells allowed the cells to continue cell division indefinitely and avoid telomere shortening provides cogent experimental evidence in support of the telomere hypothesis of replicative senescence. The cells that harbour the new gene are karyotypically normal and show no signs of transformation (Greten et al., 1998; Jiang et al., 1999), although it should be noted that human epithelial cells that express hRAS and SV40 large T antigen in addition to hTERT are, in fact, tumorogenic (Hahn et al., 1999).

In T cells, the relationship of telomeres, telomerase and replicative senescence is more complex than in fibroblasts and epithelial cells. Lymphocytes differ from most other normal somatic cells in that, under certain circumstances, they exhibit telomerase activity levels that are as high as those observed in tumor cells (Bodnar et al., 1996). Telomerase activity has also been documented in developing T cells within the thymus and in lymphoid organs (Weng et al., 1997b). Following stimulation with mitogens or a combination of antibodies to CD3 and CD28, T cells become strongly positive for telomerase activity (Bodnar et al., 1996; Weng et al., 1997a). Interestingly, this activity seems to be enhanced by CD28 signalling, since inhibiting the TCR-mediated pathway lowers telomerase levels but does not totally abrogate the enzyme activity (Weng et al., 1997a). Conversely, telomerase activity is blocked when CD28 binding is inhibited, even in the presence of strong TCR signalling (Hathcock et al., 1998; Valenzuela and Effros, 2002).

Our own studies have recently documented a striking divergence in the telomerase kinetics between CD4 and CD8 T cells. Using the long-term culture model described above, our research showed that primary exposure of CD4 and CD8 T cell subsets to stimulation with APCs results in comparable kinetics and magnitude of telomerase up-regulation. Following APC-induced upregulation, telomerase declined to undetectable levels within approximately 3 weeks in both subsets. Repeated exposure to APC results in telomerase upregulation. However, by the fourth antigenic encounter, telomerase activity was reduced to background levels in the CD8 subset, whereas high levels of telomerase activity were maintained in the CD4 subset. In addition, our analysis showed that the decline in telomerase inducibility in antigen-specific human T cells parallels loss of CD28 expression (Valenzuela and Effros, 2002). Overall, our findings imply that telomerase functions to increase replicative potential of human T cells by stabilizing telomere length during initial rounds of proliferation associated with primary antigenic stimulation, but has progressively reduced effects in subsequent stimulations as a larger proportion of antigen-specific CD28− T cells accumulate.

6. Aging in vivo

The proportion of T cells lacking CD28 expression is dramatically increased in elderly persons. Effective co-stimulation is critical to the outcome of antigen recognition and signal transduction induced by the T cell receptor (TCR). The major T cell specific costimulatory molecule is CD28 (Lenschow et al., 1996), and this molecule is expressed on nearly 100% of human T cells at birth (Azuma et al., 1993). However, over the lifespan there is a progressive accumulation of memory CD8 T cells that are CD28-negative, with some elderly persons having more than 50% of their total CD8 T cell pool being CD28-negative (Boucher et al., 1998; Effros, 2000; Looney et al., 1999). This change in the composition of the T cell memory pool is of fundamental importance, since CD28 co-stimulation has unique and diverse effects on T cell activation. CD28 has been implicated in a multitude of critical T cell functions, including lipid raft formation, IL2 gene transcription, apoptosis, stabilization of cytokine mRNA, and cell adhesion

(Holdorf et al., 2000; Mueller, 2000; Sansom, 2000; Shimizu et al., 1992; Sozou and Kirkwood, 2001). Indeed, in discussing the CD28 molecule, it has been noted that "it is truly remarkable that so many functions may be mediated by a diminutive cytoplasmic tail of only 41 amino acids" (Rudd, 1996). Thus, the absence of CD28 alone, independent of other characteristics of these putatively senescent cells, will cause them to differ in profound ways from CD28+ T cells.

The presence of large proportions of CD28− T cells within the memory CD8 T cell compartment may also indirectly influence other memory T cells of unrelated specificities. It is now well-established that homeostatic mechanisms control the proportions of memory and naïve T cells, and that the CD4 and CD8 subsets may be independently regulated (Freitas and Rocha, 2000; Rocha et al., 1989). Thus, if 50% of the memory CD8 T cell pool consists of CD28− T cells, this reduces by one-half the available "space" for replication-competent, more functional CD28+ memory CD8 T cells. Moreover, CD28− T cells are often present as part of oligoclonal expansions (Posnett et al., 1994; Schwab et al., 1997), a feature that would cause a reduction in the overall spectrum of antigenic specificities within T cell pool in the elderly.

It cannot be overemphasized that the types of experimental models used to analyse the role of CD28 in other contexts, such as analysis of CD28 knockout mice (Suresh et al., 2001), or determining the outcome of withholding CD28 stimulation from Jurkat cells, are not relevant to the analysis of those human T cells whose permanent loss of CD28 expression occurs in response to repeated/ chronic antigenic stimulation. Indeed, CD28− T cells, whether they arise in cell culture or are isolated ex vivo, exhibit a constellation of additional characteristics which make them quite distinct from knockout cells that are modified exclusively in their CD28 gene expression. In particular, evidence from both cell culture studies as well as experiments on cells isolated ex vivo indicate that CD28− T cells are the progeny of cells that were originally CD28+ (Posnett et al., 1999), suggestive of an internally regulated mechanism that has led to suppression of CD28 gene expression.

7. Potential effect of CD8 replicative senescence on immune function

The putatively senescent CD8 T cells that are present in vivo have the potential to contribute to diminished immune function in a variety of ways. First, CD8+ CD28− T cells isolated ex vivo are unable to proliferate (like their cell culture counterparts), even in response to signals that bypass cell surface receptors, such as PMA and ionomycin (Effros et al., 1996). This observation is consistent with extensive research on replicative senescence in a variety of cell types documenting the irreversible nature of the proliferative block, and its association with upregulation of cell cycle inhibitors and p53-linked checkpoints (Campisi, 2001).

If the CD8+CD28− T cells present in elderly persons are virus-specific, as we propose, their inability to undergo the requisite clonal expansion in response to encounter with their antigen will compromise the immune control over that particular virus. Indeed, the emergence of latent infections, such as VZV (shingles) and EBV (some lymphomas), as well the reduced control over acute infection with repeatedly encountered viruses (e.g. influenza) are well-documented clinical findings in elderly persons (Effros, 2001). Second, since CD28 ligation enhances the binding affinity of T cells to endothelial cells (Shimizu et al., 1992), T cells lacking CD28 may be altered in their trafficking patterns between tissue and blood. Third, if the putatively senescent T cells present in vivo produce high levels of IL-6 and TNF-α like their in vitro counterparts, their presence in vivo may be contributing to the well-documented pro-inflammatory milieu present in many elderly persons. Indeed, enhanced inflammation is now believed to play a role in many of the diseases of aging that had not been previously considered immune-mediated pathologies (Duenwald, 2002). Fourth, the resistance of CD8 + CD28− T cells tested ex vivo to apoptosis (Posnett et al., 1999) leads to their persistence, which, in turn, affects that quality and composition of the total memory pool, as discussed above.

A final aspect of senescent T cells that could have broad physiological consequences relates to the role of stress in the aging process. T cells that undergo replicative senescence in culture show transcriptional down-regulation of the *hsp70* gene in response to heat shock (Effros et al., 1994b), and T cells from elderly persons show attenuation in the molecular chaperone system hsp70, in the steroid binding hsp90, and the chaperonin hsp60 (Rao et al., 1999). These immune cell changes may contribute to the well-documented reduction in ability to respond to stress that characterizes organismic aging (Muravchick, 1998).

8. Memory cells in humans

Formal demonstration that the process of replicative senescence is actually occurring in vivo during human aging represents one aspect of the more general, challenging problem of analysing overall memory lymphocyte dynamics in humans. Studies in mice have utilized transgenic T cells, congenic cell transfer systems and longitudinal analysis of a population of T cells resulting from a single immunization to extrapolate estimates on lifespans and dynamics of memory T cells (Rufer et al., 2001). However, it is doubtful that most of the cell dynamics estimates derived from mouse studies are necessarily applicable to humans, particularly because most humans, especially older ones, have encountered many more antigens than the average laboratory mouse, so that competition for space will be more intense (Rufer et al., 2001). In addition, the nature and duration of the stimulus and the concomitant responses to other antigens will also play a role. Moreover, a difference in telomere size, telomerase activity, and spontaneous transformation rates between mice and humans precludes the use of the murine model to study the specific process of replicative senescence (Akbar et al., 2000). Nonetheless, experiments in mice have demonstrated that even infection with a single virus induces multiple rounds of cell divisions yielding abundant numbers of CD8 T cells that are non-proliferative (Voehringer et al., 2001), reminiscent of findings in chronic HIV infection in humans (Effros et al., 1996).

A variety of approaches to analyse memory T cells in the human system have been employed, and although each of these approaches has its shortcomings, collectively they are yielding an increasingly comprehensive understanding of human memory T cell biology. For example, much of the data on the kinetics of human T cells has been derived from studies on how the T cell compartment is restored after it has been destroyed by disease or eradicated by therapy, a somewhat unphysiological situation involving massive expansion in a lymphopenic environment (Mackall et al., 1995; Walker et al., 1998). Even so, information on the differences in telomere kinetics between donor and recipient memory T cells derived from these studies underscores the complexity of the T cell pool. Longitudinal studies on humans over a period of years can also provide information on the lifespan and phenotypic characteristics of memory T cells, particularly in the context of a specific antigenic stimulus. For example, telomere analysis of tetramer-binding CD8 T cells isolated during acute and chronic EBV infection has provided novel insights into T cell turnover (Maini et al., 1999; Plunkett et al., 2001) but even this elegant system cannot ensure that the same population is being studied over time.

Our own approach to analysis of memory T cells focuses on the effect of repeated antigenic encounter on the dynamics and function of these cells. Like the other approaches, cell culture analysis has its confounding issues, such as the absence of the physiological milieu, but the strength of this model system is that it allows analysis of the same population of antigen-specific T cells over time. Significantly, the cell culture model we have developed has thus far been remarkably accurate in mirroring various T cell phenotypes present in vivo in situations of known chronic antigenic stimulation (Effros et al., 1996; Effros, 2000). It is only by the identification of additional phenotypic, functional and genetic markers as cells progress in cell culture to senescence that one can more adequately address the fundamental issue of the relevance of the replicative senescence process to the in vivo aging of the immune system.

9. Concluding remarks

The process of replicative senescence was originally proposed to be intrinsically linked to species lifespan, possibly even as the *cause* of organismic aging (Campisi, 2001). It is now clear that this view was simplistic, and that replicative senescence influences lifespan in other ways, such as by hindering certain physiological processes or hampering particular organ system function. Senescent human fibroblasts, for example, stimulate pre-malignant and malignant, but not normal, epithelial cells to proliferate in culture and to form tumors in mice, thereby potentially contributing to age-related cancers (Krtolica et al., 2001). In addition, limited proliferation of fibroblasts may affect wound healing (Campisi, 1998). In the case of T cells, it could be argued that the number of population doublings (PD) achievable by each T cell is so large that the finite replicative lifespan would not be biologically meaningful in vivo. Thus, an average lifespan of 35 PDs, which, if all daughter cells continue to grow unchecked, will result in over 10^{10} cells, seems at first glance to be more than sufficient (Effros and Pawelec, 1997). However, this may not be the case, because T-cell expansion occurs in waves of proliferation followed by apoptotic mechanisms to eliminate excess cells. Thus, for certain T cells, such as those responding to latent or repeatedly encountered viruses, it is theoretically possible that the finite proliferative limit may be reached, particularly by old age. The in vivo data on CD28 expression and telomere length suggests that this, in fact, may be the case for some memory CD8 T cells. Together with the associated functional alterations, such as suppressive influence on the activity of other CTL (Looney et al., 1999) and the altered cytokine patterns, the putatively senescent CD8 T cells have the potential to profoundly influence health and longevity of the elderly.

Acknowledgements

The research reviewed in this chapter was funded in part by NIH AG 10415, the UCLA Center on Aging, and the Plott Endowment. The author holds the Elizabeth and Thomas Plott Endowed Chair in Gerontology.

References

Akbar, A.N., Soares, M.V., Plunkett, F.J., Salmon, M., 2000. Differential regulation of CD8+ T cell senescence in mice and men. Mech. Ageing Dev. 121, 69–76.

Azuma, M., Phillips, J.H., Lanier, L.L., 1993. CD28– T lymphocytes: antigenic and functional properties. J. Immunol. 150, 1147–1159.

Beverly, B., Kang, S., Lenardo, M., Schwartz, R., 1992. Reversal of in vitro T cell clonal anergy by IL-2 stimulation. Int. Immunol. 4, 661–671.

Bodnar, A., Quellette, M., Frolkis, M., Holt, S.E., Chiu, C.P., Morin, G.M., Harley, C.B., Shay, J.W., Linsteiner, S., Wright, W.E., 1998. Extension of life-span by introduction of telomerase into normal human cells. Science 279, 349–352.

Bodnar, A.G., Kim, N.W., Effros, R.B., Chiu, C.P., 1996. Mechanism of telomerase induction during T cell activation. Exp. Cell Res. 228, 58–64.

Boucher, N., Defeu-Duchesne, T., Vicaut, E., Farge, D., Effros, R.B., Schachter, F., 1998. CD28 expression in T cell aging and human longevity. Exp. Gerontol. 33, 267–282.

Campisi, J., 1998. The role of cellular senescence in skin aging. J. Invest. Dermatol. Symp. Proc. 3, 1–5.

Campisi, J., 2001. From cells to organisms: can we learn about aging from cells in culture? Exp. Gerontol. 36, 607–618.

Duenwald, M., 2002. Body's Defender Goes on the Attack. NY Times, D1–D9.

Effros, R.B., 1997. Loss of CD28 expression on T lymphocytes: a marker of replicative senescence. Dev. Compar. Immunol. 21, 471–478.

Effros, R.B., 1998. Replicative senescence in the immune system: impact of the Hayflick Limit on T cell function in the elderly. Am. J. Hum. Gen. 62, 1003–1007.

Effros, R.B., 2000. Costimulatory mechanisms in the elderly. Vaccine 18, 1661–1665.

Effros, R.B., 2001. Immune system activity. In: Masoro, E., Austad, S. (Eds.), Handbook of the Biology of Aging. Academic Press, San Diego, pp. 324–350.

Effros, R.B., Boucher, N., Porter, V., Zhu, X., Spaulding, C., Walford, R.L., Kronenberg, M., Cohen, D., Schachter, F., 1994a. Decline in CD28+ T cells in centenarians and in long-term T cell cultures: a possible cause for both in vivo and in vitro immunosenescence. Exp. Gerontol. 29, 601–609.

Effros, R.B., Zhu, X., Walford, R.L., 1994b. Stress response of senescent T lymphocytes: reduced hsp70 is independent of the proliferative block. J. Gerontol. 49, B65–B70.

Effros, R.B., Allsopp, R., Chiu, C.P., Wang, L., Hirji, K., Harley, C.B., Villeponteau, B., West, M., Giorgi, J.V., 1996. Shortened telomeres in the expanded CD28–CD8+ subset in HIV disease implicate replicative senescence in HIV pathogenesis. AIDS/Fast Track 10, F17–F22.

Effros, R.B., Pawelec, G., 1997. Replicative senescence of T lymphocytes: does the Hayflick Limit lead to immune exhaustion? Immunol. Today 18, 450–454.

Freitas, A.A., Rocha, B., 2000. Population biology of lymphocytes: the flight for survival. Annu. Rev. Immunol. 18, 83–111.

Greider, C.W., Blackburn, E.H., 1996. Telomeres, telomerase and cancer. Sci. Am. 274, 92–97.

Greten, T.F., Slansky, J.E., Kubota, R., Soldan, S.S., Jaffee, E.M., Leist, T.P., Pardoll, D.M., Jacobson, S., Schneck, J.P., 1998. Direct visualization of antigen-specific T cells: HTLV-1 Tax11-19- specific CD8(+) T cells are activated in peripheral blood and accumulate in cerebrospinal fluid from HAM/TSP patients. Proc. Natl. Acad. Sci. USA 95, 7568–7573.

Hahn, W.C., Counter, C.M., Lundberg, A.S., Beijersbergen, R.L., Brooks, M.W., Weinberg, R.A., 1999. Creation of human tumour cells with defined genetic elements [see comments]. Int. Immunol. 400, 464–468.

Harley, C., Futcher, A.B., Greider, C., 1990. Telomeres shorten during ageing of human fibroblasts. Int. Immunol. 345, 458–460.

Hathcock, K.S., Weng, N.P., Merica, R., Jenkins, M.K., Hodes, R., 1998. Antigen-dependent regulation of telomerase activity in murine T cells. J. Immunol. 160, 5702–5706.

Hayflick, L., 1992. Aging, longevity, and immortality in vitro. Exp. Gerontol. 27, 363–368.

Holdorf, A.D., Kanagawa, O., Shaw, A.S., 2000. CD28 and T cell co-stimulation. Rev. Immunogenet. 2, 175–184.

Jiang, X.R., Jimenez, G., Chang, E., Frolkis, M., Kusler, B., Sage, M., Beeche, M., Bodnar, A.G., Wahl, G.M., Tlsty, T.D., Chiu, C.P., 1999. Telomerase expression in human somatic cells does not induce changes associated with a transformed phenotype. Nat. Genet. 21, 111–114.

Krtolica, A., Parrinello, S., Lockett, S., Desprez, P.Y., Campisi, J., 2001. Senescent fibroblasts promote epithelial cell growth and tumorigenesis: a link between cancer and aging. Proc. Natl. Acad. Sci. USA 98, 12072–12077.

Lenschow, D.J., Walunas, T.L., Bluestone, J.A., 1996. CD28/B7 system of T cell costimulation. Annu. Rev. Immunol. 14, 233–258.

Looney, R.J., Falsey, A., Campbell, D., Torres, A., Kolassa, J., Brower, C., McCann, R., Menegus, M., McCormick, K., Frampton, M., Hall, W., Abraham, G.N., 1999. Role of cytomegalovirus in the T cell changes seen in elderly individuals. Clin. Immunol. 90, 213–219.

Mackall, C.L., Fleisher, T.A., Brown, M.R., Andrich, M.P., Chen, C.C., Feuerstein, I.M., Horowitz, M. E., Magrath, I.T., Shad, A.T., Steinberg, S.M., 1995. Age, thymopoiesis, and CD4+ T-lymphocyte regeneration after intensive chemotherapy [see comments]. N. Engl. J. Med. 332, 143–149.

Maini, M.K., Soares, M.V., Zilch, C.F., Akbar, A.N., Beverley, P.C., 1999. Virus-induced CD8+ T cell clonal expansion is associated with telomerase up-regulation and telomere length preservation: a mechanism for rescue from replicative senescence. J. Immunol. 162, 4521–4526.

Mueller, D.L., 2000. T cells: a proliferation of costimulatory molecules. Curr. Biol. 10, R227–R230.

Muravchick, S., 1998. The aging process: anesthetic implications. Acta Anaesthesiol. Belgica 49, 85–90.

Perillo, N.L., Naeim, F., Walford, R.L., Effros, R.B., 1993. The in vitro senescence of human lymphocytes: failure to divide is not associated with a loss of cytolytic activity or memory T cell phenotype. Mech. Ageing Dev. 67, 173–185.

Perillo, N.L., Walford, R.L., Newman, M.A., Effros, R.B., 1989. Human T lymphocytes possess a limited in vitro lifespan. Exp. Gerontol. 24, 177–187.

Plunkett, F.J., Soares, M.V., Annels, N., Hislop, A., Ivory, K., Lowdell, M., Salmon, M., Rickinson, A., Akbar, A.N., 2001. The flow cytometric analysis of telomere length in antigen-specific CD8+ T cells during acute Epstein–Barr virus infection. Blood 97, 700–707.

Posnett, D.N., Edinger, J.W., Manavalan, J.S., Irwin, C., Marodon, G., 1999. Differentiation of human CD8 T cells: implications for in vivo persistence of CD8+ CD28- cytotoxic effector clones. Int. Immunol. 11, 229–241.

Posnett, D.N., Sinha, R., Kabak, S., Russo, C., 1994. Clonal populations of T cells in normal elderly humans: the T cell equivalent to "benign monoclonal gammopathy". J. Exp. Med. 179, 609–618.

Rao, D.V., Watson, K., Jones, G.L., 1999. Age-related attenuation in the expression of the major heat shock proteins in human peripheral lymphocytes. Mech. Ageing Dev. 107, 105–118.

Reiser, J.B., Darnault, C., Guimezanes, A., Gregoire, C., Mosser, T., Schmitt-Verhulst, A.M., Fontecilla-Camps, J.C., Malissen, B., Housset, D., Mazza, G., 2000. Crystal structure of a T cell receptor bound to an allogeneic MHC molecule. [Comment In: Nat. Immunol. 2000 Oct;1(4):277–8 UI, 21177378]. Nat. Immunol. 1, 291–297.

Rocha, B., Dautigny, N., Pereira, P., 1989. Peripheral T lymphocytes: expansion potential and homeostatic regulation of pool sizes and CD4/CD8 ratios in vivo. Eur. J. Immunol. 19, 905–911.

Rudd, C.E., 1996. Upstream–downstream, CD28 cosignaling pathways and T cell function. Immunity 4, 527–534.

Rufer, N., Helg, C., Chapuis, B., Roosnek, E., 2001. Human memory T cells: lessons from stem cell transplantation. Trends Immunol. 22, 136–141.

Sansom, D.M., 2000. CD28, CTLA-4 and their ligands: who does what and to whom? Immunology 101, 169–177.

Schwab, R., Szabo, P., Manavalan, J.S., Weksler, M.E., Posnett, D.N., Pannetier, C., Kourilsky, P., Even, J., 1997. Expanded CD4+ and CD8+ T cell clones in elderly humans. J. Immunol. 158, 4493–4499.

Shay, J., Wright, W., 1996. Telomerase activity in human cancer. Curr. Opin. Oncol. 8, 66–71.

Shimizu, Y., Van Seventer, G., Ennis, E., Newman, W., Horgan, K., Shaw, S., 1992. Crosslinking of the T cell-specific accessory molecules CD7 and CD28 modulates T cell adhesion. J. Exp. Med. 175, 577–582.

Smith, J.R., Pereira-Smith, O.M., 1996. Replicative senescence: implications for in vivo aging and tumor suppression. Science 273, 63–66.

Sozou, P.D., Kirkwood, T.B., 2001. A stochastic model of cell replicative senescence based on telomere shortening, oxidative stress, and somatic mutations in nuclear and mitochondrial DNA. J. Theor. Biol. 213, 573–586.

Spaulding, C.C., Walford, R.L., Effros, R.B., 1997. Elevated serum TNFα and IL-6 levels in old mice are normalized by caloric restriction. Mech. Ageing Dev. 93, 87–94.

Spaulding, C.S., Guo, W., Effros, R.B., 1999. Resistance to apoptosis in human CD8+ T cells that reach replicative senescence after multiple rounds of antigen-specific proliferation. Exp Gerontol. 34, 633–644

Suresh, M., Whitmire, J.K., Harrington, L.E., Larsen, C.P., Pearson, T.C., Altman, J.D., Ahmed, R., 2001. Role of CD28-B7 interactions in generation and maintenance of CD8 T cell memory. J. Immunol. 167, 5565–5573.

Valenzuela, H.F., Effros, R.B., 2002. Divergent telomerase and CD28 expression patterns in human CD4 and CD8 T cells following repeated encounters with the same antigenic stimulus, Clin. Immunol. 105, 117–125.

Voehringer, D., Blaser, C., Brawand, P., Raulet, D.H., Hanke, T., Pircher, H., 2001. Viral infections induce abundant numbers of senescent CD8 T cells. J. Immunol. 167, 4838–4843.

Walker, R.E., Carter, C.S., Muul, L., Natarajan, V., Herpin, B.R., Leitman, S.F., Klein, H.G., Mullen, C.A., Metcalf, J.A., Baseler, M., Falloon, J., Davey, R.T. Jr., Kovacs, J.A., Polis, M.A., Masur, H., Blaese, R.M., Lane, H.C., 1998. Peripheral expansion of pre-existing mature T cells is an important means of CD4+ T-cell regeneration HIV-infected adults. Nat. Med. 4, 852–856.

Weng, N.P., Palmer, L.D., Levine, B.L., Lane, H.C., June, C.H., Hodes, R.J., 1997a. Tales of tails, regulation of telomere length and telomerase activity during lymphocyte development, differentiation, activation, and aging. Immunol. Rev. 160, 43–54.

Weng, N.P., Levine, B.L., June, C.H., Hodes, R.J., 1997b. Regulation of telomerase RNA template expression in human T lymphocyte development and activation. J. Immunol. 158, 3215–3220.

Advances in
Cell Aging and
Gerontology

Cultured T cell clones as models for immunosenescence

Graham Pawelec

Center for Medical Research, University of Tübingen, ZMF, Waldhörnlestr 22, D-72072 Tübingen, Germany
Correspondence address: Tel.: +49-7071-2982805; fax: +49-7071-295567.
E-mail: graham.pawelec@uni-tuebingen.de

Contents

Abbreviations

CD: cluster of differentiation; CE: cloning efficiency; (C) PD: (cumulative) population doublings; ELISA: enzyme-linked immunosorbent assay; IL: interleukin; MFI: mean fluorescence intensity; MHC: major histocompatibility complex; TCC: T cell clone; TCR: T cell receptor; TNF-α: tumor necrosis factor-α.

1. Introduction

The elderly suffer from an increased susceptibility to infectious disease and to certain types of cancer (Remarque and Pawelec, 1998; Falsey, 2000). It is hypothesised that ageing of the immune system, particularly the T cell compartment contributes to this state of affairs due to immunosenescence (Adibzadeh et al., 1996). We further hypothesise that because repeated intermittent or chronic antigen exposure may lead to T cell clonal exhaustion (reflecting the inability of T cell clones to divide indefinitely), this process plays a part in the compromised immunity of the elderly,

Advances in Cell Aging and Gerontology, vol. 13, 295–307

who have accumulated a lifetime's exposure to infectious agents, autoantigens and cancer antigens (Effros and Pawelec, 1997). Therefore, this type of clonal exhaustion caused by antigen-stimulated proliferative stress, coupled with decreased or negligible thymic output and ageing of naive cells, may contribute to age-associated immuno-deficiency. Defining age-associated alterations in T cell immunity will allow identifica-tion of mechanisms responsible for dysregulated immunity in these conditions, and facilitate development of strategies to combat such changes (Pawelec et al., 2002b).

Evidence is beginning to accumulate that immunosenescence does indeed influence disease susceptibility and outcome, but diagnostic methods for assessing this state are imperfect. It has therefore been very difficult to make a definitive statement on the relevance of immunosenescence to clinical outcome. Moreover, immunogerontological studies have been rendered difficult by the constant problem of distinguishing between the effects of underlying disease and the effects of ageing. It is not even clear whether this is possible, despite the sporadic employment of the SENIEUR protocol over the last 20 years for the purpose of identifying "perfectly" healthy donors and comparing them with "average" people (Ligthart et al., 1984). A basic problem remaining to be investigated in this context is ageing of the immune system reflected in replicative senescence and biomarkers thereof, and immunosenes-cence caused by other, additional factors. The former would be expected to occur under conditions of chronic antigenic stimulation, i.e. pathological conditions not directly associated with chronological age, rendering a distinction between age and disease impossible; whereas the latter would be independent of replicative senes-cence. This implies that we need to identify biomarkers of replicative senescence in T cells, as well as markers of ageing not associated with the degree of previous cell division, should these exist. To this end, subsets of T cells from donors of different ages and conditions need to be investigated for putative markers of proliferation (thus far, few are available, and not definitively proven, i.e. T cell receptor excision circle assessment [which is suitable only for assessing a small number of divisions], telomere lengths [confounded by telomerase expression in T cells], residual prolif-erative capacity [demanding to standardise]) compared to the other markers dis-cussed here. Factors correlating with proliferative history are needed. These would allow a distinction to be made between the two major sets of causes of ageing in the immune system: namely, either dependent on or independent of replicative senescence.

A number of cross-sectional studies over the years has suggested that maintenance of an intact immune system, particularly the T cell component, is associated with successful ageing (Roberts-Thomson et al., 1974; Murasko et al., 1988; Wayne et al., 1990; Franceschi et al., 1995). The proportion of the oldest old (> 85) who remain in good health and with well-functioning immune systems is higher than that of normal elderly individuals of around 70 years, the majority of whom show immune dysfunc-tion (and the majority of whom die before becoming very much older). As long-itudinal studies are now beginning to show more clearly (Wikby et al., 1994, 1998; Ferguson et al., 1995), a major predictive factor for longevity and clinical good performance in the very old is indeed a well-functioning immune system, which in addition, is not obsessed with recognition of very limited numbers of certain viral epitopes, and with a correspondingly decreased antigen-recognition repertoire

(Looney et al., 1999; Olsson et al., 2000; Wikby et al., 2002). There is thus reason to believe that: (1) identification of risk factors for ageing in the T cell system would be diagnostically useful in predicting which elderly persons would be most at risk from immune-influenced mortality, and taking suitable precautions using existing techniques (e.g. vaccination, supplementation); (2) that manipulating the immune system to normalise function and delay or abrogate immunosenescence using novel technologies might well prolong the healthy and productive life of the individual.

The ability to culture human T cells for extended periods provided that intermittent stimulation and a source of growth factors is present enables model longitudinal and intervention studies to be established in vitro (Adibzadeh et al., 1995). Additionally, confounding factors such as changing proportions of different cell subsets within the overall population can be relatively easily avoided by working with monoclonal lines. The lifespan of different human T clones (TCC) is very variable, but always finite. Biological parameters can be constantly monitored over this entire finite lifespan by sampling cells from these in vitro cultures. Interventions to delay or prevent replicative senescence can be conveniently screened. Thus, cultured TCC represent useful models for that component of immunosenescence which is associated with proliferative stress-induced replicative senescence of T cells (Pawelec et al., 1997). This final chapter reviews our experience with human CD4$^+$ T cells in this context.

2. Longevity of T cell clones

The techniques for generating and maintaining human T cell clones have been in existence, essentially unchanged, for more than 20 years now (Bach et al., 1979; Pawelec and Wernet, 1980). Although T cells can be maintained for extended periods in tissue culture, as was soon noted, they were not immortal in vitro and had finite, often relatively short, lifespans (see Duquesnoy and Zeevi, 1983; Pawelec et al., 1983; Effros and Walford, 1984; Effros and Walford, 1987; Effros et al., 1994; Grubeck-Loebenstein et al., 1994; Mariani et al., 1990; McCarron et al., 1987; Perillo et al., 1989; Perillo et al., 1993). Despite the enormous numbers of studies over the years, it is remarkable that, unlike in the mouse, no single case of spontaneous in vitro immortalisation has ever been convincingly documented. On the other hand, properly quantitative studies on clonal longevities of cultured human T cells remain extremely rare. Most earlier data come from work done with uncloned, bulk T cell cultures, and relatively few studies actually quantified growth capacity in terms of population doublings (PD). Therefore, questions such as the relationship between T cell cloning capacity and age and health status of the donor have been difficult or impossible to answer. McCarron et al. (1987) reported T cell longevities of uncloned cultures ranging from 62 to 172 PD, as calculated from their figures of cell expansion factors. They observed no differences in longevity of T cell cultures derived from neonatal, young adult or old adult donors. However, as these lines were not cloned, the final longevity of the cultures was determined by the longest-lived clones. Only cloning experiments can reveal the average lifespans of all the cells in a population and not merely the longest-lived, as well as the variation in

Table 1
Longevities of human T cell clones under standard culture conditions

Origin	CE%	n, clones/ expts	Percentage of clones reaching population size			Longest- lived (PD)
			20 PD (10^6)	30 PD (10^9)	40 PD (10^{12})	
CD34+ (peripheral)	55	533/6	31	17	6	60
CD34+ (cord)	43	94/2	29	15	Nyt	57
CD3+ (young)	47	1355/15	47	24	15	170
CD3+ (old SENIEUR)	52	116/2	55	22	16	72
CD3+ (old OCTO)	33	32/3	48	28	19	61
CD3+ (CML)	49	35/1	60	35	14	15

CE, cloning efficiency (calculated from percentage of wells positive in cloning plates). Longevity is expressed as a percentage of established clones (i.e. those counted as positive in calculating the CE) which survive to 20, 30 or 40 PD. Origins: CD34+, positively-selected hematopoietic stem cells from peripheral or cord blood; CD3+, normal peripheral T cells; young, apparently healthy donors under 30 years; old SENIEUR, donors over 90 years, in excellent health according to the criteria of the SENIEUR protocol; old OCTO, donors over 80 years participating in the longitudinal OCTO study, selected only by survival; CML, a middle-aged donor with chronic myelogenous leukemia in chronic phase treated with interferon-α.

longevity from clone to clone. This is likely to be more physiologically relevant than simply establishing the longevity of one rare cell present in the original population. Sequential loss of T cells due to different longevities would parallel shrinkage of the T cell repertoire in vivo, a phenomenon which is observed to occur, at least in CD8$^+$ T cells. McCarron et al. (1987) also examined longevity of a small number of T cell clones in their seminal paper. Their findings for TCC were rather different from their results in uncloned populations: those derived from neonates averaged 52 PD, those derived from young adults (20–30 years) managed 40 PD, but those from the elderly (70–90 years) only 32 PD. These appear to be the only data of this type available in the literature, apart from our own. Our results have been summarised and updated previously (Pawelec et al., 2000, 2002a). Here, we again update and extend the "life table" to summarise findings so far on the longevities of human T cell clones derived from various different sources (Table 1).

2.1. T cell clones from young healthy donors

Inspection of the data in Table 1 makes it immediately apparent that although individual T cell clones might show very varied longevities, the overall patterns for T cells of quite different origins are remarkably similar. First, let us consider the bulk of the data on cloning of peripheral T cells from young healthy donors, the usual source of material for this type of experiment. These cells can be cloned with a high efficiency (cloning efficiency, CE, nearly 50%). Essentially all of the T cell clones (TCC) obtained are CD4+ not CD8+, suggesting that these cloning conditions are favourable for T cell outgrowth, but only of CD4+ cells. As the original cells to be cloned consisted of a mixture of CD4 and CD8 cells in these experiments, the CE of 50% therefore implies that almost all CD4+ cells are initially capable of forming clones (defined as generation of ca. 1000 cells from one initial cell).

We can therefore conclude that the CD4 T cell repertoire is present more or less in its entirety in the starting clonal populations under these conditions. However, after 20 PD, when the clone size has increased from 10^3 to 10^6, about half of the clones have already been lost (mean of 53% for cloning of young donor CD3 cells, Table 1). At 30 PD, another half have been lost so that only one quarter of the originally clonable cells is still present. However, at 40 PD (which theoretically represents a really large clone size of 10^{12} cells), although more clones have been lost, 15% of the original starting clonal population remains. Finally, the longest-lived clone in all of these experiments achieved 170 PD, exceeding the number of PD often accepted as indicating immortalisation (Tang et al., 2001; Mathon et al., 2001). However, it is important to note that this clone was in fact not immortalised, and still retained a finite lifespan, dying at this point. These results indicate constant attrition of the T cell population at the clonal level, but with retention of perhaps up to 5% of the original CD4 repertoire up to 40 PD and with retention of very rare clones for considerably longer.

2.2. T cell clones from other sources

How do these longevity data compare with those on T cell clones derived from the old, the young, the sick? Our starting hypothesis was that T cells generated in situ from CD34 + hematopoietic progenitors would have greater longevities due to their lack of a previous proliferative history (as T cells at least). Reciprocally, we anticipated that CE of cells from the elderly would be reduced as would their clonal expansion capacity, and that this would be more prevalent in less healthy donors than in those rigorously selected for good health. Finally, we predicted that T cells from situations of chronic in vivo antigenic stress, such as in cancer, would also have lower CE and longevities for the same reasons. As can be seen from Table 1, none of these expectations was fulfilled.

First, cloning efficiencies. It is not the case that CE are lower in the elderly than in the young, and neither do they depend on the state of health of the elderly (although there may be a tendency towards slightly lower CE in the donors from the Swedish study, who were not specifically selected for good health, Table 1). They are equally high when the origin of the cells to be cloned is from a situation of antigenic stress (here, CML). They are also similar in the extra-thymic T cell differentiation cultures employed. Bearing in mind that the system selected for CD4 cells, and that the CE are so high that we cannot be dealing with rare contaminants, it has to be concluded that CE of the T cells from these markedly different sources are nonetheless not markedly different. So are their longevities different? Old cells with a proliferative history in vivo might manage the ca. 10 PD needed to be counted as a clone, but not the 20, 30 or more as an established TCC. However, it is clear from the collected data in Table 1 that this is not the case either. Very similar proportions of TCC achieve 20, 30 and 40 PD regardless of their origin. In addition, the maximum longevity of the longest-lived clone is also similar and very close to the "Hayflick Limit" of 50–60 PD. Exceptional clones, such as that from the young donor which achieved 170 PD, have not been seen yet, but this may merely be due to the smaller number of clones studied. Overall, it seems that CE and longevities of T cells in culture are essentially identical regardless of whether the cells to be cloned are derived from progenitors, young or old donors, healthy or sick.

How should we interpret these results? The simplest interpretation is that CD4 cells do in fact behave similarly no matter what their origin. Thus, if in vivo selection allows only fully functional CD4+ T cells to continue to exist (i.e. those capable of the clonal expansion required for successful immune responses), these results could be explained. In this respect, there may be a major difference between CD4+ and CD8+ T cells. Clonal expansions of T cells are easily detected in the elderly, or even middle-aged, but these are essentially limited to the CD8 subset (Posnett et al., 1994), and consist predominantly of dysfunctional cells carrying receptors for latent viruses, especially CMV (Khan et al., 2002; Ouyang et al., manuscript submitted for publication). Thus, using tetramer technology, we have observed excessive numbers of CMV-specific CD8 cells which, on sorting and stimulating with specific CMV antigen, are found to contain a much smaller fraction of functional cells than those of the same specificity sorted from the young. They lack CD28 and express a negative regulatory receptor associated with an inability to divide. It has been suggested that cells such as these accumulate because of a defect in the apoptotic pathways which would normally ensure their removal from the system. Certainly, resistance to apoptosis has been demonstrated in this type of cell (Posnett et al., 1999; Spaulding et al., 1999; Globerson and Effros, 2000). They may therefore be filling the "immunological space" with dysfunctional cells, and their removal, analogous to the situation in a mouse model (Zhou et al., 1995; Hsu et al., 2001) might "rejuvenate" the system. CD4+ cells, on the other, may become more, not less, resistant to apoptosis as they age (Pawelec et al., 1996) and this may explain the difficulty in demonstrating the kind of clonal expansions seen in CD8 cells. It may also mean that the CD4 cell selection process is more effective and leads to a retention only of properly functional cells in the elderly, which can still be cloned with the same efficiency and have the same longevity as those obtained from young donors.

3. Changes in surface phenotype during culture

Notwithstanding the above considerations, the in vitro clonal culture model is informative for longitudinal studies of age-related changes to CD4 cells under chronic antigenic stress. Because of the constraints of the cloning procedure, "young" cells are already at least ca. 22, 23 PD, but at this stage, we still have a good representation of the starting repertoire (see discussion above). However, those clones that can be studied over an extended time do of course represent only the small fraction of maybe 10% which are capable of that degree of longevity. One of the simplest analytical techniques to study these cells over time is to use flow cytometry with monoclonal antibodies to molecules expressed on the cell surface. We have examined a large range of different surface markers and found relatively few which change with age (i.e. with increasing PD). One of the most reproducible

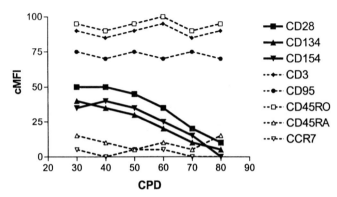

Fig. 1. Representative surface staining pattern on a CD4+ TCC. Cloned cells were cryopreserved 7 days after subculturing over the lifespan of the clone, then thawed at different ages (CPD) and tested at the same time in immunofluorescence with antibodies detecting the markers shown. The density of surface expression is given as corrected mean fluorescence intensity (cMFI) which takes the background staining and autofluorescence of the cells into account. This common pattern of age-associated surface molecule alterations includes decreasing levels of CD28, CD134 and CD154, with stable levels of CD3 (marking the TCR), CD45RO and CD95, together with maintained low or negative staining with CD45RA and the CCR7 chemokine receptor.

remains CD28. Because the study populations are monoclonal, results are not expressed as percent of cells which are positive, but as the density of the molecule at the surface. Although we have performed more accurate quantification experiments with beads it is usually sufficient to present results as median fluorescence values after background correction. This does not provide a readout of the actual number of molecules bound by antibody, but within an experiment, different populations can be directly compared for expression level because the fluorescence intensity correlates with the number of binding sites. The results so far can be summarised as follows: although different clones show different expression patterns, again there are no obvious differences depending on the origin of the cells to be cloned. The most common pattern of age-associated alterations is shown representatively in Fig. 1, although this is by no means universal. Age-associated alterations most frequently observed in these TCC involve a reduction of the level of expression of the costimulatory receptor CD28, as well as the putative costimulatory receptors CD134 and CD154, whereas despite the increased susceptibility to apoptosis, CD95 remains constant. The cells have a memory effector phenotype (CD45RA-negative, CCR7-negative, CD45RO$^+$) as would be expected from chronically-stimulated effector cells. The level of the TCR also remains stable, suggesting that these cells may retain the ability to recognise and respond to antigen. This typical pattern is occasionally not observed; most commonly expression of CD134 and/or CD154 is retained with age. However, the most consistent pattern is CD28 reduction

CD28 re-expression and autocrine TNF-α production

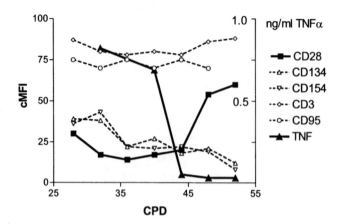

Fig. 2. CD28 re-expression by a CD4+ TCC. Rare clones such as this one (clone 401–2) recover CD28 expression at higher CPD (left-hand scale). This may correlate with decreasing autocrine TNF-α production (right hand scale) measured in ELISA as ng/ml of TNF-α produced per million cells per 24 h period per ml culture medium after mitogenic stimulation.

which is essentially universal in the ageing TCC. Nonetheless, in certain clones, although they show this age-associated decrease at first, CD28 re-expression can occur later (Fig. 2). We have correlated this CD28 re-expression with a decreased ability of the TCC to secrete TNF-α, at least in certain clones (Fig. 2). This is consistent with the observation that TNF-α downregulates CD28 expression (Bryl et al., 2001) and with out earlier observations that TNF-α can directly inhibit some TCC (Pawelec et al., 1989).

4. Changes in function during culture

Major functional changes follow from the decreased level of expression of costimulatory receptors, CD28 as defined here, and in all likelihood, others yet to be investigated. The most dramatic changes are in the patterns of cytokines secreted, commonly resulting in decreased levels of IL 2 and increased levels of IL 10 (Pawelec et al., 1997). Because the expression of the TCR is maintained and antigen-specific signalling still occurs, it is likely that these differences are caused by differences in the delivery of costimuli to the T cells, ever-increasing numbers of which are now being defined (Table 2). The balance of these, and their function, will determine the consequences of TCR ligation, apparently more so more than this antigen-specific "signal one" itself (Pawelec et al., 1997). Not very much is known about the regulatory mechanisms for CD28 expression, let alone for the expression of these other positive and negative costimulatory receptors. This is likely to be an area of intensive investigation by immunogerontologists over the next few years.

Table 2
CD4 T cell costimulation – hypothetical scheme with minimum number of ligand/receptor pairs

Ligand	Receptor	Signal	Main effect
MHC + antigen	TCR	1	Anergy/apoptosis
CD80	CD28	2.1	Activation (IL 2, IL 4)
CD86	CD28	2.2	Activation (IL 2, IL 4)
CD80/86	CD152	−2.1	De-activation ($\downarrow$ IL 2, IL 4)
B7RP1	ICOS	2.3	Activation (IL 4, IL 10, TNF-α)
PDL1, PDL2	PD-1	−2.2	De-activation ($\downarrow$IL 2, IL 10, IFN-γ)
B7H3	?	2.4?	Activation (IFN-γ)
CD40	CD154	2.5?	Activation (?)
OX-40L	CD134	2.6?	Activation (?)
4-1BB ligand	CD137	2.7?	Activation (?)
CD70	CD27	2.8?	Activation (?)

Signalling via the TCR provides an obligatory signal 1; costimulation via summation of positive and negative costimulatory signals as shown determines the final outcome of antigen contact (activation, anergy or apoptosis).

5. Relevance in vivo

As mentioned in Section 2 above, should clonal longevities in vivo parallel the in vitro model, the consequence would be repertoire shrinkage. This is because although newly generated T cells leave the thymus throughout life, due to thymic involution, by the far the greatest fraction of T cells in the adult has been generated prior to puberty (Jamieson et al., 1999). Because of this age-associated thymic involution, the output of new T cells in adulthood and old-age is severely limited under natural conditions. Even under the non-natural stress of peripheral T cell reconstitution after, say, chemotherapy, the majority of T cells is derived from pre-existing cells, rather than recent thymic emigrants (Mackall et al., 1995). Therefore, the concept has arisen that most of the T cell repertoire in the elderly represents what was already present in early life. As discussed above, particularly in that major subset of T cells which recognises endogenous antigen together with MHC class I molecules, and which are specialised to lyse virus-infected cells (CD8 + cytotoxic lymphocytes, CTL) one finds clonal expansions of virus-recognising cells in the elderly (Posnett et al., 1994). In the other major T cell compartment, the CD4+, MHC class II-restricted T helper (Th) cells, such clonal expansions are far less marked but can be detected in both experimental animals and humans (Grunewald and Wigzell, 1996; Mosley et al., 1998). This scenario implies that given a limited number of T cells at the beginning of life, and inadequate replenishment during ageing, a situation of age-associated "immunodeficiency" may eventually arise, due to the clonal exhaustion of the relevant antigen-specific T cells. Such a hypothesis is based on the knowledge that the number of T cells activated by antigenic challenge is generally small and that a successful immune response demands an extensive and dramatic increase in the numbers of cells carrying that antigenic specificity by clonal expansion. When the source of antigen is destroyed, the excess lymphocytes are destroyed too, and on antigen re-challenge, the process of clonal expansion must be repeated. Because normal somatic cells possess a finite capacity for cell division, "clonal exhaustion" caused by repetitive antigenic challenge (e.g. in infection) may occur. Recent evidence regarding CD8 cells in infectious mononucleosis has resulted in a direct estimate of the number of PDs required for successful control of the infection (Maini et al., 1999a,b). This number (28 PD) agrees very well with previous theoretical estimates (Effros and Pawelec, 1997) and suggests that clonal exhaustion may play a real and significant role even in primary infections, at least in CD8 cells. It is unclear what effects are seen in CD4 cells.

6. Intervention to extend longevity and restore function

There is still a great deal of argument about the impact of culture conditions on cellular "replicative senescence" (usually studied with fibroblasts). TCC are clearly very susceptible to alterations in culture conditions, perhaps helping to account for the difficulties many labs experience in trying to maintain them for extended periods. We are in the process of optimising culture conditions for TCC and of analysing biomarkers of culture ageing in these cells. One of the most likely reasons for cessation of growth may have to do with increasing levels of oxidative DNA damage sustained by the cells under standard culture conditions, i.e. in air, with 20% oxygen, triggering growth arrest and apoptosis (Hyland et al., 2001). Reducing oxygen tension may be a sensible "remediation" approach, which is currently being explored. This, together with the use of anti-oxidants and other agents, as well as improved culture vessels and nutrition-delivery technology, may improve culture conditions sufficiently to observe thus far indistinguishable differences in behaviour of T cells from different sources in these experiments. The European Union-supported collaborative network "Immunology and Ageing in Europe, ImAginE" and the focussed project "T cells in Ageing, T-CIA" are presently attempting to resolve these questions.

7. Conclusions

Much evidence points to infectious disease as the major cause of morbidity and mortality in the elderly and it seems very likely that immune dysfunction contributes to this state of affairs. Although innate and humoural immunity are relatively less affected by ageing, it seems that the T cell compartment shows marked age-associated alterations, both in subset composition and function of the cells within subsets. Longitudinal ex vivo studies suggest that immune parameters which are predominantly T cell-related can be clustered to yield an immune risk phenotype (IRP) predictive of mortality in the very elderly. At least some of these changes in T cells, those to do with chronic antigenic stimulation, can be modelled in clonal cultures in vitro. These cultures represent valuable tools for studying an important aspect of T cell immunosenescence. Interventions to selectively target changes identified in cultured T cells may be applicable in vivo to improve health and quality of life of the elderly, and reduce health care costs.

Acknowledgements

The experimental work described here was supported by grants from the DFG (Pa 361/7-1), and the VERUM Foundation, and was performed under the aegis of EU projects coordinated by GP [Concerted Action EUCAMBIS (BMHI-CT1994-1209, 1995–1998), Thematic Network ImAginE (QLK6-CT1999-02031, 2000–2003) and the RTD project T-CIA (QLK6-CT2002-02283, 2002–2005)].

References

Adibzadeh, M., Pohla, H., Rehbein, A., Pawelec, G., 1995. Long-term culture of monoclonal human T lymphocytes: models for immunosenescence? Mech. Ageing Dev. 83, 171–183.

Adibzadeh, M., Mariani, E., Bartoloni, C., Beckman, I., Ligthart, G., Remarque, E., Shall, S., Solana, R., Taylor, G.M., Barnett, Y., Pawelec, G., 1996. Lifespans of T lymphocytes. Mech. Ageing Dev. 91, 145–154.

Bach, F.H., Inouye, H., Hank, J.A., Alter, B.J., 1979. Human T-lymphocyte clones reactive in primed lymphocyte typing and cytotoxicity. Nature 281, 307–309.

Bryl, E., Vallejo, A.N., Weyand, C.M., Goronzy, J.J., 2001. Down-regulation of CD28 expression by TNF-alpha. J. Immunol. 167, 3231–3238.

Duquesnoy, R., Zeevi, A., 1983. Immunogenetic analysis of the HLA complex with alloreactive T cell clones. Hum. Immunol. 8, 17–23.

Effros, R.B., Pawelec, G., 1997. Replicative senescence of T lymphocytes: does the Hayflick Limit lead to immune exhaustion? Immunol. Today 18, 450–454.

Effros, R.B., Walford, R.L., 1984. T cell cultures and the Hayflick limit. Hum. Immunol. 9, 49–65.

Effros, R.B., Walford, R.L., 1987. Neonatal T cells as a model system to study the possible *In vitro* senescence of lymphocytes. Exp Gerontol 22, 307–316.

Effros, R.B., Zhu, X.M., Walford, R.L., 1994. Stress Response of Senescent T Lymphocytes – Reduced Hsp70 is Independent of the Proliferative Block. J Gerontol 49, B65–B70.

Falsey, A.R., 2000. Epidemiology of infectious disease. In: Grimly Evans, J., Williams, T.F., Beattie, B.L., Michel, J.-P., Wilcock, G.K. (Eds.), Oxford Textbook of Geriatric Medicine, 2nd ed. OUP, Oxford, pp. 55–64.

Ferguson, F.G., Wikby, A., Maxson, P., Olsson, J., Johansson, B., 1995. Immune parameters in a longitudinal study of a very old population of Swedish people: a comparison between survivors and nonsurvivors. J. Gerontol. Ser. A-Biol. Sci. Med. 50, B378–B382.

Franceschi, C., Monti, D., Sansoni, P., Cossarizza, A., 1995. The immunology of exceptional individuals: the lesson of centenarians. Immunol. Today 16, 12–16.

Globerson, A., Effros, R.B., 2000. Ageing of lymphocytes and lymphocytes in the aged. Immunol. Today 21, 515–521.

Grubeck-Loebenstein, B., Lechner, H., Trieb, K., 1994. Long-term in vitro growth of human T cell clones: can postmitotic 'senescent' cell populations be defined? Int. Arch. Allergy Immunol. 104, 232–239.

Grunewald, J., Wigzell, H., 1996. T-cell expansions in healthy individuals. Immunologist 4, 99–103.

Hsu, H.C., Shi, J., Yang, P., Xu, X., Dodd, C., Matsuki, Y., Zhang, H.G., Mountz, J.D., 2001. Activated CD8(+) T cells from aged mice exhibit decreased activation-induced cell death. Mech. Ageing Dev. 122, 1663–1684.

Hyland, P., Barnett, C., Pawelec, G., Barnett, Y., 2001. Age-related accumulation of oxidative DNA damage and alterations in levels of p16(INK4a/CDKN2a), p21(WAF1/CIP1/SDI1) and p27(KIP1) in human CD4+ T cell clones in vitro. Mech. Ageing Dev. 122, 1151–1167.

Jamieson, B.D., Douek, D.C., Killian, S., Hultin, L.E., Scripture-Adams, D.D., Giorgi, J.V., Marelli, D., Koup, R.A., Zack, J.A., 1999. Generation of functional thymocytes in the human adult. Immunity 10, 569–575.

Khan, N., Shariff, N., Cobbold, M., Bruton, R., Ainsworth, J.A., Sinclair, A.J., Nayak, L., Moss, P.A., 2002. Cytomegalovirus seropositivity drives the CD8 T cell repertoire toward greater clonality in healthy elderly individuals. J. Immunol. 169, 1984–1992.

Ligthart, G.J., Corberand, J.X., Fournier, C., Galanaud, P., Hijmans, W., Kennes, B., Müller-Hermelink, H.K., Steinmann, G.G., 1984. Admission criteria for immunogerontological studies in man: the SENIEUR protocol. Mech. Ageing Dev. 28, 47–55.

Looney, R.J., Falsey, A., Campbell, D., Torres, A., Kolassa, J., Brower, C., McCann, R., Menegus, N., McCormick, K., Frampton, M., Hall, W., Abraham, G.N., 1999. Role of cytomegalovirus in the T cell changes seen in elderly individuals. Clin. Immunol. 90, 213–219.

Mackall, C.L., Fleisher, T.A., Brown, M.R., Andrich, M.P., Chen, C.C., Feuerstein, I.M., Horowitz, M. E., Magrath, I.T., Shad, A.T., Steinberg, S.M., Wexler, L.H., Gress, R.E., 1995. Age, thymopoiesis, and CD4+ T-lymphocyte regeneration after intensive chemotherapy. N. Engl. J. Med. 332, 143–149.

Maini, M.K., Casorati, G., Dellabona, P., Wack, A., Beverley, P.C.L., 1999a. T-cell clonality in immune responses. Immunol. Today 20, 262–266.

Maini, M.K., Soares, M.V.D., Zilch, C.F., Akbar, A.N., Beverley, P.C.L., 1999b. Virus-induced CD8(+) T cell clonal expansion is associated with telomerase up-regulation and telomere length preservation: a mechanism for rescue from replicative senescence. J. Immunol. 162, 4521–4526.

Mariani, E., Roda, P., Mariani, A.R., Vitale, M., Degrassi, A., Papa, S., Faccini, A., 1990. Age-associated changes in CD8+ and CD16+ cell reactivity: clonal analysis. Clin. Exp. Immunol. 81, 479–484.

Mathon, N.F., Malcolm, D.S., Harrisingh, M.C., Cheng, L.L., Lloyd, A.C., 2001. Lack of replicative senescence in normal rodent glia. Science 291, 872–875.

McCarron, M., Osborne, Y., Story, C., Dempsey, J.L., Turner, R., Morley, A., 1987. Effect of age on lymphocyte proliferation. Mech. Ageing Dev. 41, 211–218.

Mosley, R.L., Koker, M.M., Miller, R.A., 1998. Idiosyncratic alterations of TCR size distributions affecting both CD4 and CD8 T cell subsets in aging mice. Cell. Immunol. 189, 10–18.

Murasko, D.M., Weiner, P., Kaye, D., 1988. Association of lack of mitogen induced lymphocyte proliferation with increased mortality in the elderly. Aging: Immunol. Infect. Dis. 1, 1–23.

Olsson, J., Wikby, A., Johansson, B., Lofgren, S., Nilsson, B.O., Ferguson, F.G., 2000. Age-related change in peripheral blood T-lymphocyte subpopulations and cytomegalovirus infection in the very old: the Swedish longitudinal OCTO immune study. Mech. Ageing Dev. 121, 187–201.

Pawelec, G., Wernet, P., 1980. Restimulation properties of human alloreactive cloned T cell lines Dissection of HLA-D-region alleles in population studies and in family segregation analysis. Immunogenetics 11, 507–519.

Pawelec, G., Schneider, E.M., Wernet, P., 1983. Human T-cell clones with multiple and changing functions: indications of unexpected flexibility in immune response networks. Immunol. Today 4, 275–278.

Pawelec, G.P., Rehbein, A., Schaudt, K., Busch, F.W., 1989. IL-4-responsive human helper T cell clones are resistant to growth inhibition by tumor necrosis factor-α. J. Immunol. 143, 902–906.

Pawelec, G., Sansom, D., Rehbein, A., Adibzadeh, M., Beckman, I., 1996. Decreased proliferative capacity and increased susceptibility to activation-induced cell death in late-passage human CD4(+) TCR2(+) cultured T cell clones. Exp. Gerontol. 31, 655–668.

Pawelec, G., Rehbein, A., Haehnel, K., Merl, A., Adibzadeh, M., 1997. Human T cell clones as a model for immunosenescence. Immunol. Rev. 160, 31–43.

Pawelec, G., Mariani, E., Bradley, B., Solana, R., 2000. Longevity in vitro of human CD4+ T helper cell clones derived from young donors and elderly donors, or from progenitor cells: age-associated differences in cell surface molecule expression and cytokine secretion. Biogerontology 1, 247–254.

Pawelec, G., Barnett, Y., Mariani, E., Solana, R., 2002a. Human CD4+ T cell clone longevity in tissue culture: lack of influence of donor age or cell origin. Exp. Gerontol. 37, 265–270.

Pawelec, G., Ouyang, Q., Colonna-Romano, G., Candore, G., Lio, D., Caruso, C., 2002b. Is human immunosenescence clinically relevant? Looking for "immunological risk phenotypes". Trends Immunol. 23, 330–332.

Perillo, N.L., Naeim, F., Watford, R.L., Effros, R.B., 1993. The Invitro Senescence of Human T-Lymphocytes – Failure to Divide Is Not Associated with a Loss of Cytolytic Activity or Memory T-Cell Phenotype. Mech Ageing Dev 67, 173–185.

Perillo, N.L., Walford, R.L., Newman, M.A., Effros, R.B., 1989. Human T lymphocytes possess a limited *In vitro* life span. Exp. Gerontol. 24, 177–187.

Perillo, N.L., Naeim, F., Walford, R.L., Effros, R.B., 1993. The in vitro senescence of human T-lymphocytes – failure to divide is not associated with a loss of cytolytic activity of memory T-cell Phenotype. Mech. Ageing Dev. 67, 173–185.

Posnett, D.N., Sinha, R., Kabak, S., Russo, C., 1994. Clonal populations of T cells in normal elderly humans—the T cell equivalent to benign monoclonal gammapathy. J. Exp. Med. 179, 609–618.

Posnett, D.N., Edinger, J.W., Manavalan, J.S., Irwin, C., Marodon, G., 1999. Differentiation of human CD8 T cells: implications for in vivo persistence of CD8(+)CD28(−) cytotoxic effector clones. Int. Immunol. 11, 229–241.

Remarque, E., Pawelec, G., 1998. T cell immunosenescence and its clinical relevance in man. Rev. Clin. Gerontol. 8, 5–25.

Roberts-Thomson, I.C., Whittingham, S., Youngchaiyud, U., Mackay, I.R., 1974. Ageing, immune response and mortality. Lancet 2, 368–370.

Spaulding, C., Guo, W., Effros, R.B., 1999. Resistance to apoptosis in human CD8+ T cells that reach replicative senescence after multiple rounds of antigen-specific proliferation. Exp. Gerontol. 34, 633–644.

Tang, D.G., Tokumoto, Y.M., Apperly, J.A., Lloyd, A.C., Raff, M.C., 2001. Lack of replicative senescence in cultured rat Oligodendrocyte precursor cells. Science 291, 868–871.

Wayne, S.J., Rhyne, R.L., Garry, P.J., Goodwin, J.S., 1990. Cell-mediated immunity as a predictor of morbidity and mortality in subjects over 60. J. Gerontol. 45, 45–48.

Wikby, A., Johansson, B., Ferguson, F., Olsson, J., 1994. Age-related changes in immune parameters in a very old population of Swedish people: a longitudinal study. Exp. Gerontol. 29, 531–541.

Wikby, A., Maxson, P., Olsson, J., Johansson, B., Ferguson, F.G., 1998. Changes in CD8 and CD4 lymphocyte subsets, T cell proliferation responses and non-survival in the very old: the Swedish longitudinal OCTO-immune study. Mech. Ageing Dev. 102, 187–198.

Wikby, A., Johansson, B., Olsson, J., Löfgren, S., Nilsson, B.-O., Ferguson, F.G., 2002. Expansion of peripheral blood CD8 T lymphocyte subpopulations and an association with cytomegalovirus positivity in the elderly: the Swedish NONA-immune study. Exp. Gerontol. 37, 445–453.

Zhou, T., Edwards, C.K., Mountz, J.D., 1995. Prevention of age-related T cell apoptosis defect in CD2-fas-transgenic mice. J. Exp. Med. 182, 129–137.

List of Contributors

Anders Wikby,
Boo Johansson and
Frederick Ferguson

Address correspondence to: Professor Anders Wikby
University College of the Health Sciences
Department of Natural Science and Biomedicine
P.O. Box 1038, Barnapsgatan 39, SE-551 11 Jönköping
Sweden Tel.: +46-381-35101 Fax: +46-381-17620
E-mail: Anders.Wikby@hhj.hj.se

Felicia A. Huppert,
Eleanor M. Pinto,
Kevin Morgan and
Carol Brayne

Address correspondence to: Dr. Carol Brayne
University of Cambridge, Department of Community
Medicine, Institute of Public Health, Forvie Site,
Robinson Way, Cambridge CB2 2SR
England, Tel.: +44-1223-330-334
Fax: +44-1223-330-330
E-mail: carol.brayne@medschl.cam.ac.uk

Giuseppina Candore,
Giuseppina Colonna-Romano,
Domenico Lio and
Calogero Caruso

Address correspondence to: Professor Calogero
Caruso, University of Palermo, Section for Cellular
and Molecular Pathology, Department of
Pathobiology, Corso Tukory 211, I-90134
Palermo, Italy, Tel.: +39-91-655-5911
Fax: +39-91-655-5933
E-mail: Marcoc@mbox.unipa.it

Ameila Globerson

The Multidisciplinary Center for Research on Ageing
Ben Gurion University of the Negev, IL-84105
Beer Sheva, Israel, Tel.: +972-8-934-3995
Fax: +972-8-934-4165
E-mail: Lcglober@weizmann.weizmann.ac.il

Jeffrey Pido-Lopez and
Richard Aspinall

Address correspondence to: Dr. Richard Aspinall
University of London, Imperial College
Department of Immunology, 369 Fulham Road
London SW10 9NH, England
Tel.: +44-181-746-5993, Fax: +44-181-746-5997
E-mail: R.aspinall@ic.ac.uk

Tamas Fulop,
Katsuiku Hirokawa,
Gilles Dupuis,
Anis Larbi and
Graham Pawelec

Address correspondence to: Professor Tamas Fulop, Jr.
Université de Sherbrooke, Centre de recherche en
gérontologie et gériatrie, Institut Universitaire de
Gériatrie, 1036, Belvedere Sud, Sherbrooke, Québec
Canada J1H 4C4, Tel.: +1-819-829-7131

Fax: + 1-819-829-7141
E-mail: tfulop@courrier.usherb.ca

Javier G. Casado,
Olga DelaRosa,
Esther Peralbo,
Raquel Tarazona and
Rafael Solana

Address correspondence to: Professor Rafael Solana
University of Córdoba, Department of Immunology
Hospital Reina Sofia, Ave. Menedez Pidal s/n
ES-14004 Córdoba, Spain, Tel.: + 34-957-217532
Fax: + 34-957-217532, E-mail: Rsolana@uco.es

Erminia Mariani and
Andrea Facchini

Address correspondence to: Dr. Erminia Mariani
Istituti Ortopedici Rizzoli
Laboratorio di Immunologia e Genetica, Via di
Barbiano 1/10, I-40136 Bologna, Italy
Tel.: + 39-51-6366 803, Fax: + 39-51-6366 807
E-mail: Marianie@alma.unibo.it

Julie McLeod

University of the West of England, Faculty of Applied
Sciences, Coldharbour Lane, Bristol BS16 1QY
England, Tel.: + 44-117-344-2531
Fax: + 44-117-976-3871
E-mail: Julie.mcleod@uwe.ac.uk

Sudhir Gupta

University of California, Irvine, CA 92697, USA
Tel.: + 1-949-824-5818, Fax: + 1-949-824-4362
E-mail: sgupta@uci.edu

Owen A. Ross,
Martin D. Curran,
Derek Middleton,
Brian P. McIlhatton,
Paul Hyland, Orla Duggan,
Kathryn Annett,
Christopher Barnett and
Yvonne Barnett

Address correspondence to: Dr. Yvonne Barnett
University of Ulster, Department of Biomedical
Sciences, Cromore Road, Coleraine BT52 1SA
Northern Ireland, Tel.: + 44-1265-324-627
Fax: + 44-1265-324-965
E-mail: Yvonne.barnett@ulst.ac.uk

Daniela Frasca,
Luisa Guidi and Gino Doria

Address correspondence to: Dr. Daniela Frasca
University of Miami School of Medicine
Department of Microbiology and Immunology
1600 NW 10th Ave, Miami, FL 33136, USA
Tel.: + 1-305-243-6040, Fax: + 1-305-243-5522
E-mail: DFrasca@med.miami.edu

T. G. Sourlingas and
K.E. Sekeri-Pataryas

Address correspondence to: Dr. Kalliope
Sekeri-Pataryas, National Center for Scientific

Research "Demokritos", Institute of Biology
P.O. Box 60228, Ag. Parakevi, GR-15310 Athens
Greece, Tel.: +30-1-651-3021
Fax: +30-1-651-0594
E-mail: Ksek@mail.demokritos.gr

Klaus-Helge Ibs,
Philip Gabriel and
Lothar Rink

Address correspondence to: Professor Lothar Rink
Institute of Immunology, RWTH-Aachen
University Hospital, Pauwelsstr. 30, D-52074 Aachen
Germany, Tel.: +49-241-808-0208
Fax: +49-241-808-2613
E-mail: LRink@klinikum.rwth-aachen.de

Eugenio Mocchegiani,
Robertina Giacconi,
Mario Muzzioli and
Catia Cipriano

Address correspondence to: Dr. Eugenio Mocchegiani
Italian National Research Centers on Ageing
Immunology Center, Via Birarelli 8, I-60121 Ancona
Italy, Tel.: +39-71-800-4216, Fax: +39-71-206 6791
E-mail: E.mocchegiani@inrca.it

Rita B. Effros

University of California at Los Angeles
Department of Pathology and Laboratory Medicine
10833 Le Conte Avenue, Los Angeles, CA 90095-1732
USA, Tel.: +1-310-825-0748, Fax: +1-310-206-0657
E-mail: Reffros@mednet.ucla.edu

Graham Pawelec

Center for Medical Research, University of Tübingen
ZMF, Waldhörnlestr 22, D-72072 Tübingen
Germany, Tel.: +49-7071-2982805
Fax: +49-7071-295567
E-mail: graham.pawelec@uni-tuebingen.de

Advances in
Cell Aging and Gerontology
Series Editor: Mark P. Mattson
URL: http://www.elsevier.nl/locate/series/acag

Aims and Scope:

Advances in Cell Aging and Gerontology (ACAG) is dedicated to providing timely review articles on prominent and emerging research in the area of molecular, cellular and organismal aspects of aging and age-related disease. The average human life expectancy continues to increase and, accordingly, the impact of the dysfunction and diseases associated with aging are becoming a major problem in our society. The field of aging research is rapidly becoming the niche of thousands of laboratories worldwide that encompass expertise ranging from genetics and evolution to molecular and cellular biology, biochemistry and behavior. ACAG consists of edited volumes that each critically review a major subject area within the realms of fundamental mechanisms of the aging process and age-related diseases such as cancer, cardiovascular disease, diabetes and neurodegenerative disorders. Particular emphasis is placed upon: the identification of new genes linked to the aging process and specific age-related diseases; the elucidation of cellular signal transduction pathways that promote or retard cellular aging; understanding the impact of diet and behavior on aging at the molecular and cellular levels; and the application of basic research to the development of lifespan extension and disease prevention strategies. ACAG will provide a valuable resource for scientists at all levels from graduate students to senior scientists and physicians.

Books Published:

Lightning Source UK Ltd.
Milton Keynes UK
04 December 2010

163811UK00007B/65/A